GENETICS - RESEARCH AND ISSUES

MICROARRAYS

PRINCIPLES, APPLICATIONS AND TECHNOLOGIES

GENETICS - RESEARCH AND ISSUES

Additional books in this series can be found on Nova's website under the Series tab.

Additional e-books in this series can be found on Nova's website under the e-book tab.

GENETICS - RESEARCH AND ISSUES

MICROARRAYS

PRINCIPLES, APPLICATIONS AND TECHNOLOGIES

JAMES V. ROGERS
EDITOR

NOVA BIOMEDICAL

New York

Library of Congress Cataloging-in-Publication Data

ISBN: 978-1-62948-669-7

Library of Congress Control Number: 2013955545

Published by Nova Science Publishers, Inc. † New York

Contents

Preface

In April 2003, the completion of the Human Genome Project was the capstone to a scientific journey that began decades earlier when Francis Crick and James Watson unraveled the mystery of the human genetic code. The sequencing of the human genome opened the flood gates of ideas and imagination to explore the structure and function of specific genes. Empowered researchers applied the fundamental principles of biology's central dogma to advance areas of scientific investigation ranging from genetics and evolution to molecular medicine, health, and disease.

The rapid advance of science and technology yielded the development of the microarray technique. Deciphering genetic sequences propelled the use of microarrays from a new fad in genome research to a ubiquitous tool in biological investigations. Turning genes on or off by the thousands critically orchestrates everyday biological processes and the amount of data generated from a single microarray experiment can be overwhelming, time-consuming, and tedious. By combining technological advances in microarrays with statistical and bioinformatics software, investigators have become armed with some of the most powerful means of biological problem-solving. Statistical reduction enables researchers to sort through the transcriptome and generate useable lists of significantly-modulated genes highlighting changes that chance alone cannot explain.

The underlying theme of this book is the use and application of microarray technology throughout the life sciences. This compilation of chapters provides an overview of how microarrays can be used to address diverse complex biological questions. In this collection, authors present information on using microarrays to unlock molecular mechanisms and gene expression patterns associated with research areas ranging from biodefense, pathological changes, and monitoring antimicrobial resistance genes to diagnostics, marine biodiversity, and dermal toxicology.

By using microarray technologies and specialized software, gene-expression profiles can provide insight into the development or evaluation of prophylactic and/or therapeutic treatments that target specific molecular pathways or genes. These data can ultimately lead to effective analyses using a molecular-systems biology approach that can feed into translational and predictive research. Due to the abundance of microarray data available in the open scientific literature, researchers can apply statistical approaches to identify biomarkers and perform risk assessments without the need of experimentation. In this book, authors utilized previously published data for the functional prediction and understanding of molecular processes, the identification of reference genes using meta-analysis, and the development of a phenotypic anchoring model for drug-induced liver injury.

Biomarker-based discovery and risk assessment are important to provide mechanistic insight into factors as such as genetic background, diet, and lifestyle modulate the risk of disease. The continual development and refinement of improved molecular techniques to predict, treat, and prevent disease is critical to improving human health. Researchers must identify key components indicative of individuals or populations that pose an increased risk of infection or disease in order to achieve the goal of highly-individualized medical interventions (e.g., personalized medicine). For example, one chapter identifies an 18-gene signature that differentiates treatment responders and non-responders to interferon therapy. Based on the expression levels of this gene signature, a positive prediction value higher than 96% was observed for accurately predicting whether an HCV patient will or will not respond to interferon therapy. Moreover, another chapter describes the use microarrays to uncover mechanistic details underlying Alzheimer's disease pathology that track to cholesterol and lipid metabolism while highlighting the interactions between cholesterol and amyloid-β peptide.

Although using microarrays for monitoring changes in mRNA transcripts has dominated the application of this cutting-edge tool, other approaches for using microarrays include expression analysis of non-coding RNAs. Both microRNAs and long non-coding RNAs are involved in many biological processes by regulating transcriptional activity and subsequent processes. In this book, three chapters discuss the expression, annotation, and function of microRNAs and long non-coding RNAs to characterize macrophage microvesicle function, cutaneous inflammatory disorders, and elucidate their potential involvement in pathophysiological roles.

Almost every cell type within an organism contains the same genetic information locked up within DNA, but what makes these cells different from one another is based on the number, type, and magnitude of genes that are activated. The transcription of DNA to messenger RNA and subsequent translation into proteins is what leads to the unique phenotype exhibited by each cell, which could ultimately be normal or abnormal. To help differentiate one cell type from another, or differences among tissues, DNA microarrays have advanced the understanding of the genome with respect to structure and function. The implementation of microarrays and companion statistical and informatics tools is continually evolving and helping scientists discover new and exciting aspects of life science that will advance systems biology to a new level of understanding of how biological organisms function and respond to internal and external stimuli as well as provide insight into personalized or predictive medicine.

In: Microarrays: Principles, Applications and Technologies
Editor: James V. Rogers
ISBN: 978-1-62948-669-7

Chapter 1

Microarray Technology for Category 'A' Bioterrorism Agent Research: Current Status and Perspectives

***Lauren Huettner*[1,2] *and Eric M. Vela*[2*]**
[1]Division of Veterinary Public Health, College of Public Health,
The Ohio State University, Columbus, OH, US
[2]Battelle, Columbus, OH, US

Abstract

Since the terrorist attacks and subsequent anthrax letters in the early 2000s, the United States has been on heightened alert for potential bioterrorist attacks, or the deliberate release of viruses, bacteria, or other agents used to cause morbidity and mortality in people, animals, or plants. Although many infectious pathogens could be employed in the event of an attack, certain pathogens known as category A bioterrorism agents pose the greatest risk to the public and national security, and could be potentially life-threatening if weaponized. Current pathogen detection and diagnostic methods are not ideal, providing results with poor speed and accuracy, and host-pathogen interaction molecular profiles have yet to be fully characterized. Microarray technology provides unique resources for the rapid and accurate detection and diagnosis of these pathogens, as well as a platform to allow researchers to create genetic profiles for epidemiological studies and study host-pathogen interactions for the development and testing of vaccines, therapeutics and prophylaxis.

Introduction

Microarray technology enables the study of interactions between hundreds to thousands of molecules simultaneously through the analysis of a very small amount of analytes and

[*] Corresponding Author (velae@battelle.org).

poses a unique way of characterizing how cells and organisms adapt to changes in the external environment [1]. Molecules are differentially expressed during various cellular pathways and mechanisms constantly occurring within an organism, and changes in the environment result in changes in molecular expression. Microarrays are used to analyze differences in gene expression levels, small nucleotide polymorphisms, protein and carbohydrate levels, as well as antibody expression. This method is not considerably different from other hybridization techniques, such as Northern and Southern blotting, but the unique aspect of microarray technology is the incredible amount of information that is obtained at one time. Historically, technology only allowed researchers to focus on a minimal number of analytes, but microarray technology allows for the concurrent study of a myriad of genes. The simultaneous and parallel determination of the expression of numerous analytes that microarray technology allows renders it a powerful laboratory tool [2]. Although microarray technology is still developing as a technique and must overcome many barriers, microarrays are quickly becoming successful, reliable and mature genomic tools for both biomedical and clinical research, along with epidemiological and clinical practices. Microarrays have revolutionized the current scientific research approaches involving various diseases, including heart disease, metabolic disease, mental illness, cancer biology, infectious disease, and bio- or agroterrorism by introducing a new platform for disease detection, diagnostics, the effects of pathogen infection, and vaccine and therapeutic drug development.

Microarray technology consists of a compilation of microscopic features affixed onto a small surface support, where probes with target molecules produce qualitative or quantitative data after being hybridized with test material. The type of array can be distinguished by characteristics including the nature of the probe, the solid-surface support used, and target detection [3]. For example, DNA microarrays are created by arranging minuscule amounts of thousands of gene sequences on a single microscope slide (a database of over 40,000 gene sequences is available to researchers). Once a gene is activated, the cell begins to copy specific segments of that gene and generate messenger RNA (mRNA) that is present in various cells and can be collected and labeled by using reverse transcriptase (RT) enzyme and fluorescent nucleotides. The process allows for the generation and labeling of complementary DNA (cDNA) to the respective mRNA. Once the cDNA is generated, it can be collected and placed on DNA microarray slides for hybridization commonly involving red or green fluorescent dyes. Specialized scanners can be used to measure the fluorescent intensity for each area on the microarray slide. An active gene produces more molecules of mRNA, resulting in more labeled-cDNA, which can in turn be hybridized to the DNA on the microarray slide to generate various levels of fluorescent intensity. Thus, less active genes produce fewer mRNA resulting in a lower level of fluorescent intensity. Measurement of fluorescence intensity signifies expression levels of a particular gene when compared to controls. For host-pathogen interactions, changes in gene expression occurs in various cells and tissues in response to a disease state, and the changes in gene expression are consistent when the host organism has been exposed to a specific infectious disease pathogen. These changes in gene expression may be specific to the pathogen and can be utilized as potential biomarkers to identify a clinical profile associated with the infectious disease, otherwise known as a "disease signature." Microarray technology offers a unique method to study host-pathogen interactions at a global level and can be used as a tool to delineate molecular mechanisms of tissue responses in host organisms infected with various pathogens. These technologies may also offer the possibility of identifying new approaches to diagnostics, detection of disease, and development of new vaccines, therapeutics, and prophylaxis.

With the looming threats of infectious agents being used as bioweapons, it is clear that there is a strong need for better detection, diagnostic, and drug development tools. Since the terrorist attacks and subsequent anthrax letters in the early 2000s, the United States has been on heightened alert for potential bioterrorist attacks. Bioterrorism is defined by the Centers for Disease Control and Prevention (CDC) as the deliberate release of viruses, bacteria, or other agents used to cause illness or death in people, animals, or plants [4]. Bioterrorism agents are classified into categories, A, B, or C agents, each defined by how easily the pathogen can spread and the severity of morbidity and mortality that they may cause. Category A agents, the highest priority agents, pose the greatest risk to the public and national security due to four main factors: (1) the ease of person to person transmissibility; (2) infection may result in high mortality rates; (3) outbreaks of the pathogen may cause public panic; and (4) the pathogens require special action for public health preparedness [5]. These pathogenic agents include, but are not limited to *Bacillus anthracis* (anthrax), *Clostridium botulinum* toxin (botulism), *Yersinia pestis* (plague), Variola major (smallpox), *Francisella tularensis* (tularemia), and certain viral hemorrhagic fevers including Ebola virus, Marburg virus, Lassa virus and Machupo virus (Table 1) (6-9). Studies show that The United States is inadequately prepared to face further biological attacks (10). This can be partially attributed to the lack of government organization, but can also be due to the lack of appropriate vaccines, prophylaxis, therapeutics, and unperfected diagnostic tools. Although fairly new technology, microarrays have been increasingly utilized in scientific studies and the use of the technology in studies pertaining to category A bioterrorism agents has increased greatly since the 2001 attacks (Figure 1). Microarrays can help bridge the gap in the insufficient scientific knowledge of these deadly pathogens and have the potential to be key contributors in the development of new detection and diagnostic methods, and technologies aimed at developing new drugs, along with the determination of drug efficacy.

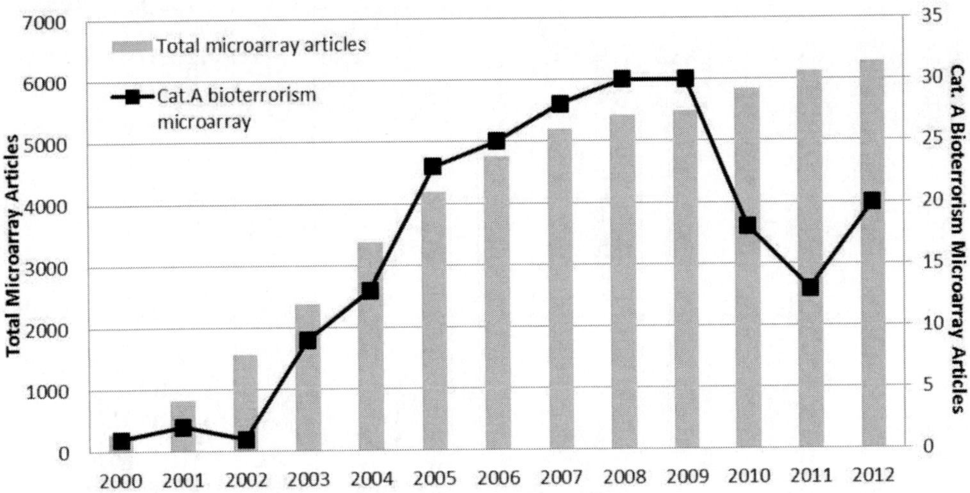

Figure 1. Microarray publications. The number of manuscripts published utilizing microarray technology (bars) and the number of microarray publications that have category A bioterrorism applications (line) are pictured.

Table 1. Category A bioterrorism agents and the main aspects of their clinical disease in the event of a bioweapon attack

Agent	Disease	Route of exposure	Incubation	Initial Clinical Symptoms	Diagnosis	Vaccine	Treatment	Fatality Rate with Treatment
Bacillus anthracis	Anthrax	Inhalational	Usually < 1 week	non-specific; flu-like	Culture	Yes	Antimicrobial therapy	75%
Clostridium botulinum	Botulism	Foodborne	12-36 hours post ingestion	non-specific; flu-like	Culture	No	Botulinum antitoxin	5-15%
Variola major	Smallpox	Inhalational	7-17 days	non-specific; flu-like followed by pustular rash	Lesion evaluation, PCR	Yes	Supportive	30% for classical; Near 100% for hemorrhagic or flat forms
Francisella tularensis	Tularemia	Inhalational	1-14 days	non-specific; flu-like	Culture	No	Antimicrobial therapy	2%
Yersinia pestis	Plague	Inhalational	1-6 days	non-specific; flu-like followed by rapidly developing pneumonia	Culture	No	Antimicrobial therapy	11%
Ebola Virus	Ebola hemorrhagic fever	Inhalational	2-21 days	non-specific; flu-like	ELISA, PCR, virus isolation	No	Supportive	49-81%
Marburg Virus	Marburg hemorrhagic fever	Inhalational	5-10 days	non-specific; flu-like	ELISA, PCR, virus isolation	No	Supportive	23-90%
Machupo Virus	Bolivian hemorrhagic fever	Inhalational	4-21 days	non-specific; flu-like	ELISA	No	Ribavirin, Supportive	18-22%
Lassa Virus	Lassa fever	Inhalational	1-3 weeks	non-specific; flu-like	ELISA	No	Ribavirin, Supportive	15-20%

Detection and Diagnostic Tools

The US Homeland Security Presidential Directive 21 defines biosurveillance as "the process of active data-gathering with appropriate analysis and interpretation of biosphere data that might relate to disease activity and threats to human or animal health—whether infectious, toxic, metabolic, or otherwise, and regardless of intentional or natural origin—in order to achieve early warning of health threats, early detection of health events, and overall situational awareness of disease activity"[5]. Conventional biosurveillance relies on the combined use of an automated system to evaluate case and suspect case reporting, biosensors, statistical surveillance, and data visualization of pre-diagnostic and diagnostic data for the earliest possible detection of biological events that may be indicative of a public health emergency. Alongside being able to rapidly and efficiently detect a potentially dangerous agent in the environment and various samples, the accurate diagnosis of individuals that have possibly been exposed to such agents is equally, if not more important. Being able to quickly distinguish an infection of a bioterrorism agent from any other pathogen is critical due to the fact that these agents have the ability to cause major disease, even death in a short period of time, and lead to a presentation of non-specific flu-like clinical symptoms, mostly clinically indistinguishable from the majority of diseases in the stages of early infection. Until recently, the use of traditional pathogen detection and diagnostic methods, such as biochemical assays, polymerase chain reaction (PCR), enrichment culture, serology, and microscopy, have been limiting due to their specific laboratory and time requirements, making them inefficient in addressing potential bioterrorism concerns [11]. Novel, more technologically advanced molecular detection and diagnostic methods, such as microarrays, provide the potential for rapid, highly sensitive and specific pathogen identification [12, 13] (Table 2).

Table 2. Microarrays used for the detection and diagnosis of category A bioterrorism agents and the main findings and conclusions from the studies

Microarray used	Category A agent(s) studied	Main findings and conclusion	Reference
DNA			
Oligonucleotide	*C. botulinum*	The array provides a rapid approach for the simultaneously detection of botulinum neurotoxin genes and determining the relatedness of type A *C. botulinum* strains isolated from different sources associated with a botulism outbreak.	10
	Y. pestis, B. anthracis	The oligonucleotide microarray can specifically detect and identify *Y. pestis* and *B. anthracis* from similar species in milk samples.	5
	Variola major	DNA from 52 samples of various ORV species were tested and accurately identified. The new microarray method is a valuable tool for the rapid and accurate detection and differentiation of various OPV species, including VARV species.	13
	Variola major, Ebola virus, Marburg virus	The presence of the viruses was confirmed after microarray analysis of nasopharyngeal aspirates, blood, urine, and tissue samples from persons infected.	17

Table 2. (Continued)

Microarray used	Category A agent(s) studied	Main findings and conclusion	Reference
Resequencing	*B. anthracis*	The arrays represent a "one reaction" genotyping technology that possesses the ability to discriminate highly similar isolates and other strains from *B. cereus* s.l.Clade 1.	12
	B. anthracis, Y. pestis, F. tularensis	The array platform can identify *B. anthracis, F. tularensis, Y. pestis* and distinguish them from similar species in a single assay.	14
	Ebola Zaire, Lassa Virus	The assay detects and identifies closely related strains of filoviruses of the Ebola Zaire group, or the Machupo and Lassa arenaviruses , but is also able to differentiate near neighbors of the same agent types.	15
Whole genome	Lassa virus	There are strong transcriptional changes in the immune response to LASV exposure, but this was paired with a lack of pro-inflammatory cytokine response. The results provide a foundation for biomarker identification for clinical diagnosis.	16
Multi-genome	*B. anthracis*	The sources of genome diversity and aspects of genome evolution of members of the *B. cereus* group, but especially *B. anthracis*, were elucidated, improving diagnostics and enabling the identification of biomarkers in the future.	11
Carbohydrate			
Multiplex (polysaccharide)	*B. anthracis, F. tularensis*	Antibodies resulting from infection of *B. anthracis* and *F. tularensis* were about to be detected and differentiated using this single platform.	19
Protein			
Multiplex (antibody-functionalized barcoded resins (BCRs))	*B. anthracis, F. tularensis*	Anthrax PA, *F. tularensis* lipopolysaccharide and CD14 antigens were accurately identified and quantified in tetraplex assays, showing potential for the use of BCRs as multiples antigen detection and identification on a microarray format.	21
ELISA based antibody	*C. botulinum*	Simultaneous detection of BoNT serotypes A, B, C, D, E, and F was achieved with high sensitivity providing a rapid detection tool for *C. botulinum* toxins in various samples.	22

DNA Arrays

DNA microarrays have been shown to be an effective high-throughput detection and diagnostic tool for the identification of pathogenic microorganisms in clinical, environmental, food, and water samples and have received considerable attention due to the ability to simultaneously analyze a very large number of nucleic acid sequence targets and detect multiple genetic targets or genomes from multiple pathogens on a single slide [14]. The assays are characterized based on the type of DNA fragments that are used on the array, such as complementary DNA (cDNA) or oligonucleotides [2]. DNA microarray technology allows for the entire genome of one or multiple organisms to be studied at one time with relatively simple techniques.

Foodborne Detection

The public food supply is considered a vulnerable target for bioterrorism attacks, and events related to the deliberate contamination of food using conventional pathogens, such as Escherichia coli O157:H7 and Salmonella, but also nonconventional foodborne pathogens, such as weaponized category A agents [11]. Early detection of disease inducing pathogens in the food supply is critical in the event of an attack or a potential natural outbreak in order to spare the susceptible public populations from a potentially life threatening disease and to spare the food industry from economic havoc due to loss of productivity and business, legal issues, cost of medical treatments for the infected, and recall and destruction of contaminated products [15]. Conventional pathogen detection methods such as culture methods, polymerase chain reaction (PCR) and immunology-based methods may take hours to days to produce a definitive result [16]. Microarray technology offers a high throughput technique to detect multiple foodborne pathogens in food samples and poses as a useful laboratory tool to rapidly detect agents and eliminate contaminated products, preventing public consumption of harmful food items. Oligonucleotide DNA microarrays have been utilized in the detection of a variety of category A bioterrorism agents that have the potential to contaminate the public food supply, including *B. anthracis*, *C. botulinum*, and *Y. pestis*. A focused oligonucleotide microarray format was used by investigators to provide a rapid approach for neurotoxin gene detection and preliminary determination of the relatedness of strains isolated from different food sources associated with a *C. botulinum* outbreak [17]. The focused microarray provides useful information beyond the toxin type of the strain involved with a particular outbreak, namely epidemiological associations in an outbreak investigation [17]. Similarly to the previous study, a new generation low density oligonucleotide microarray was created to allow the simultaneous detection and identification of *Y. pestis* and *B. anthracis* to be used in a foodborne application [11]. The microarray was designated as the "Y-PESTIS/B-ANTHRACIS 4x2K Array" and contains 533 and 1,707 probes for *Y. pestis* and *B. anthracis*, respectively, that target the pathogen's virulence genes [11]. The specific targeting of these genes provides information on the virulence potential of strains implicated, rendering the microarray especially useful in the detection and profiling of potentially life threatening biothreat agents.

Pathogen Differentiation and Clinical Diagnosis

The Bacillus cereus sensu lato group includes multiple species including *B. cereus*, *B. thuringiensis*, *B. mycoides*, *B. pseudomycoides*, *B. weihenstephanensis* and *B. anthracis*, the causative agent of anthrax. These species share 99% sequence identity and belong to one phylogenetic group [18]. A multi-genome DNA microarray was utilized to investigate *B. cereus* group genomic diversity and to elucidate events associated with the emergence of *B. anthracis* [18]. The study found indications that the evolution of *B. anthracis* was associated with the acquisition and/or maintenance of a limited but specific set of chromosomally-encoded virulence-associated genes along with the genes encoded on the virulence plasmids. The data obtained from the study allows for the better understanding of events associated with the evolution of *B. anthracis* and will help to improve diagnostics for the pathogen. Similarly, a customized resequencing DNA microarray was used to distinguish infectious *B. anthracis* from closely related B. cereus sensu lato strains utilizing isolated genomic DNA from each strain and results were compared with multi-locus sequencing typing (MLST) data, the traditional method of DNA genotyping [19]. The resulting data suggested that *B. anthracis* resequencing microarray technology is the most effective and valuable tool for genotyping Clade 1 *B. cereus* strains along with strains within the *B. anthracis* lineage, offering advantages over MLST. Other methods of species identification, including amplified fragment length polymorphism analysis and multi-locus sequence typing, are numerous, but have a lower genetic resolution sensitivity level [19].

An oligonucleotide DNA microarray method was developed for simultaneous detection and identification of six species of Orthopoxvirus (OPV) including Variola (VARV), Monkeypox (MPXV), Cowpox, Camelpox, Vaccinia (VACV), and Ectromelia (ECTV) viruses (20). The method allows for the discrimination of OPV species from other pathogens that cause infections with clinical manifestations similar to OPV infections. Being able to detect and differentiate OPV species is especially important in biodefense due to the fact that the majority of the world population does not have protective immunity to the viruses after the discontinuation of mass vaccination and is susceptible to deadly infection caused by various OPV, mainly VARV, the causative agent of smallpox. Other methods to discriminate OPV include differentiation of viral protein profiles using Western blotting and other serological assays, which require production, isolation, and unambiguous typing of live virus, which can prove difficult. PCR-based methods are also utilized, allowing for rapid and sensitive detection, but do not have output capacity comparable to microarray technology [20].

The respiratory pathogen microarray version 1 (RPM v.1) chip, a resequencing DNA microarray, was utilized under blinded conditions to detect and discriminate different strains of B. anthracis, Y. pestis, and F. tularensis in clinical samples [21]. Resequencing arrays were developed to combine the multiplexing power of standard microarrays with the discrimination capabilities of sequence analysis, while they offer the capability of having a single test that detects and discriminates between a target pathogen and its genetically similar pathogens. The RPM v.1 chip was primarily designed to identify common respiratory pathogens and has eleven targets devoted to the detection of six CDC category A bioterrorism agents that may present with nonspecific, flu-like symptoms at early stages of infection. The study demonstrated the potential of resequencing arrays to detect and identify specific sequences from a variety of inactivated biothreat organisms [21]. Similarly, a resequencing microarray (a broad-range resequencing pathogen microarray [RPM] for detection of tropical and emerging infectious agents [TEI] including biothreat agents: RPM-TEI v 1.0) was used to detect and differentiate multiple hemorrhagic fever-inducing filoviruses and arenaviruses, including Ebola Zaire and Lassa virus (LASV), respectively, along with multiple other infectious agents with high specificity and sensitivity [22]. LASV was further studied by utilizing a whole genome microarray to elucidate temporal host response in the peripheral blood mononuclear cells (PBMCs) of non-human primates following aerosol exposure to the virus [23]. The data obtained allowed the investigators to further elucidate the host immune responses during LASV infection and provided a foundation for LASV specific biomarker identification, a potential method of clinical diagnosis. A microarray platform, the GreeneChip microarray system, was designed for pathogen surveillance and discovery of multiple viral and bacterial infectious pathogens, including OPVs, Ebola virus (EBOV), and Marburg virus (MARV) [24]. The GreeneChip technology introduces sample preparation and labeling methods that enhance sensitivity. The new array has been used in an outbreak investigation when other methods failed to detect a microorganism in a fatal hemorrhagic fever case and has been validated with cultured viral isolates and tested with blood, respiratory, urine, and tissue samples containing the viral pathogens [24].

Carbohydrate Arrays

The use of carbohydrate microarray technology for the detection of bacterial and viral infections is growing in interest. The technology holds great promise as a high-throughput means of detecting the interactions of proteins with diverse oligosaccharide sequences of glycoproteins, glycolipids and polysaccharides [25]. With increasing awareness of post-genome glycosylation as a potentially important form of post-translational modification of proteins, carbohydrate microarray technology is especially needed for the detection and characterization of carbohydrate ligands, which are largely diverse and unable to be cloned in the case of oligosaccharide ligands. Because of these limitations, conventional methods are not ideal and high-throughput microarray technology is being utilized to analyze carbohydrate-protein interactions to elucidate mechanisms of infection, inflammation, and immunity [25]. A microarray platform was developed in 2008 using bacterial 'signature' carbohydrates associated with *B. anthracis* and *F. tularensis* infection [26]. The trisaccharide derived from *F. tularensis* was converted to glycosylamines by reductive amination and was then immobilized on epoxide-functionalized glass slides along with anthrose, a *B. anthracis* specific carbohydrate, and the corresponding tetrasaccharides attached to a linker containing –NH2 group. The microarray was used to specifically detect antibodies in the serum of melioidosis patients, or animals immunized with the causative bacteria of tularemia or anthrax [26].

Protein Arrays

Protein microarrays, or protein chips, are miniaturized, parallel assay systems that contain small amounts of purified proteins in a high-density format [27]. Protein microarray technology offers a post-genomic strategy to investigate host-pathogen interactions by identifying multiple protein-molecule interactions as well as immunogenic characteristics of the host at any stage of disease. There are two types of protein microarrays: analytical and functional. Analytical protein microarrays are composed of well-characterized biomolecules with specific binding activities, such as antibodies, to analyze the components of complex biological samples or to determine whether a sample contains a specific protein of interest and have been used for protein expression profiling, biomarker identification, clinical diagnosis, and environmental/food safety analysis. Functional protein arrays are used for drug target identification/validation and studies of protein interaction, biochemical activity, and immune responses [27]. A multiplex assay based on antibody-functionalized barcoded resins (BCRs) was used to detect antibodies against human pathogens *F. tularensis* lipopolysaccharide (LPS) and *B. anthracis* protective antigen (PA) toxin, demonstrating an alternative platform for multiplex immunoassays capable of detecting antigens [28]. Enzyme-linked immunosorbent assay (ELISA)-based protein microarrays allow for the simultaneous analysis of up to fifty proteins in a small sample volume within a single experiment, making the assays ideal for high-throughput protein analyses. ELISA-based protein antibody microarrays have been developed for the simultaneous and sensitive detection of *C. botulinum* neurotoxin (BoNT) serotypes A, B, C, D, E, and F [29]. The current gold standard of *C. botulinum* detection, the standard mouse bioassay, is outperformed as a result of the microarray being optimized with an even higher sensitivity than that of the bioassay. The assays can specifically detect and differentiate between individual BoNT serotypes and was evaluated in both clinical and food samples. These simultaneous and highly sensitive assays have the potential for high-throughput measurements of BoNT serotypes A, B, C, D, E, and F from a range of environmental, food, biodefense, and clinical samples [29].

Elucidating Disease Pathogenesis for the Purpose of Drug Development

Understanding the mechanisms of host and agent pathology during a diseased state is essential when developing new vaccines and therapeutics. The discovery of novel molecular mechanisms for drug intervention targets allows for the manufacturing of more effective drug products to prevent or combat life threatening diseases by being able to elucidate novel molecular drug targets.

Microarray technology has allowed for the elucidation of host and agent pathology in multiple category A bioterrorism agents and proves as a promising tool for the development of new vaccines, prophylaxis, and therapeutic drugs for these agents, some of which currently have none available (Table 3).

Table 3. Microarrays used for pathogen-host response profiling of category A bioterrorism agents for the purpose of drug development and the main findings and conclusions from the studies

Microarray used	Category A agent(s) studied	Main findings and conclusion	Reference
DNA			
Whole genome	*B. anthracis*	LT directly targets human peripheral monocytes and causes multiple aberrant gene responses normally associated with defects in monocyte's normal signaling transduction pathways and function.	28
	B. anthracis	The expression of 295 unigenes encoding for proteins functiong in both innate and adaptive immunity were significantly increased in PA-stimulated vs. non-stimulated PBMCs and no significant decrease in gene expression was observed.	29
	Variola major	58 early infection biomarkers (EIB) were expressed in PBMCs from macaques infected with two Variola strains. The expression levels were correlated with disease progression and severity, suggesting a pathogen-specific re-organization of the gene expression and protein interaction networks occurs during infection.	30

Microarray used	Category A agent(s) studied	Main findings and conclusion	Reference
	F. tularensis	Novel host pathways potentially associated with Francisella infection were identified. Transcriptional profiling in vitro can be used as a reliable tool to identify potential biomarkers that conform to human tularemia.	32
	F. tularensis	Whole-genome transcriptional profiling revealed that *F. tularensis* (Ft*fabI)* and *de novo* fatty acid biosynthetic genes are transcriptionally active during infection. Ft*fabI* inhibitor is a relevant drug target for individuals infected with *F. tularensis* bacterium.	34
	F. tularensis	Elevated cdGMP in *F. novicida* stimulates biofilm formation and inhibits virulence, suggesting that differences in human virulence between *F. novicida* and *F. tularensis* may be due in part to the absence of cdGMP signaling in *F. tularensis.*	36
Oligonucleotide	*F. tularensis*	Several host response genes' expression levels were reduced, including the pro-inflammatory signaling pathway when compared to less virulent *F. tularensis* strains. Several new potentially important factors in host cell subversion were identified and may serve as novel targets for drug discovery.	33
cDNA	*F. tularensis*	The relative expression of a limited number of genes differed significantly between live virus strain (LVS) wild-type and Δ*iclR* strains. IclR does not contribute significantly to the virulence or pathogenesis of *F. tularensis* LVS or SchuS4.	31
	Ebola virus, Marburg virus	EBOV and MARV have the ability to modulate host cell gene expression to antagonize the activity of the innate antiviral response and that REBOV has the reduced ability to block expression of interferon-relates gene expression.	37
Porcine	Ebola virus	After infection, a series of proinflammatory events occur, immune cells infiltrate into the lungs and lead to severe respiratory distress, providing the background structure to further study the mechanisms of antiviral response evasion caused by certain viral hemorrhagic fevers	38
Murine	*F. tularensis*	The data confirmed down-regulation of the toll-like receptor pathway and the induction of IFN-g inducible genes after challenge with virulent *F. tularensis*, SchuS4. Pathway analysis indicates a novel role for cellular junctions and an epithelial response to infection, and respiratory epithelium seems to play an important role in host response.	39
	F. tularensis	F. tularensis LVS infection applies an oxidative stress on the target cells and many of the host-defense mechanisms appear to be intended to counteract this stress.	40
	Y. pestis	The data provides new information on the underlying transcrptional mechanism in mice to *Y. pestis* infection.	41
Protein			
Antibody	Variola major	Host-response profiles of subjects inoculated with MVA and Dryvax were comparable, particularly with respect to responses to membrane proteins that might play a role in virus neutralization.	42

Table 3. (Continued)

Microarray used	Category A agent(s) studied	Main findings and conclusion	Reference
	Variola major	MVA has an immunogenicity comparable to that of Dryvax, making MVA a potential vaccination alternatinve to Dryvax.	43
	Variola major	The generation of B cell hybridomas resulted in monoclonal antibodies against a variety of VACV antigens, data that could contribute to the mapping of B cell epitopes in VACV that contribute to immune protection and the generation of novel drugs.	44
	Variola major	Proteins recognized by antibodies from all vaccinees were less than ten percent of the total VACV proteome.	45
	Variola major	Antibodies from vaccination recognized a small number of proteins shared with MPXV and humoral responses unique to MPXV antigens were involved in recovery from infection with the virus.	46
	F. tularensis	Thirty-one previously unreported immunoreactive antigens and data suggesting that a killed vaccine can provide protection against highly virulent *F. tularensis* strains were discovered.	49
	F. tularensis	Fourteen anti-Francisella hybridoma antibodies were characterized for the potential of the vaccine strain to be used as immunotherapy of tularemia infections.	50
	F. tularensis	Signals measuring strong antigen-antibody binding activity can be computationally analyzed to determine infection using the array technology.	51
ELISpot antibody	*Y. pestis*	Both humoral and cellular responses against *Y. pestis* in humans were reported. Antibody to F1, LcrV and YopD can remain in recovered patients for ten or more years and specific memory T cell responses to F1 and LcrV could not be deleted in those patients 4 to 6 years post-infection.	48
Novel			
Tissue	Lassa virus	Used to characterize the in-vivo expression profile of CD317 from infected cells in culture by using non-transformed tissues as a framework for future studies. CD317 is a possible target for immunotherapy.	52
Skin electroporation	Variola major	Developed to deliver a synthetic, multivalent, highly concentrated, DNA vaccine to macaques that provides protection from a highly pathogenic MPXV challenge resulting in significantly improved efficacious DNA vaccine platform that produced very similar immune responses to that of live viral infection.	53

DNA Arrays

A whole genome DNA microarray was utilized to elucidate the roles and contributions of monocytes during the host immune system collapse due to *B. anthracis* infection by allowing for the study of lethal toxin (LT) impairment on human peripheral monocytes [30]. The microarray analysis of the mechanisms behind the rapid onset of death in anthrax victims provides the potential for new targets for controlling inflammation and enhancing host defense. Similarly, the Human HGU113 Plus 2.0 whole genome oligonucleotide array was employed for the transcriptional analysis of genes that encode for proteins associated with immune activation in protective antigen-stimulated PBMCs from vaccinated non-human primates [31]. The transcriptional expression patterns further explored with the microarray can be utilized to demonstrate the potential relationship between the development of an innate and adaptive immune response to PA stimulation, providing more information to aid in the evaluation of vaccines against anthrax infection [31].

Also using a whole genome microarray data paired with molecular annotation, a new computational analysis technique was presented to identify infection biomarkers potentially specific to smallpox infection [32]. Using molecular profiles previously collected from cynomolgus macaque PBMCs, which can allow for the identification of pathogen-specific biomarker before clinical symptoms appear, it was concluded that the new method of analysis uncovered a set of genes associated with disease severity and progression independent of the Variola strain or the type of exposure used to infect the animals. The analysis also elucidated the possibility of VARV being able to complete key biological processes by utilizing specific host gene products [32]. The novel discoveries regarding VARV pathogenesis and host response have important implications in vaccine development and efficacy testing, along with the development of prophylaxis and therapeutic countermeasures.

DNA microarrays have been increasingly utilized for the disease characterization of *F. tularensis*. The contribution of IclR, to the intracellular growth, virulence, and gene regulation of human pathogenic *F. tularensis* subspecies was investigated using a cDNA array [33]. Ic1R is a presumed transcriptional regulator that is required for full virulence of the animal pathogen, *F. tularensis* subspecies *novicida* U112. Analysis of the genes identified in the microarray suggested that the majority of genes affected by IclR have differences in sequence between the studied subspecies. PBMCs were once again used to study gene expression patterns in humans infected with the *F. tularensis* live vaccine strain using a whole human genome Affymetrix gene chip [34]. Similarly, oligonucleotide microarrays were used to analyze gene regulation patterns in PBMCs and ultimately the host response after infection with the highly virulent Type A. *F. tularensis subsp. tularensis* [35]. A murine model of infection for *F. tularensis* was utilized to create a whole-genome transcriptional profile with DNA microarrays from infected tissues [36]. The data obtained from the study revealed that *F. tularensis* FabI (FtFabI) inhibitor is a relevant drug target for individuals infected with *F. tularensis* bacterium [36]. A whole-genome NimbleGen microarray was used to elucidate differences in human virulence between *F. novicida*, a rare human pathogen considered to be a subspecies of *F. tularensis* on the basis of DNA similarity [37], and *F. tularensis*. The microarray analyses revealed that elevated cdGMP in *F. novicida* stimulates biofilm formation and inhibits virulence, which may suggest that differences in human virulence between *F. novicida* and *F. tularensis* may be due in part to the absence of cdGMP signaling in *F. tularensis* [38]. Murine DNA microarray technology was utilized by multiple investigators to further elucidate host response mechanisms employed during *F. tularensis* infection [39, 40]. The first study uncovered evidence that epithelium tissue in the lung may play a key role in the host response to infection [39], where the second study discovered that when mice were infected with *F. tularensis* live vaccine strain (LVS), the pathogen employed an oxidative stress upon the target cells and many of the host-defense mechanisms appear to be intended to counteract the stress [40]. The data obtained from these multiple microarray studies identify novel host-response mechanisms potentially associated with *F. tularensis* infection and indicates the use of gene expression profiling for biomarker selection in an active ongoing *F. tularensis* infection which could be used as novel vaccine, prophylactic, and therapeutic drug targets.

Gene profiling for the purpose of identifying key molecular mechanisms during the host response of mice infected with *Y. pestis* was similarly performed using a murine genome microarray [41]. The study uncovered important data related to key biological processes that occur during early stages of *Y. pestis* infection, including 534 significantly altered probe sets pertaining to immune function, apoptosis, cell cycling, protein degradation, and cytoskeletal processes. The resulting data can be used alongside previously recorded data for the development of vaccines and therapeutic drugs in the future.

Human cDNA microarrays were used to study the effects of filovirus infection on host cell gene expression by characterizing the responses in human liver cells infected with EBOV-Zaire (ZEBOV), EBOV-Reston (REBOV), and MARV [42]. The investigators found that Ebola virus and MARV have the ability to modulate host cell gene expression to antagonize the activity of the innate antiviral response and that REBOV has the reduced ability to block expression of interferon-related genes [42]. Also, two experiments were performed, aimed at further illuminating the mechanisms involved in the cellular infiltration of ZEBOV into the lungs and to identify the cells that harbor the ZEBOV antigen in 5-6 week old pigs [43]. A porcine 4x44K microarray was utilized, proving that a series of proinflammatory events occur, immune cells infiltrate into the lungs and lead to severe respiratory distress. These findings provide the background structure to further study the mechanisms of antiviral response evasion caused by certain viral hemorrhagic fevers, leading to new targets for therapeutic interventions.

Protein Arrays

The use of protein microarray technology is becoming increasingly abundant in infectious disease pathogenesis and drug development studies. Detection of antibodies expressed during a diseased state are the main focus of these microarrays, allowing the investigator to better understand host immune responses. Protein array technology was utilized to study host immune response by detecting dominant antigens found in the human serum profiles after being inoculated with Modified Vaccinia virus Ankara (MVA), an attenuated strain of VACV, and Dryvax, a lyophilized calf lymph virus and until recently, the only licensed smallpox virus in the United States [44]. The host response profiles of the human subjects inoculated with MVA and Dryvax were found to be comparable, particularly with respect to responses to membrane proteins that might be expected to mediate virus neutralization. The study was also intended to prioritize antigens for the development of immunoassays (ELISAs) that could be used to monitor vaccination and to produce data that will aid in the development of novel MVA-based smallpox vaccines [44]. Using a protein microarray, it was determined that despite the mutations and deletions in MVA, the modified virus has an immunogenicity that is comparable to that of Dryvax, suggesting that MVA may be a useful vaccination alternative to Dryvax [45]. Similarly, host antibody responses in subjects that received the VACV immunization for smallpox were evaluated using a protein microarray [46]. The investigators focused on generating and characterizing B cell hybridomas in immunized mice. These hybridomas resulted in monoclonal antibodies against a variety of VACV antigens, which could be used to map B cell epitopes in the smallpox vaccine that contribute to immune protection, and could contribute to novel vaccine, prophylactic, and therapeutic development [46]. Human antibody responses to the VACV vaccine were measured with a protein microarray developed to be used as a method for the development of replacement vaccines for use by immunocompromized or low-responding individuals [47]. It was reported that the proteins recognized by antibodies from all vaccinees were less than ten percent of the total VACV proteome. This new microarray technology will potentially allow for a full inventory of VACV proteins required for optimal protection against smallpox, leading to new research to elucidate possible protein-protein interactions, immune responses and other proteome based interactions during VACV infection, providing more information for the development of potentially safer, better defined vaccines.

Continuing to measure antibody responses in vaccinees of the smallpox vaccine, VACV, antibody responses to MPXV infection were compared to human antibody responses after smallpox vaccination by using a protein microarray in order to identity of antigens within the MPXV proteome contributing to immune responses [48]. The resulting data suggested that antibodies from vaccination recognized a small number of proteins shared with MPXV and humoral responses unique to MPXV antigens were involved in recovery from infection with the virus. MPXV, along with VARV, is a poxvirus in the genus orthopoxvirus and is known to cause smallpox-like disease in humans along with non-human primates [48]. Vaccinating with VACV is effective in the protection of a subject against smallpox, but it is also known to be effective in protecting that subject against MPXV infection as well [49]. Studying the proteome of MPXV alongside VACV can provide important clues to formulate better vaccines and drug treatments for those who are infected with, or live in an endemic area for MPXV, further proving protein microarray technology to be a valuable laboratory and public health tool.

The protective immune mechanisms against the bacterium *Y. pestis* have been explored to identify mechanisms to aid in the critical understanding of developing effective vaccines [50]. The humoral and memory cellular immune responses to the F1 and LcrV proteins in subjects with *Y. pestis* infection were investigated using a protein microarray along with an enzyme-linked immunosorbent spot assay (ELISpot) in order to further elucidate the host immune response to *Y. pestis* infection [50]. It was discovered that although antibodies to F1, LcrV and YopD could be detected up to ten years post exposure, specific memory T cell responses to F1 and LcrV could not be detected after four years post infection, illuminating the need for an effective live attenuated *Y. pestis* vaccine. Three separate investigators used protein microarray technology to study host immune responses associated with *F. tularensis* infection [51-53]. Using *F. tularensis* live virus strain (LVS) injected into mice, immunodominant antigens associated with *F. tularensis* infection were identified [51]. They discovered 31 previously unreported immunoreactive antigens, and data suggesting that a killed vaccine can provide protection against highly-virulent strains of *F. tularensis* was generated. Such information that can give insight into the protective immune response will potentially be used in the future by developing studies to elucidate novel host immune responses in regards to the newly discovered antigens. These future studies will potentially have an impact in the design of non-living vaccines for *F. tularensis* and other intracellular pathogens [51]. Similarly, mice challenged with the *F. tularensis* LVS were used to characterize 14 anti-*Francisella* hybridoma antibodies for the potential of the vaccine strain to be used as immunotherapy of tularemia infections [52]. Lastly, protein microarray chips were probed with sera from individuals infected with *F. tularensis* to identify a serodiagnostic antigen set by using antibody responses to protein markers [53].

Novel Arrays

Novel microarray technologies are being developed to be able to provide more in depth information on pathogen host responses and to further elucidate mechanisms and targets for drug therapies. A novel tissue microarray was used to characterize the in-vivo expression profile of CD317, an intrinsic immunity factor that restricts the release of enveloped viruses, in this case LASV, from infected cells in culture [54]. In order to do so, non-transformed tissues were used to provide a framework for future studies of the biological functions associated with CD317 in humans, focusing on its importance as an antiviral factor, along with using CD317 as a target for immunotherapy [54]. Similarly, a novel minimally invasive skin electroporation microarray was developed for the purpose of delivering a synthetic, multivalent, highly concentrated, DNA vaccine to macaques that provides protection from a highly pathogenic MPXV challenge [55]. The result was a significantly improved efficacious DNA vaccine platform that produced immune responses that mimicked live viral infection. The data collected likely has relevance for vaccine design against complex human and animal pathogens, including VACV. Capturing interactions between pathogen and human hosts at the proteomic scale will prove increasingly more important to understanding the emergence of new infectious diseases and for devising methods for medical intervention.

Genotyping and Expression Profiling

Understanding disease-related biology using a systems biology approach has been a key factor in being able to discern gene networks and the cellular pathways that are homologous with disease onset. Delineating detailed maps of these pathways and networks allow for new and better research to be applied towards better elucidating disease pathogenesis in various animal models and in the human along with being able to better understand the epidemiological aspects of the disease causing agent. Increasing molecular epidemiological studies utilizing microarray technology have been performed in recent years to monitor the mutations of category A bioterrorism agent genomes, study the phylogeny of pathogen strains throughout the world, and understand the evolution of the different clades of the specified pathogens (Table 4). The utilization of microarray technology has allowed for the genotyping and expression profiling of multiple category A bioterrorism agents and their closely related pathogen species in order to better understand the phylogeny and transmission history of these agents.

DNA Arrays

A high density comparative genomic hybridization (CGH) microarray was used to clarify the organization of genes within *C. botulinum* type A strains that carried an unexpressed

BoNT/B gene [56]. Gene organization and nucleotide sequences of both the type A and B BoNT gene clusters were analyzed due to the importance of type A *C. botulinum* in human disease and the prevalence of type A strains having unexpressed type B neurotoxin genes [56]. The study contributed to the knowledge pertaining to the origin of BoNT-gene clusters and the evolution of clostridial strains that produce botulinum neurotoxin [56]. Similarly, the use of microarrays for the detection and determination of pathogen biomarkers has been explored for *C. botulinum* strains [57]. ATCC 3502 microarrays based on the *C. botulinum* ATCC 3502 genome were used to analyze 32 Nordic group I *C. botulinum* type B strains isolated from various sources. The analysis revealed that the two predominating Nordic group I *C. botulinum* type B clusters have different resistance to toxic compounds, reflecting variable evolutionary responses to environmental stress. These findings are important for future epidemiologic definitions of *C. botulinum* and infectious forms of human botulism [57].

Table 4. Microarrays used for genotyping and expression profiling of category A bioterrorism agents and the main findings and conclusions from the studies

Microarray used	Category A Agent studied	Main findings and conclusions	Reference
DNA			
CGH	*C. botulinum*	In type A(B) strain NCTC 2916, the neurotoxin gene is of type BoNT/A1 within a gene cluster that has identical organization to that found in BoNT/A2 type strains. This data may be significant in establishing the origin of the BoNT-gene cluster.	54
Whole genome	*C. botulinum*	The two predominating Nordic group I *C. botulinum* type B clusters have different resistance to toxic compounds, reflecting variable evolutionary responses to environmental stress	55
	C. botulinum	Elevated CO2 levels had little effect on the neurotoxin gene expression, growth, or neurotoxin formation of *C. botulinum* ATCC 3502, suggesting that CO2 might not be as effective against foodborne bacteria as once believed.	56
Oligonucleotide	*C. botulinum*	A unique strain from the Southern Hemisphere that produces a novel type E botulinum neurotoxin subtype was profiled, expanding the understanding of the global distribution of *C. botulinum* type E strains. The type E toxin gene cluster may be able to insert into other strains with a more diverse genetic background.	58
MIP	*F. tularensis*	Phylogenetic analyses suggest a North American origin for the studied B-radiation clade and multiple dispersal events between North America and Eurasia, indicating a complex transmission history for *F. tularensis* subsp. holarctica.	57

A whole genome microarray was employed to quantify the effect of carbon dioxide (CO_2) concentration on the growth, relative gene expression and neurotoxin production by *C. botulinum* type A1 strain ATCC 3502 [58]. CO_2 is commonly used in modified atmosphere packaging of food as an antimicrobial. The study found that the elevated CO_2 levels had little effect on the neurotoxin gene expression, growth, or neurotoxin formation of *C. botulinum*

ATCC 3502, suggesting that CO_2 might not be as effective against foodborne bacteria as once believed [58]. Microarray technology has been more recently applied to the epidemiology field, where it is being used to determine pathogen phylogeography and transmission histories. Both Molecular Inversion Probe (MIP) arrays and high-density microarray single-nucleotide polymorphism (SNP) genotyping were utilized to further investigate the phylogenetic structure of *F. tularensis* and to designate new groups within the pathogen [59]. Similarly, focused oligonucleotide DNA microarray technology was used to characterize *C. botulinum* type E strain, a previously unreported isolate found in Argentinian soil in 1995, and the first type E strain originating from the Southern hemisphere [60]. Using DNA microarray technology to expand the understanding of the global distribution of category A agent strains allows for the better understanding of agent mechanisms and their ability to adapt to different environments.

Conclusion

Microarray technologies are a powerful laboratory tool. Possessing the ability to simultaneously detect and identify thousands of microscopic analytes, including genetic material, proteins, and carbohydrates, microarrays have just started opened up a realm of possibilities in the clinical and research world, but the potential power that the technology holds has yet to be realized. With the threat of a potential bioterrorism attack involving one or multiple bioterrorism agents, it is important to have tools that aid in the rapid and accurate detection and diagnosis of these pathogens, along with greater details into the host-agent pathogenesis mechanisms for the development and evaluation of vaccines, therapeutic, and prophylactic drugs. It is clear that microarray technology has proven useful as a laboratory tool when dealing with these dangerous agents, being utilized by many investigators to elucidate many molecular aspects of disease pathogenesis and for the detection of the pathogens or their secondary products. Ideally, looking forward, microarray technology will evolve into a more independent, self-contained, field deployable point of care tool for the purposes of rapid, low-cost and accurate detection and diagnosis of infectious pathogens in various samples that will ultimately impact clinical medicine, epidemiology, and biodefense research.

References

[1] Broadhead ML, Clark JC, Dass CR, Choong PF. Microarray: an instrument for cancer surgeons of the future? *ANZ J Surg* 2010 Jul;80(7-8):531-6.

[2] Villasenor-Park J, Ortega-Loayza AG. Microarray technique, analysis, and applications in dermatology. *J Invest Dermatol* 2013 Apr;133(4):e7.

[3] Miller MB, Tang YW. Basic concepts of microarrays and potential applications in clinical microbiology. *Clin Microbiol Rev* 2009 Oct;22(4):611-33.

[4] Centers for Disease Control and Prevention. Bioterrorism Overview. Centers for Disease Control and Prevention Emergency Preparedness and Response. 2013. Ref Type: Online Source

[5] Morse SS. Public health surveillance and infectious disease detection. *Biosecur Bioterror* 2012 Mar;10(1):6-16.
[6] CDC Emergency Risk Communication Branch (ERCB), Division of Emergency Operations (DEO), Office of Public Health Preparedness and Response (OPHPR). Bioterrorism Agents Fact Sheets. 2013. Ref Type: Online Source
[7] King JW. Ebola Virus. 2013. Ref Type: Online Source
[8] Pathogen Regulation Directorate PHAoC. Machupo Virus: Pathogen Safety Data Sheet. 2010. Ref Type: Online Source
[9] Bolivian hemorrhagic fever. 2013. Ref Type: Online Source
[10] Danzig R. A decade of countering bioterrorism: incremental progress, fundamental failings. *Biosecur Bioterror* 2012 Mar;10(1):49-54.
[11] Goji N, Macmillan T, Amoako KK. A New Generation Microarray for the Simultaneous Detection and Identification of Yersinia pestis and Bacillus anthracis in Food. *J Pathog* 2012;2012:627036.
[12] Lipkin WI. Microbe hunting. *Microbiol Mol Biol Rev* 2010 Sep;74(3):363-77.
[13] Relman DA. Microbial genomics and infectious diseases. *N Engl J Med* 2011 Jul 28;365(4):347-57.
[14] Seidel M, Niessner R. Automated analytical microarrays: a critical review. *Anal Bioanal Chem* 2008 Jul;391(5):1521-44.
[15] Shin GW, Hwang HS, Chung B, Jung GY. Recent developments in CE-based detection methods for food-borne pathogens. *Electrophoresis* 2010 Jul;31(13):2137-53.
[16] Velusamy V, Arshak K, Korostynska O, Oliwa K, Adley C. An overview of foodborne pathogen detection: in the perspective of biosensors. *Biotechnol Adv* 2010 Mar;28(2):232-54.
[17] Raphael BH, Joseph LA, McCroskey LM, Luquez C, Maslanka SE. Detection and differentiation of Clostridium botulinum type A strains using a focused DNA microarray. *Mol Cell Probes* 2010 Jun;24(3):146-53.
[18] Papazisi L, Rasko DA, Ratnayake S, Bock GR, Remortel BG, Appalla L et al. Investigating the genome diversity of B. cereus and evolutionary aspects of B. anthracis emergence. *Genomics* 2011 Jul;98(1):26-39.
[19] Zwick ME, Kiley MP, Stewart AC, Mateczun A, Read TD. Genotyping of Bacillus cereus strains by microarray-based resequencing. *PLoS One* 2008;3(7):e2513.
[20] Ryabinin VA, Shundrin LA, Kostina EB, Laassri M, Chizhikov V, Shchelkunov SN et al. Microarray assay for detection and discrimination of Orthopoxvirus species. *J Med Virol* 2006 Oct;78(10):1325-40.
[21] Taitt CR, Malanoski AP, Lin B, Stenger DA, Ligler FS, Kusterbeck AW et al. Discrimination between biothreat agents and 'near neighbor' species using a resequencing array. *FEMS Immunol Med Microbiol* 2008 Dec;54(3):356-64.
[22] Leski TA, Lin B, Malanoski AP, Wang Z, Long NC, Meador CE et al. Testing and validation of high density resequencing microarray for broad range biothreat agents detection. *PLoS One* 2009;4(8):e6569.
[23] Malhotra S, Yen JY, Honko AN, Garamszegi S, Caballero IS, Johnson JC et al. Transcriptional profiling of the circulating immune response to lassa virus in an aerosol model of exposure. *PLoS Negl Trop Dis* 2013 Apr;7(4):e2171.

[24] Palacios G, Quan PL, Jabado OJ, Conlan S, Hirschberg DL, Liu Y et al. Panmicrobial oligonucleotide array for diagnosis of infectious diseases. *Emerg Infect Dis* 2007 Jan;13(1):73-81.

[25] Feizi T, Fazio F, Chai W, Wong CH. Carbohydrate microarrays - a new set of technologies at the frontiers of glycomics. *Curr Opin Struct Biol* 2003 Oct;13(5):637-45.

[26] Parthasarathy N, Saksena R, Kovac P, Deshazer D, Peacock SJ, Wuthiekanun V et al. Application of carbohydrate microarray technology for the detection of Burkholderia pseudomallei, Bacillus anthracis and Francisella tularensis antibodies. *Carbohydr Res* 2008 Nov 3;343(16):2783-8.

[27] Tao SC, Chen CS, Zhu H. Applications of protein microarray technology. *Comb Chem High Throughput Screen* 2007 Sep;10(8):706-18.

[28] Blais DR, Alvarez-Puebla RA, Bravo-Vasquez JP, Fenniri H, Pezacki JP. Multiplex pathogen detection based on spatially addressable microarrays of barcoded resins. *Biotechnol J* 2008 Jul;3(7):948-53.

[29] Zhang Y, Lou J, Jenko KL, Marks JD, Varnum SM. Simultaneous and sensitive detection of six serotypes of botulinum neurotoxin using enzyme-linked immunosorbent assay-based protein antibody microarrays. *Anal Biochem* 2012 Nov 15;430(2):185-92.

[30] Chauncey KM, Lopez MC, Sidhu G, Szarowicz SE, Baker HV, Quinn C et al. Bacillus anthracis' lethal toxin induces broad transcriptional responses in human peripheral monocytes. *BMC Immunol* 2012;13:33.

[31] Rogers J, Casbohm S, Giannunzio L, Sinnott L, Rust S, Brys A et al. Transcriptional analysis of protective antigen-stimulates PBMC from non-human primates vaccinated with the anthrax vaccine absorbed. *Vaccine* 2006 Jan 25;24(2006):3609-17.

[32] Valdivia-Granda WA, Kann MG, Malaga J. Transcriptional interactions during smallpox infection and identification of early infection biomarkers. *Pac Symp Biocomput* 2007;100-11.

[33] Mortensen BL, Fuller JR, Taft-Benz S, Kijek TM, Miller CN, Huang MT et al. Effects of the putative transcriptional regulator IclR on Francisella tularensis pathogenesis. *Infect Immun* 2010 Dec;78(12):5022-32.

[34] Paranavitana C, Pittman PR, Velauthapillai M, Zelazowska E, Dasilva L. Transcriptional profiling of Francisella tularensis infected peripheral blood mononuclear cells: a predictive tool for tularemia. *FEMS Immunol Med Microbiol* 2008 Oct;54(1):92-103.

[35] Butchar JP, Cremer TJ, Clay CD, Gavrilin MA, Wewers MD, Marsh CB et al. Microarray analysis of human monocytes infected with Francisella tularensis identifies new targets of host response subversion. *PLoS One* 2008;3(8):e2924.

[36] Kingry LC, Cummings JE, Brookman KW, Bommineni GR, Tonge PJ, Slayden RA. The Francisella tularensis FabI enoyl-acyl carrier protein reductase gene is essential to bacterial viability and is expressed during infection. *J Bacteriol* 2013 Jan;195(2):351-8.

[37] McLendon MK, Apicella MA, Allen LA. Francisella tularensis: taxonomy, genetics, and Immunopathogenesis of a potential agent of biowarfare. *Annu Rev Microbiol* 2006;60:167-85.

[38] Zogaj X, Wyatt GC, Klose KE. Cyclic di-GMP stimulates biofilm formation and inhibits virulence of Francisella novicida. *Infect Immun* 2012 Dec;80(12):4239-47.

[39] David J, Gates AJ, Griffiths GD, Lukaszewski RA. Gene expression following low dose inhalational Francisella tularensis (SchuS4) exposure in Balb/c mice and the potential role of the epithelium and cell adhesion. *Microbes Infect* 2012 Apr;14(4):369-79.

[40] Andersson H, Hartmanova B, Ryden P, Noppa L, Naslund L, Sjostedt A. A microarray analysis of the murine macrophage response to infection with Francisella tularensis LVS. *J Med Microbiol* 2006 Aug;55(Pt 8):1023-33.

[41] Rogers J, Choi Y, Giannunzio L, Sabourin P, Bornman D, Blosser E et al. Transcriptional responses in spleens from mice exposed to *Yersinia Pestis* CO92. *Microbial Pathogeneisis* 2007 Apr 12;43(2007):67-77.

[42] Kash JC, Muhlberger E, Carter V, Grosch M, Perwitasari O, Proll SC et al. Global suppression of the host antiviral response by Ebola- and Marburgviruses: increased antagonism of the type I interferon response is associated with enhanced virulence. *J Virol* 2006 Mar;80(6):3009-20.

[43] Nfon CK, Leung A, Smith G, Embury-Hyatt C, Kobinger G, Weingartl HM. Immunopathogenesis of Severe Acute Respiratory Disease in Zaire ebolavirus-Infected Pigs. *PLoS One* 2013;8(4):e61904.

[44] Hermanson G, Chun S, Felgner J, Tan X, Pablo J, Nakajima-Sasaki R et al. Measurement of antibody responses to Modified Vaccinia virus Ankara (MVA) and Dryvax((R)) using proteome microarrays and development of recombinant protein ELISAs. *Vaccine* 2012 Jan 11;30(3):614-25.

[45] Davies DH, Wyatt LS, Newman FK, Earl PL, Chun S, Hernandez JE et al. Antibody profiling by proteome microarray reveals the immunogenicity of the attenuated smallpox vaccine modified vaccinia virus ankara is comparable to that of Dryvax. *J Virol* 2008 Jan;82(2):652-63.

[46] Meng X, Zhong Y, Embry A, Yan B, Lu S, Zhong G et al. Generation and characterization of a large panel of murine monoclonal antibodies against vaccinia virus. *Virology* 2011 Jan 20;409(2):271-9.

[47] Schmid K, Keasey SL, Pittman P, Emerson GL, Meegan J, Tikhonov AP et al. Analysis of the human immune response to vaccinia by use of a novel protein microarray suggests that antibodies recognize less than 10% of the total viral proteome. *Proteomics Clin Appl* 2008 Oct;2(10-11):1528-38.

[48] Keasey S, Pugh C, Tikhonov A, Chen G, Schweitzer B, Nalca A et al. Proteomic basis of the antibody response to monkeypox virus infection examined in cynomolgus macaques and a comparison to human smallpox vaccination. *PLoS One* 2010;5(12):e15547.

[49] Jezek Z, Szczeniowski M, Paluku KM, Mutombo M. Human monkeypox: clinical features of 282 patients. *J Infect Dis* 1987 Aug;156(2):293-8.

[50] Li B, Du C, Zhou L, Bi Y, Wang X, Wen L et al. Humoral and cellular immune responses to Yersinia pestis infection in long-term recovered plague patients. *Clin Vaccine Immunol* 2012 Feb;19(2):228-34.

[51] Eyles JE, Unal B, Hartley MG, Newstead SL, Flick-Smith H, Prior JL et al. Immunodominant Francisella tularensis antigens identified using proteome microarray. *Proteomics* 2007 Jun;7(13):2172-83.

[52] Lu Z, Roche MI, Hui JH, Unal B, Felgner PL, Gulati S et al. Generation and characterization of hybridoma antibodies for immunotherapy of tularemia. *Immunol Lett* 2007 Oct 15;112(2):92-103.

[53] Sundaresh S, Randall A, Unal B, Petersen JM, Belisle JT, Hartley MG et al. From protein microarrays to diagnostic antigen discovery: a study of the pathogen Francisella tularensis. *Bioinformatics* 2007 Jul 1;23(13):i508-i518.

[54] Erikson E, Adam T, Schmidt S, Lehmann-Koch J, Over B, Goffinet C et al. In vivo expression profile of the antiviral restriction factor and tumor-targeting antigen CD317/BST-2/HM1.24/tetherin in humans. *Proc Natl Acad Sci U S A* 2011 Aug 16;108(33):13688-93.

[55] Hirao LA, Draghia-Akli R, Prigge JT, Yang M, Satishchandran A, Wu L et al. Multivalent smallpox DNA vaccine delivered by intradermal electroporation drives protective immunity in nonhuman primates against lethal monkeypox challenge. *J Infect Dis* 2011 Jan 1;203(1):95-102.

[56] Rodriguez JM, Collins MD, East AK. Gene organization and sequence determination of the two botulinum neurotoxin gene clusters in Clostridium botulinum type A(B) strain NCTC 2916. *Curr Microbiol* 1998 Apr;36(4):226-31.

[57] Lindstrom M, Hinderink K, Somervuo P, Kiviniemi K, Nevas M, Chen Y et al. Comparative genomic hybridization analysis of two predominant Nordic group I (proteolytic) Clostridium botulinum type B clusters. *Appl Environ Microbiol* 2009 May;75(9):2643-51.

[58] Artin I, Mason DR, Pin C, Schelin J, Peck MW, Holst E et al. Effects of carbon dioxide on growth of proteolytic Clostridium botulinum, its ability to produce neurotoxin, and its transcriptome. *Appl Environ Microbiol* 2010 Feb;76(4):1168-72.

[59] Vogler AJ, Birdsell D, Price LB, Bowers JR, Beckstrom-Sternberg SM, Auerbach RK et al. Phylogeography of Francisella tularensis: global expansion of a highly fit clone. *J Bacteriol* 2009 Apr;191(8):2474-84.

[60] Raphael BH, Lautenschlager M, Kalb SR, de Jong LI, Frace M, Luquez C et al. Analysis of a unique Clostridium botulinum strain from the Southern hemisphere producing a novel type E botulinum neurotoxin subtype. *BMC Microbiol* 2012;12:245.

In: Microarrays: Principles, Applications and Technologies
Editor: James V. Rogers
ISBN: 978-1-62948-669-7

Chapter 2

Similarity Threshold Selection Tools for Gene Co-Expression Networks

Liliana López-Kleine[*] and Luis Guillermo Leal
Department of Statistics, Universidad Nacional de Colombia,
Sede Bogotá, Colombia

Abstract

Microarray data are often used for the construction of co-expression networks, which, in turn are used for functional prediction and understanding of molecular processes among others. Based on the hypothesis that expression of genes with similar function is highly correlated, a similarity measure is used to construct a gene pair-wise correlation matrix as a first step in gene co-expression network reconstruction. Then, an adjacency matrix is deduced assigning a link between genes on the network when the similarity measure exceeds a previously chosen threshold. Several methods have been proposed in order to choose this threshold objectively, such as methods based on random matrix theory, methods based on topological features of the graph such as the clustering coefficient and methods based on spectral graph theory. These methods assume that random networks and biological networks have identifiable different structures. In the present chapter some of these methods, their usefulness, data-quality dependence and their assumptions will be reviewed. Moreover, their limitations when applied to publicly available plant microarray data and improvements on one method that make it more flexible and less data-quality dependent, will be presented.

Introduction

The study of gene relationships based on gene co-expression and their effect on biological function and phenotype is a focal point in systems biology. Microarray data and more recently RNA-sequencing data are often used for the construction of gene co-expression

[*] Corresponding author (llopezk@unal.edu.co).

networks, which, in turn are used for functional prediction and interpretation of gene relationships.

Based on the hypothesis that the expression of genes with similar function is highly correlated, a similarity measure is used to construct a gene pair-wise correlation matrix as a first step in gene co-expression network reconstruction. In co-expression networks, nodes represent gene products and edges indicate a significant correlation of expression between a gene pair (co-expression).

Many methods for construction of co-expression networks compare gene expression measurements from samples across multiple experimental conditions using a gene expression profile similarity measurement. The most common is the Pearson's correlation coefficient. When the behavior of input data is not linear, mutual information functions (MI) can be calculated to determine the relationships among genes [1]. Once a statistical method has been calculated a gene by gene similarity matrix is obtained. After construction of the similarity matrix, a threshold must be determined to decide whether the relationship is significant and therefore biologically meaningful, or due to noise. Then, based on the threshold, an adjacency matrix is constructed assigning an edge between genes on the network when the similarity measure exceeds the previously chosen threshold. Values in the similarity matrix below the threshold are set to zero and the other values to one. The result is an adjacency matrix in which each value of one represents an edge in the final co-expression network.

The selection of an expression threshold has an effect on conclusions and functional predictions derived from the resulting network. The simplest methodology that has been applied to select a threshold is the use of an arbitrary threshold [2]. Nevertheless, this method is not objective and requires previous biological knowledge. Slightly more objective, but still risky due to the possibility of declaring noisy relationships as significant, is the method of retaining only a top percentage of the strongest relationships [3].

Many authors have proposed and applied methods based on statistics of the pair wise correlation: permutation testing [4], linear regression[5], rankbased methods [6, 7], Fisher's test of homogeneity [8], Partial Correlation and Information Theory (PCIT) [9], Weighted Gene Co-expression Network Analysis (WGCNA) [10–12], or use of additional information in supervised machine learning methods [13]. Other methods, such as the number of maximal cliques, correlation of control spots with expressed genes, Bonferroni correction of p-values, and statistical power have been reviewed and tested by Borate et al. [14]. Several methods based on the adjacency matrix data structure have been proposed in order to choose this threshold objectively.

The most popular methods are based on random matrix theory [15, 16], on topological features of the graph such as the clustering coefficient [17, 18] and also based on spectral graph theory [9]. These methods share the assumption that random networks and biological networks have identifiable and different structures.

In this chapter, we will review several methods for threshold definition and then focus primarily on methods for knowledge independent threshold definition based on the adjacency matrix structure, which are based on the hypothesis that random graphs have different and identifiable structures in comparison with real biological networks. Moreover, we will discuss these assumption and its implications applying RMT and clustering coefficient methods on publicly available gene expression data.

Methods for Knowledge-independent Threshold Definition

Spectral Graph Theory

The number of connected components of a network can be identified using the spectrum of the network [19]. Ding et al. [20] observed that "nearly-disconnected" portions can also be identified by examination of the eigenvector associated with the smallest nonzero eigenvalue of the network, often called the Fiedler vector. The ability to find the nearly-disconnected pieces allows identifying those nodes sharing a cluster in the network.

Perkins et al. [9] propose a method for threshold selection based on spectral graph theory. Eigenvalues and eigenvectors are computed for a transformation of the adjacency matrix of the network constructed at various threshold values. From these transformations (spectra of the graphs), basic spectral clustering methods are used to examine the set of gene-gene relationships and select a threshold dependent upon the community structure of the data. The authors show that the largest components of gene co-expression networks often contain the majority of the network nodes. The exception occurs at high thresholds where the network becomes very sparse. It can be shown that the multiplicity of the zero eigenvalue is equal to the number of connected components in the graph. Therefore, when analyzing only the spectrum of the largest component, the smallest eigenvalue will be equal to zero while the remaining eigenvalues will be nonzero. The smallest nonzero eigenvalue is called the algebraic connectivity of the component. Lower connectivity values indicate the presence of nearly-disconnected components [20]. As the threshold is increased, loosely connected components of the graph will increase. Perkins et al. [9] propose selecting a threshold value that maximizes the number of these components and thus minimizes the number of edges connecting the pieces. This method is based on the assumption that biological networks have subgraphs. Nevertheless, when not enough microarray information is used and an unstructured graph is being constructed, this assumption could not be met [9].

Partial Correlation and Information Theory (PCIT)

Reverter and Chan [12] present the PCIT algorithm for the reconstruction of gene co-expression networks that combines the concept partial correlation coefficient with information theory to identify significant gene to gene associations defining edges in the reconstruction of networks. The properties of PCIT are examined in the context of the topology of the reconstructed network, including connectivity structure, clustering coefficient, and sensitivity. Their results show that PCIT algorithm is more sensitive than established methods and capable of detecting functionally validated gene–gene interactions coming from absolute r values as low as 0.3. These bona fide associations, which often relate to genes with low variation in expression patterns, are beyond the detection limits of conventional fixed-threshold methods, and would be overlooked by studies relying on those methods.

The proposed PCIT algorithm contains two distinct steps as follows:

i. For every trio of genes in x, y and z, the three first-order partial correlation between x and y given z coefficients are computed: $r_{xy,z} = \frac{r_{xy}-r_{xz}r_{yz}}{\sqrt{(1-r_{xz}^2)(1-r_{yz}^2)}}$. It indicates the relationship between x and y that is independent of z.

ii. For every trio of genes, and in order to obtain the local threshold for capturing significant associations, the average ratio (ε) of partial to direct correlation is computed as follows: $\varepsilon = \frac{1}{3}\left(\frac{r_{xy,z}}{r_{xy}} + \frac{r_{xz,y}}{r_{xz}} + \frac{r_{yz,x}}{r_{yz}}\right)$.

iii. Connections between genes are discarded if: $|r_{xy}| \leq |\varepsilon r_{xz}|$ and $|r_{xy}| \leq |\varepsilon r_{yz}|$. Otherwise, the association is defined as significant.

Rather than strong assumptions, this methodology depends highly on the quality of gene expression data and information contained in the other genomic data utilized. The combination of knowledge extraction from different data sources will be useful for model organisms in which a great amount of high-quality genomic data is available.

Cliques

Current approaches to threshold selection are typically statistically based, and do not fully reflect the connectivity of the data [21]. Palla et al. [22] found that a threshold based on clique size was effective at separating networks. A clique is a subgraph in which all nodes are connected to each other [23]. The number of cliques has been investigated by Borate et al. [14] as a method to determine the threshold. The authors calculated the maximal number of cliques at each correlation threshold, starting at $r = 0.99$ with decreasing in steps of 0.01. The number of maximal cliques increased due to greater connections among genes. When clique number increased two or three times the previous value, the resulting correlation was chosen as the threshold.

A disadvantage of using cliques is the computational requirement, which grows exponentially with gene number [14]. This method seems very simple and does not need particular assumptions, but depends on the fact that a structured and complex biological network is constructed. If expression data from a single experiment is used, the co-expression networks may not exhibit such structure; thus, the method could show some difficulties in application.

Random Matrix Theory (RMT)

Initially proposed by Wigner [24] for studying the spectrum of complex nuclei, RMT is a powerful approach for identifying phase transitions associated with disorder and noise in statistical physics and materials science. It has been successfully used for studying the behavior of complex systems and recently for the study of biological systems [15]. RMT based approaches have been tested with multiple correlation metrics (e.g., Pearson's and Spearman's linear correlation coefficient or MI) [15].

RMT predicts that real symmetric matrices have two universal ensembles. Based on the nearest neighbor spacing distribution (NNSD) of eigenvalues, RMT assumes NNSD follows a

Gaussian orthogonal ensemble (GOE) distribution if there exists correlation between nearest neighbor eigenvalues, but follows Poisson ensemble (Poisson distribution) if there is no correlation [25]. The GOE distribution of NNSD is assumed to exist when random or noisy relationships between elements are being represented in the analyzed symmetric matrix [26]. Therefore, deviations from GOE universal prediction can be used to distinguish non-random properties from random noise [26].

The correlation matrix of gene expressions (M) is the combination of the high correlation part Mc (nonrandom part of M), which signifies the correlation of gene expressions specified to changes in biological systems, and the weak correlation part Mr (so-called noise), which expresses random relations between gene expressions [15]: M = Mc + Mr.

Luo et al., [15] based their threshold detection methodology on the hypothesis that the NNSD of M will follow GOE and the NNSD of Mc will follow Poisson distribution. The transition of NNSD between GOE and Poisson distributions can serve as a reference point to distinguish non-random relationships between genes from random noise. This shift can be identified and can then be used as a threshold for choosing real relationships and filtering relationships due to noise. To determine this transition point, RMT tests successively correlation thresholds by beginning at a small initial correlation value and then gradually increasing this threshold, decreasing the number of non-zero values in the similarity matrix. Eigenvalues and the NNSD are determined at each iteration. The NNSD of eigenvalues is determined by sorting the eigenvalues, removing duplicates, and then calculating the differences between each adjacent eigenvalues (spacing distance). To determine the threshold that defines the transition to a Gaussian distribution, a Chi-square test is performed at each successive level in order to test goodness to fit to a Poisson distribution, while taking care not to select a local maximum [15, 16].

In the work of Luo et al. [15], this RMT methodology was applied to data from different organisms: *Shewanella oneidensis*, *Escherichia coli*, yeast, *Arabidipsis thaliana*, *Drosophila*, mouse and human. Their report also predicted the function of unknown genes, which had been verified in the scientific literature and their own experimental validation. Although these results apparently indicate that assumptions and methodology are very general, when we applied the methodology to a specific microarray data set of *A. thaliana,* the distribution of NNSD was not as expected, indicating that RMT threshold definition is not applicable to any microarray data set (see application to real world microarray data below).

Clustering Coefficient

The clustering coefficient *Ci* of a node [27] is defined as $C_i = \frac{2E_i}{k_i(k_i-1)}$, where $k_i(> 1)$ is the number of first-neighbors of node *i* and E_i is the number of edges between first neighbors. A clustering coefficient of one occurs when all the k_i nodes are connected to each other [17]. If the similarity threshold (τ_v) is set at its minimum value of zero, a fully connected network with clustering coefficient $C(\tau_v) = \frac{1}{K}\sum_{k_i>1} C_i$ of one will be obtained. As the cut-off is increased, edges will get eliminated leading to a lowering of the $C(\tau_v)$. However, at a reasonably high similarity threshold, one can expect the $C(\tau_v)$ to start increasing due to the emergence of highly disconnected subnetworks. Gupta *et al.* [17] state that the 'critical' threshold should be set at the highest similarity value below the point at which a sharp increase/decrease in the $C(\tau_v)$ is observed. This is based on the hypothesis that below this threshold, noisy relationships that do not have real biological meaning will be detected.

Elo et al. [18] proposed a method improving that of Gupta et al. [17] to automatically (not graphically) select a threshold that preserves as many valid coexpression links as possible, while simultaneously controlling the number of false detections. The procedure is based on comparing the observed clustering coefficient and its randomized counterpart as the number of connections is gradually decreased. The authors base their method on the hypothesis that when the threshold is raised, the removed links are likely to be noise as long as the difference between the observed clustering coefficient and the clustering coefficient of a random graph increases monotonically. This observation is formulated as a discrete optimization problem, where the critical threshold is found establishing the minimum τ_v in: $\tau^* = \underset{v}{min}\{\tau_v: C(\tau_v) - C_r(\tau_v) > C(\tau_{v+1}) - C_r(\tau_{v+1})\}$ over several thresholds: $\tau_v \in [0.01, 0.99]$ and $\tau_{v+1} = \tau_v + 0.01$. Here $C(\tau_v)$ denotes the clustering coefficient of the coexpression network at threshold τ_v and $C_r(\tau_v)$ that of the random network [18]. The random reference network is constructed at each threshold based on the configuration model [28]. As for the RMT approach presented in the preceding section, this methodology based on clustering coefficients needs the reference network meets the configuration model, which is not the case for all microarray data as we show below.

Application to Real-world Microarray Data

The most objective methods for threshold independent definition that do not highly depend on the data quality, as reported on previous applications, require that assumptions based on graph theory and the particular structure of biological networks are met. The strongest assumptions are required by the RMT and clustering coefficient methods, which are based on the fact that the constructed gene co-expression network differs from a random network and that these differences are identifiable.

In the following section, the RMT and clustering coefficient method are applied to publicly available *A. thaliana* microarray data in order to test the applicability of these methods. For the clustering coefficient method, we propose an improvement in order to make it more flexible and, therefore, suitable for microarray data with different characteristics.

Expression Data and Similarity Matrix

A total of 39 pathogen resistance microarray datasets from *A. thaliana* were obtained from publicly available data[1], pre-processed [29] and filtered based on differential expression [30] to conform an expression matrix (E_{nxp}) of p=560 samples and n=21,122 informative

[1] GEO DataSets repositories were queried for the experiments with accession: GSE12856, GSE13739, GSE14961, GSE15236, GSE16471, GSE16472, GSE16497, GSE17382, GSE17875, GSE19273, GSE20188, GSE21762, GSE21920, GSE26679, GSE26973, GSE28800, GSE431, GSE5513, GSE5752, GSE5753, GSE5754, GSE5755, GSE5756, GSE5757, GSE5758, GSE6176, GSE6831, GSE8319, GSE10426, GSE10713, GSE13390, GSE15880, GSE15881, GSE18757, GSE25838, GSE34081, GSE7990, GSE8877 and GSE31230.

genes. A similarity matrix ($S_{nxn} = \{s_{i,j}\}$) was calculated on expression data using the mutual information coefficient as similarity metric between pairs of genes [31]. The shrink entropy estimator and the global equal width discretization were used in conjunction to calculate the similarities ($s_{i,j}$) [32, 33].

Random Matrix Theory Method Application on Plant Microarray Data

Methodology

Equation 1 was used to calculate a correlation matrix (M_{nxn}). Similarity thresholds (τ_v) were transformed to the range of [0, 1] with 0.01 of step variation as follows.

$$M_{nxn} = [m_{i,j}] \text{ where } m_{i,j} = \begin{cases} s_{i,j} \ if \ s_{i,j} \geq \tau_v \\ 0 \ if \ s_{i,j} < \tau_v \end{cases} \quad \text{(Eq. 1)}$$

For each threshold a correlation matrix and its eigenvalues were calculated. Afterwards, eigenvalues were unfolded using the spectral unfolding technique described by Luo et al. [15]. The NNSD was assessed from the unfolded eigenvalues (Equation 2). NNSD was compared with the Wigner-Dyson distribution (for the GOE) and Poisson distribution (Equations 3 and 4). A Chi-square test with a confidence level of α=0.001 was used to test the goodness of fit to each of these distributions:

$$e_i = \lambda_{i+1}^d - \lambda_i^d \quad \text{(Eq. 2)}$$

$$P_{GOE}(e) \approx \frac{\pi e}{2} exp\left(-\frac{\pi e^2}{4}\right) \quad \text{(Eq. 3)}$$

$$P_{Poisson}(e) = exp(-e) \quad \text{(Eq. 4)}$$

In Equations 2, 3 and 4, e is the level spacing, λ_{i+1}^d and λ_i^d are two nearest unfolded eigenvalues.

Results

Histograms of level spacing (*e*) for each threshold were contrasted with theoretical density functions and a transition from the GOE model was determined. Low similarity thresholds produced eigenvalue fluctuations following the GOE (Figure 1a). As τ_v increases, less correlation between eigenvalues was observed and NNSD diverged from the GOE model (Figure 1b). For high values of τ_v the histogram of level spacing was biased to low values (Figure 1c and d), while the NNSD did not fit the GOE.

In this application, a transition from GOE to Poisson distributions was not observed. NNSD fit the GOE model (Wigner-Dyson distribution) until $\tau_v = 0.81$ when the Chi-square statistic showed a small peak (Figure 2a). However, NNSD did not follow a Poisson distribution for all the evaluated thresholds (Figure 2b). A small decrease in Chi-square values was observed at high thresholds, but not significant to conclude that *e* distributes Poisson. Therefore we conclude that the RMT threshold definition model is also data-dependent and cannot be used for any microarray data set.

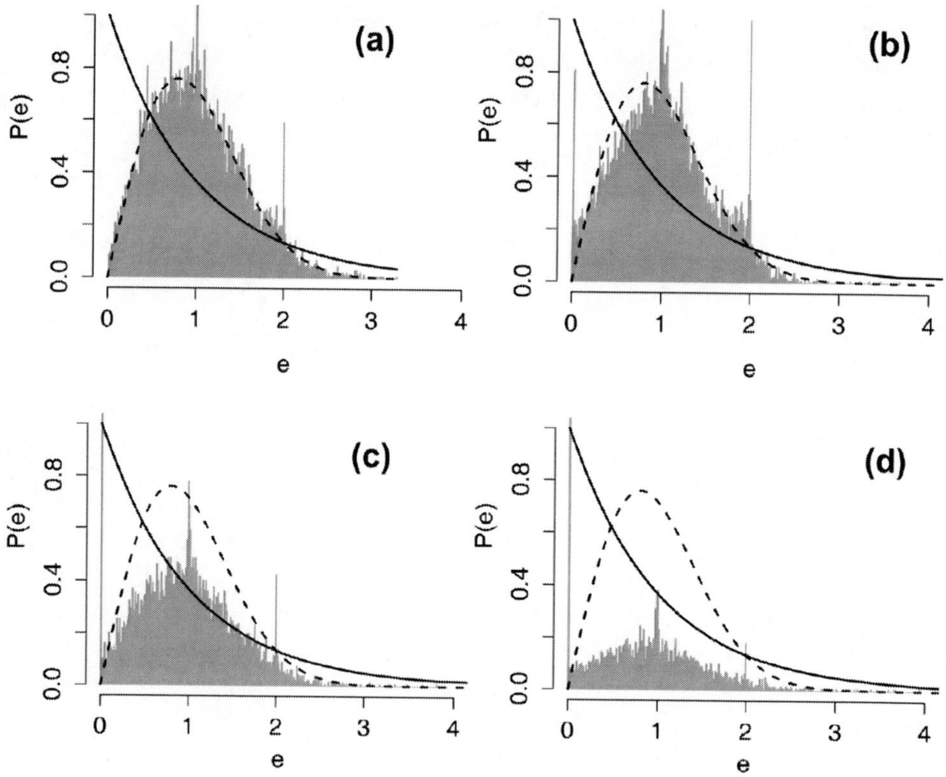

Figure 1. Histograms of level spacing (*e*) compared with Wigner-Dyson (dashed line) and Poisson (straight line) distributions. Four similarity thresholds are shown: (a) 0.60 (b) 0.81 (c) 0.85 (d) 0.89.

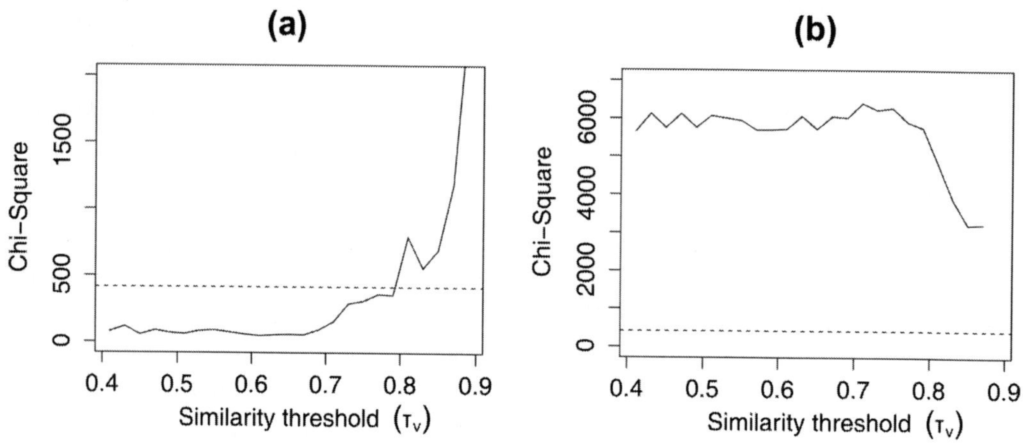

Figure 2. Chi-square statistics from goodness of fit tests that compare NNSD with (a) Wigner-Dyson and (b) Poisson distributions. Dashed horizontal lines indicate the critical value for α=0.05.

Modification of the Clustering Coefficient Method

Methodology

As mentioned above, the network clustering coefficient for a given threshold ($C(\tau_v)$) is compared with a clustering coefficient expected for a random network ($C_r(\tau_v)$) [18]. In this application, we have modified the original optimization problem proposed by Elo et al. [18], and tested two additional rules to find the similarity threshold (τ^*): i) The first local maximum of absolute difference between clustering coefficients (Equation 5). ii) The first local maximum of the variation coefficient of the node degree ($CVK(\tau_v)$) (Equation 6).

$$\tau^* = \min_{v}\{\tau_v : |C(\tau_v) - C_r(\tau_v)| > |C(\tau_{v+1}) - C_r(\tau_{v+1})|\} \quad \text{(Eq. 5)}$$

$$\tau^* = \min_{v}\{\tau_v : CVK(\tau_v) > CVK(\tau_{v+1})\} \quad \text{(Eq. 6)}$$

We implemented the following algorithm to compare the accuracy of both rules:

a) Three groups of 100 simulated networks (SNs) were created and represented by their adjacency matrices using the igraph library [34] in R [35]. Groups I, II and III have the same clustering coefficient ($C(\tau_v) = 0.02$) and degree distribution ($P(k) \sim k^{-2}$), but different variation coefficients $CVK(\tau_v)$ (2.8, 1.0, and 3.1, respectively).
b) A similarity matrix was created for every adjacency matrix.
   i. A similarity array of n^2 similarities ($s_{i,j}$) following a normal distribution was assessed.
   ii. The array was sorted in descending order.
   iii. The first n_E higher similarities were randomly assigned to pairs of nodes connected in the SN. Consequently, a theoretical threshold τ_t^* for the SN was found as the the n_E-*th* element in the array.
   iv. Remaining similarities lower than τ_t^* were assigned randomly to non-connected pairs.
c) An observed threshold τ^* was calculated using each rule (Equations 5 and 6). Curves $|C(\tau_v) - C_r(\tau_v)|$ Vs τ_v and $CVK(\tau_v)$ Vs τ_v were smoothened prior to identification of thresholds.
d) The absolute error between thresholds (η) was used to compare the theoretical and calculated thresholds of each SN, $\tau_{t,i}^*$ and τ_i^* respectively (Equation 7).

$$\eta = \sum_{i=1}^{100} |\tau_i^* - \tau_{t,i}^*| \quad \text{(Eq. 7)}$$

Results

The method presented by Elo et al. [18] and two additional decision rules proposed in this chapter (see Equations 5 and 6) for the similarity threshold definition were compared on the same *A. thaliana* microarray data described above. Figure 3a shows the inequality $C_r(\tau_v) > C(\tau_v)$ when $\tau_v > 0.84$, which means that the *A. thaliana* network constructed on these particular microarray dataset is not as clustered as the realizations of many randomized networks with the same degree distribution. It is also evident that $(C(\tau_v) - C_r(\tau_v)) < 0$; therefore, the decision rule by Elo *et al.* [18] is not suitable for the microarray data analyzed here. In contrast, the absolute difference between clustering coefficients (Figure 3a) and the coefficient of variation (Figure 3b) curves show a constant increase until reaching$\tau_v = 0.91$.

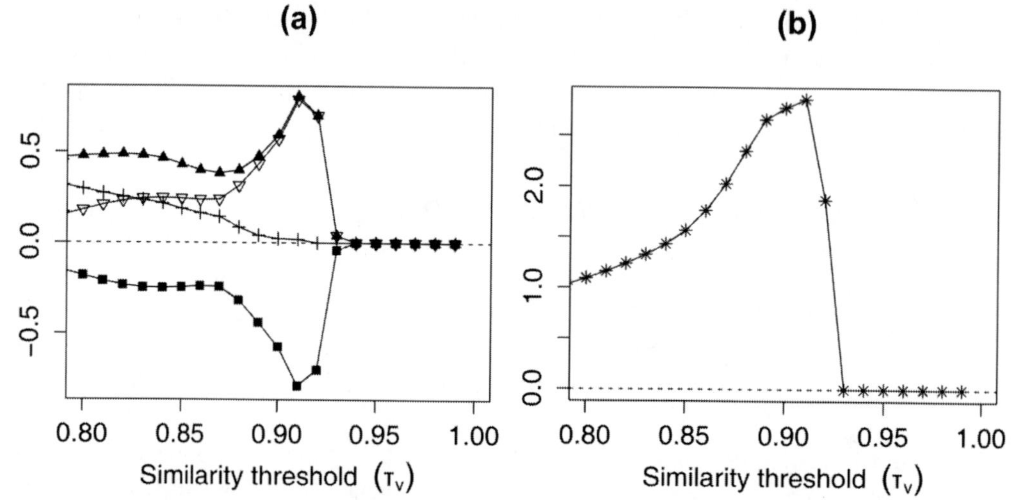

Figure 3. Comparison of methodologies for threshold selection in *A. thaliana* similarity matrix. In plot (a), (▼) indicates the expected clustering coefficient for a random network $C_r(\tau_v)$, (+) indicates the network clustering coefficient $C(\tau_v)$, (■) indicates the $(C(\tau_v) - C_r(\tau_v))$ difference and (▲)the $|C(\tau_v) - C_r(\tau_v)|$ difference. In plot (b), the coefficient of variation of node degree $(CVK(\tau_v))$ is shown.

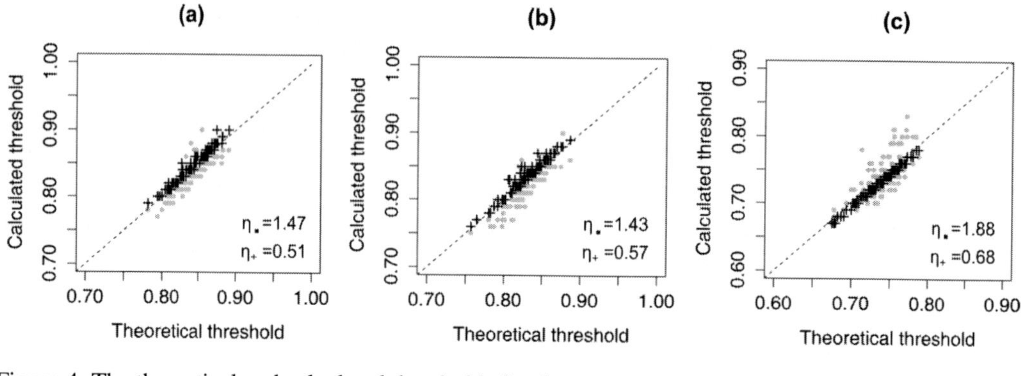

Figure 4. The theoretical and calculated thresholds for three groups of 100 simulated networks. (a) group I, (b) group II, (c) group III. In plots (a, b and c): (•) indicates the thresholds calculated by Equation 5 in $|C(\tau_v) - C_r(\tau_v)|$ Vs τ_v curves. (+) indicates the thresholds calculated by Equation 6 in $CVK(\tau_v)$ Vs τ_v curves. $\eta_{\cdot}$ and η_+ are the absolute error for each group of calculated thresholds (Equation 7).

These decision rules we propose in this chapter showed a precision for similarity threshold selection as shown by simulations (Figure 4). For all the three groups of SNs, the calculated threshold was close to the theoretical threshold. The rule based on coefficients of variation shows the lowest deviations from theoretical thresholds. Furthermore, its performance is adequate for those SNs with higher or lower $CVK(\tau_v)$ than 2.8, which is the $CVK(\tau_v)$ for *A. thaliana* network constructed at $\tau_v = 0.91$ in figure 3b.

Random networks are characterized by high $CVK(\tau_v)$, as Newman (2003) defined $C_r(\tau_v)$ in function of $CVK(\tau_v)$. The opposite situation, low $CVK(\tau_v)$, is more frequent in scale-free networks. We recreated both scenarios in the simulations we undertook. The SNs in group II have the lowest $CVK(\tau_v)$. Higher $CVK(\tau_v)$ are to be found in groups I and III. Consequently, we conclude that both decision rules given in Equations 5 and 6, could be used for networks having different topologies, not only with randomized structures similar to the here constructed *A. thaliana* network, but also with scale-free networks.

Conclusion

In this chapter, we review several methods for knowledge independent threshold selection and expose their usefulness and limitations. The most objective methods such as those based on RMT or clustering coefficient, require the strongest assumptions. Therefore, their usefulness and applicability can be impaired when assumptions are not met. Nevertheless, it is possible to modify these methods in order to adapt them and make them less data-dependent. Here, we propose a modification to the clustering coefficient method that provides a more universal approach regarding assumptions, thereby improving the previously proposed method.

References

[1] Wang H, Wang Q, Li X, Shen B, Ding M, Shen Z. Towards patterns tree of gene coexpression in eukaryotic species. *Bioinformatics* 2008;24:1367–73.

[2] Freeman TC, Goldovsky L, Brosch M, van Dongen S, Mazière P, Grocock RJ et al. Construction, visualisation, and clustering of transcription networks from microarray expression data. *PLoS Comput Biol* 2007;3:2032–42.

[3] Ala U, Piro RM, Grassi E, Damasco C, Silengo L, Oti M et al. Prediction of human disease genes by human-mouse conserved coexpression analysis. *PLoS Comput Biol* 2008;4:17.

[4] Carter SL, Brechbühler CM, Griffin M, Bond AT. Gene co-expression network topology provides a framework for molecular characterization of cellular state. *Bioinformatics* 2004;20:2242–50.

[5] Persson S, Wei H, Milne J, Page GP, Somerville CR. Identification of genes required for cellulose synthesis by regression analysis of public microarray data sets. *Proc Natl Acad Sci U S A* 2005;102:8633–8.

[6] Stuart JM, Segal E, Koller D, Kim SK. A gene-coexpression network for global discovery of conserved genetic modules. *Science* 2003;302:249–55.

[7] Wolfe CJ, Kohane IS, Butte AJ. Systematic survey reveals general applicability of “guilt-by-association” within gene coexpression networks. *BMC Bioinformatics* 2005;6:227.

[8] Nayak RR, Kearns M, Spielman RS, Cheung VG. Coexpression network based on natural variation in human gene expression reveals gene interactions and functions. *Genome Res* 2009;19:1953–62.

[9] Perkins AD, Langston MA. Threshold selection in gene co-expression networks using spectral graph theory techniques. *BMC Bioinformatics* 2009;10:S4.

[10] Barabási AL, Ravasza E, Vicsek T. Deterministic Scale-Free Networks. *Physica A* 2001;64:559.

[11] Barabási A-L, Oltvai ZN. Network biology: understanding the cell's functional organization. *Nat Rev Genet* 2004;5:101–13.

[12] Reverter A, Chan EKF. Combining partial correlation and an information theory approach to the reversed engineering of gene co-expression networks. *Bioinformatics* 2008;24:2491–7.

[13] Puelma T, Gutiérrez RA, Soto A. Discriminative local subspaces in gene expression data for effective gene function prediction. *Bioinformatics* 2012;28:2256–64.

[14] Borate BR, Chesler EJ, Langston MA, Saxton AM, Voy BH. Comparison of threshold selection methods for microarray gene co-expression matrices. *BMC Res Notes* 2009;2:240.

[15] Luo F, Yang Y, Zhong J, Gao H, Khan L, Thompson DK et al. Constructing gene co-expression networks and predicting functions of unknown genes by random matrix theory. *BMC Bioinformatics* 2007;8:299.

[16] Gibson SM, Ficklin SP, Isaacson S, Luo F, Feltus FA, Smith MC. Massive-scale gene co-expression network construction and robustness testing using random matrix theory. *PLoS ONE* 2013;8:e55871.

[17] Gupta A, Maranas CD, Albert R. Elucidation of directionality for co-expressed genes: predicting intra-operon termination sites. *Bioinformatics* 2006;22:7.

[18] Elo LL, Järvenpää H, Oresic M, Lahesmaa R, Aittokallio T. Systematic construction of gene coexpression networks with applications to human T helper cell differentiation process. *Bioinformatics* 2007;23:2096–103.

[19] Chung FRK. Spectral Graph Theory. In: CBMS Regional Conference Series in Mathematics. *Providence: American Mathematical Society,* 1997:1–92.

[20] Ding CHQ, He X, Zha H. A spectral method to separate disconnected and nearly-disconnected web graph components. In: Proceedings of the seventh ACM SIGKDD international conference on knowledge discovery and data mining. San Francisco: *ACM Press,* 2001:275–80.

[21] Butte AJ, Kohane IS. Mutual information relevance networks: functional genomic clustering using pairwise entropy measurements. *Pac Symp Biocomput* 2000;426:418–29.

[22] Palla G, Derényi I, Farkas I, Vicsek T. Uncovering the overlapping community structure of complex networks in nature and society. *Nature* 2005;435:814–8.

[23] Baldwin NE, Chesler EJ, Kirov S, Langston MA, Snoddy JR, Williams RW et al. Computational, integrative, and comparative methods for the elucidation of genetic coexpression networks. *J Biomed Biotechnol* 2005;2005:172–80.

[24] Wigner E. Random matrices in physics. *SIAM Rev Soc Ind Appl Math* 1967;9:1–23.

[25] Mehta ML. Random Matrices. 2nd ed. New York: *Academic Press,* 1990:1–687.

[26] Plerou V, Gopikrishnan P, Rosenow B, Amaral AN, Stanley HE. Universal and nonuniversal properties of cross correlations in financial time series eigenvalue spacings. *Phys Rev Lett* 1999;83:1471–74.

[27] Albert R, Barabasi AL. Statistical mechanics of complex networks. *Rev Mod Phys* 2001;74:1–78.

[28] Newman MEJ. Random graphs as models of networks. In: Bornholdt S, Schuster HG, ed. Handbook of Graphs and Networks. Berlin: *Wiley-VCH*, 2003:35–68.

[29] Irizarry R, Hobbs B, Collin F, Beazer-Barclay YD, Antonellis KJ, Scherf U et al. Exploration, normalization, and summaries of high density oligonucleotide array probe level data. *Biostatistics* 2003;4:249–64.

[30] Tusher VG, Tibshirani R, Chu G. Significance analysis of microarrays applied to the ionizing radiation response. *Proc Natl Acad Sci U S A* 2001;98:5116–21.

[31] Dionisio A, Menezes R, Mendes D. Mutual information: a measure of dependency for nonlinear time series. *Physica A* 2004;344:326–9.

[32] Hausser J. Improving Entropy Estimation and the Inference of Genetic Regulatory Networks. 2006. Available at: http://jean.hausser.org/content/200609061402/main.pdf. Accessed on 9 August 2013

[33] Meyer PE, Lafitte F, Bontempi G. minet: A R/Bioconductor Package for Inferring Large Transcriptional Networks Using Mutual Information. *BMC Bioinformatics* 2008;9:461.

[34] Csardi G, Nepusz T. The igraph software package for complex network research. *Inter Journal* 2006;1:1695.

[35] R Development Core Team R. R: A Language and Environment for Statistical Computing. *Vienna: R Foundation for Statistical Computing*, 2011:1–409.

In: Microarrays: Principles, Applications and Technologies
Editor: James V. Rogers
ISBN: 978-1-62948-669-7

Chapter 3

Mining Gene Expression Microarrays for Long Non-Coding RNAs

L. Zhang,[1] D. R. Wagner[1,2] and Y. Devaux*[*1]
[1]Laboratory of Cardiovascular Research,
Centre de Recherche Public de la Santé, Luxembourg, Luxembourg
[2]Division of Cardiology, Centre Hospitalier, Luxembourg, Luxembourg

Abstract

Microarrays have been widely used in gene expression research. The amount of gene expression data has exponentially increased during the past decade. A great deal of data is stored in public data repositories like Gene Expression Omnibus and ArrayExpress, which can be freely accessed and exploited. With the development of high throughput techniques such as next generation sequencing, the number of long non-coding RNAs (lncRNAs) is highly increasing, while the number of protein-coding RNAs remains stable. LncRNAs play regulatory roles at almost every stage of gene expression from epigenetic modification to regulation of post-transcriptional process. Some lncRNAs are aberrantly expressed in disease states like cancer, Alzheimer's disease, and coronary artery disease. However, the research on lncRNAs is limited by the paucity of specific analytical tools. Some lncRNAs have been found on commonly used gene expression microarrays. A computational and manual re-annotation pipeline was developed to identify lncRNAs in traditional gene expression microarrays. Using this pipeline, one can get insights into the pathophysiologic function of lncRNAs. The approach motivates mining public and private microarray gene expression datasets to investigate the function of lncRNAs in disease development and progression. Ultimately, this may lead to the discovery of novel therapeutic targets.

[*] Correspondence to Yvan Devaux, Laboratory of Cardiovascular Research, Centre de Recherche Public de la Santé, 84 rue Val Fleuri L-1526, Luxembourg, Luxembourg; Tel +352 26970300; Fax +352 26970396; Email: yvan.devaux@crp-sante.lu.

Introduction

With the development of high-throughput sequencing techniques, it has been observed that only a small proportion of mammalian genomes encode protein-coding RNAs. The number of non-coding RNAs is increasing exponentially. So far, most of the attention from the research community has been devoted to short regulatory non-coding RNAs, e.g., microRNAs.

Another group of non-coding RNAs, longer than 200 nucleotides, has recently been identified by Kapranov et al. [1] and termed long non-coding RNAs (lncRNAs). LncRNAs act at almost every stage of gene expression, from epigenetic modification to post-transcriptional regulation [2-4]. It is therefore plausible that lncRNAs may be involved in developmental processes and disease progression.

This has been demonstrated for cancer, and lncRNAs may also serve as biomarkers in certain types of cancer [5, 6]. The majority of lncRNAs were uncovered in genome-wide association studies. For instance, 8 single nucleotide polymorphisms in a lncRNA called antisense noncoding RNA in the INK4 locus (ANRIL) have been associated with susceptibility to coronary artery disease [7]. Genetic variants in another lncRNA called myocardial infarction-associated transcript (MIAT) conferred susceptibility to myocardial infarction [8]. Analytical tools to characterize lncRNA expression profiles would speed up the biomedical research on the functional roles of this novel class of non-coding RNAs.

Whole genome expression microarrays have been widely used in the past decade to associate gene expression signatures with disease states. A huge amount of expression data is either stored in servers of research laboratories or available from public repositories. But, to date, these microarray platforms have been mostly used to study protein-coding genes. As a result of genome sequencing, it appeared that only a small proportion (~2%) of mammalian genes encodes protein-coding RNAs, whereas a much greater proportion is transcribed into non-coding RNAs [9, 10].

The rapid update of genomic information stored in public databases allowed the re-annotation for lncRNAs of the probes spotted on gene expression microarrays [11]. Gene expression platforms such as Affymetrix have been used to investigate lncRNAs in neuroscience [12]. Confident that there should be much more information on lncRNAs hidden behind whole genome microarray platforms, an analytical pipeline has been developed to identify all lncRNAs represented in these platforms. In this chapter, we describe the pipeline and explain how to combine gene expression microarrays with curated lncRNA sequences to obtain insights into the expression of lncRNAs. Furthermore, we provide some basic principles to predict the function of lncRNAs.

LncRNAs in Public Databases

As a relatively new group of transcripts, there is still no specialized and comprehensive database for long non-coding RNAs. LncRNAdb is the only database dedicated to lncRNAs that have been shown to have or to be associated with biological functions in eukaryotes [13]. Of note, lncRNAdb also contains mRNA with regulatory roles [13].

The database currently holds over 150 sequences from about 60 species, among which 126 sequences are from human and 94 from mouse. Each manually curated RNA in the lncRNAdb has reference information on sequence, genomic location, structure, expression patterns, subcellular localization, conservation, and function when known.

Obviously, the quantity of RNA sequences in lncRNAdb is far from enough to re-annotate thousands of probes on microarrays. Thus, other nucleotide databases are needed. Two widely-used databases, RefSeq [14] and Ensembl [15] contain large-scale transcriptomic sequence data. RefSeq (Reference Sequence) collection provides a comprehensive, integrated, non-redundant, well-annotated set of sequences, including genomic DNA, transcripts, and proteins.

The categories of sequences in RefSeq can be distinguished by the prefix of their accession number. For example, according to RefSeq definition for accession format (http://www.ncbi.nlm.nih.gov/projects/RefSeq/key.html), prefix NM means mature messenger RNA (mRNA) transcript; NR and XR prefix mean non-coding RNA. Of note, XR prefix indicates that the transcript has not been individually reviewed, or has been provided by a genome annotation process.

Furthermore, ribosome RNA, microRNA, small nucleolar RNA and small nuclear RNA are annotated in RefSeq database and they can be distinguished from lncRNAs having a regulatory function. Ensembl is a computing system which produces and maintains automatic annotation on selected eukaryotic genomes. Apart from protein-coding RNAs, the Ensembl gene set also includes automatically-annotated non-coding RNAs (ncRNAs) (http://www.ensembl.org/info/docs/genebuild/ncrna.html): tRNA (transfer RNA), Mt-tRNA (transfer RNA located in the mitochondrial genome), rRNA (ribosomal RNA), scRNA (small cytoplasmic RNA), snRNA (small nuclear RNA), snoRNA (small nucleolar RNA), miRNA (microRNA precursor), misc_RNA (miscellaneous other RNA) and lincRNA (long intergenic non-coding RNA). According to the annotation, lncRNAs with regulatory function can be extracted from Ensembl gene set.

Moreover, several large international consortia are working on the annotation of human and mouse transcriptome, such as H-InvDB (H-Invitational Database) and FANTOM (Functional Annotation of the mouse).

The H-InvDB extensively analyses all human transcripts and provides curated annotations of human genes and transcripts that include gene structure, alternative splicing variants, non-coding functional RNAs, and protein functions. [16]. In this database, a flag "predicted coding potential" allows clarifying RNA transcripts into "protein coding" and "non-protein-coding" groups. The FANTOM consortium aims at providing the ultimate characterization of mouse transcriptome, with more than 35,000 non-coding transcripts identified [17]. FANTOM3 ncRNA ID list can be freely downloaded from ftp://fantom.gsc.riken.jp/FANTOM3/noncoding/. Table 1 shows the number of human and mouse lncRNAs (>200nt) contained in each of the above cited databases.

Table 1. Number of lncRNAs in public repertoires

	lncRNAdb	RefSeq (release 59)	Ensemble ncRNA (release 71)	H-InvDB	FANTOM3
Human	126	5,389	11,226	386	
Mouse	94	2,861	4,165		22,664
URL	http://lncrnadb.com/	ftp://ftp.ncbi.nlm.nih.gov/refseq/	http://www.ensembl.org/info/data/ftp/index.html	http://h-invitational.jp/hinv/dataset/download.cgi	ftp://fantom.gsc.riken.jp/FANTOM3/annotation/

Microarray Platforms

Set up in 1997 [18], the first genome-wide expression platform has been widely used to associate gene expression signatures with disease states. More recently, the NCode™ ncRNA expression arrays have been developed and allow simultaneous profiling more than 10000 lncRNA transcripts and 25,179 protein-coding transcripts (human). In addition, lncRNAs have been indicated by manufacturers on the latest mature commercial microarray platforms. For example, Agilent SurePrint G3 Mouse GE Array contains more than 16,000 lncRNAs and Affymetrix GeneChip® Mouse Gene ST Array has about 2,000 lncRNA transcripts. It has to be noted that in older platforms, such as the Affymetrix U133 microarray platform, some lncRNA sequences are fortuitously represented. Using such platform, Michelhaugh et al. [12] were able to show that lncRNA expression profiles are altered in the nucleus accumbens of heroin abusers. Typically, probe annotation and design is based on different source databases. The rapid update of genomic information has greatly improved microarray probe annotation and mapping. As an illustration to this point, the latest Affymetrix gene expression microarray (GeneChip® Human Gene 2.0 ST Array) was based on the 51st version of RefSeq while the current version is 59th. Dai et al. [19] indicated that a regular update of probe annotation may lead to severe changes in data interpretation. Furthermore, following annotation of ABA (Allen Brain Atlas) probes [20], Mercer et al. [21] identified 849 ncRNAs expressed in the adult mouse brain. Similarly, re-annotation of the probes in the GNF (Genomics Institute of the Novartis Research Foundation) Gene Expression Atlas data, Pang et al. [22] found over 1,000 ncRNAs expressed in human and mouse CD8+ T cells. Moreover, commercial microarrays still have some inaccuracies in probe annotation For example, about 11% of the probes of the Affymetrix GeneChip U95A are non-specific and 9% do not find perfect match in the genome [23]. Therefore, it is now apparent that previously designed gene expression microarrays can be useful for lncRNA research. Also, gene expression data stored in public data repositories like Gene Expression Omnibus [24] and ArrayExpress [25] can be freely accessed and exploited. To date, more than 28,000 series of gene expression profiles have been stored in GEO database. These public microarray repositories constitute a rich of data to study the involvement of lncRNAs in pathophysiology.

Probe Re-annotation Strategy

The core principle behind microarrays is hybridization between two nucleotide sequences. Consequently, the principal process of sequence re-annotation is alignment. Blast (Basic Local Alignment Search Tool) is one of the most popular bioinformatics tools to compare biological sequences [26, 27]. The recent packages Blast+ can be downloaded and performed in user-selected databases [28]. During the blast, the probe sequences provided by gene expression microarray manufacturers are aligned to long non-coding transcript sequences from one or several public databases (Table 1). Only probes having 100% request homology to a transcript are retained. Then, probes matching a lncRNA are aligned to protein-coding sequences and perfect matched probes are eliminated. Some probes recognise both coding and non-coding RNAs. Since non-coding RNAs are generally expressed at a lower level than coding RNAs [29], it is important to discard probes matching common parts of coding and non-coding RNA sequences (Figure 1). There are two options for probe re-annotation: *de novo* probe re-annotation and re-annotation of discriminative probes.

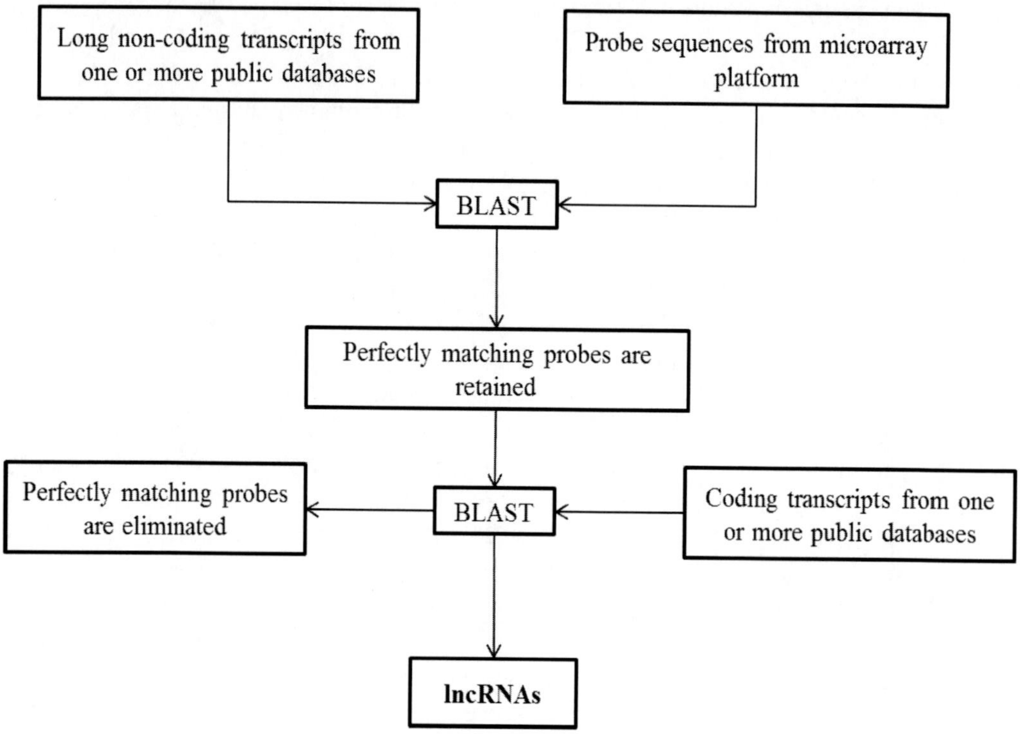

Figure 1. Microarray probe re-annotation strategy.

De Novo Probe Re-annotation

Microarray images quantification and feature annotation depend on the annotation files provided by manufacturers. These files contain information on feature position, accession number, gene symbol, and functions. Editable annotation files can be found in manufacturer's web site. For example, using the make cdfenv R package (CDF Environment Maker http://www.bioconductor.org/packages/2.12/bioc/html/makecdfenv.html), new CDF file of Affymetrix arrays could be created with the re-annotated probes. LncRNAs are easily visualized directly in analysis output. With this option, users can obtain lncRNAs expression profiles from their own microarray data or microarray data downloaded from public databases. Interestingly, this allows constructing the co-expression network of coding-non-coding genes [30], i.e., analyzing in a concomitant manner non-coding genes and protein coding genes, this is critical while analyzing the function that lncRNAs play in the regulation of protein gene expression.

Re-annotation for Discriminative Probes

Since gene dysregulation can be the source of disease development, microarrays have been mostly used to identify differentially expressed genes between two conditions, e.g., diseased vs. control. Whole probe re-annotation is time-consuming and in some cases, microarray raw data are not published in any database and only a list of differentially expressed (discriminative) probe ID is presented. Thus, the second and rapid option aims at re-annotating only discriminative microarray probes. The sequences of discriminative probes are extracted from original FASTA file provided by manufacturer, and these sequences are aligned to selected lncRNA database.

Prediction of lncRNA Function

The multiple functions of lncRNAs are far from being well-characterized and constitute an active field of investigation [31]. Here, we describe that microarrays can be used to gain insights into the role of lncRNAs in gene regulation.

Coding-Non-Coding Gene Co-Expression Network

Genes within a co-expressed module commonly have similar functions, so mining modules in a co-expression network are an efficient approach for predicting gene function [32]. Microarray based co-expression networks have generally been constructed with protein or protein coding genes. Liao et al. [30] re-annotated the expression profiles of all coding and non-coding genes expressed in the widely-used commercial mouse array (Affymetrix Mouse Genome 430 2.0 Array), and constructed a coding-non-coding gene co-expression network. Using the Markov cluster algorithm for module searching, 1,695 modules were found and, among these, 550 contained both coding and long non-coding genes. Sixty-two modules were significantly enriched for at least one gene ontology biological process term, suggesting that they have functional roles. In addition, these modules contained 218 lncRNAs functionally annotated by the most significantly enriched biological process. This study showing the presence of both coding and non-coding RNAs in the same modules suggests that these 2 types of RNAs regulate similar biological functions.

Co-Localization of Coding and Non-Coding RNAs

LncRNAs are often closely associated with protein-coding genes in the genome, which allows an easy regulation of the expression of the neighboring coding genes [33]. Ponjavic et al. [34] showed that lncRNAs expressed in the developing brain are preferentially located adjacent to protein-coding genes. Also, after analyzing data from Affymetrix Mouse 430 2.0 array, Liao et al. [30] observed that 4,105 of the 4,571 lncRNAs represented in the arrays are localized within 100 kb of any of 14,861 protein coding genes. Although the co-localization of coding and non-coding genes may evolve to have a common local regulatory architecture, this organization may also reflect the functions that many lncRNAs have in chromatin modification and in the expression of neighboring genes. Upstream lncRNAs may overlap the promoter regions of coding genes and may regulate their expression at the transcriptional level. For instance, down-regulation of the lncRNA, intergenic 10, reduced the expression of its bilateral flanking coding genes, Fank1 and Adam12 [35]. Using loss-of-function experiments, we have observed that inhibition of the lncRNA 2900055J20Rik could reduce the expression of its downstream coding gene Kctd16, encoding a potassium channel (unpublished data).Of note, more than half of mammalian coding genes have complementary non-coding antisense transcription, which is also accompanied by overlapping, intronic and bidirectional non-coding transcription [36]. For instance, the lncRNA Tsix is transcribed antisense to and initiates silencing of the expression of the other lncRNA Xist. Xist is one of the first lncRNAs identified as a silencer of the c chromosomes in females to ensure the proper amount of gene activity [37]. Thus, integrating expression and location data of both lncRNAs and coding genes in the same microarrays is an asset to study the function of lncRNAs in gene regulation.

Conclusion

Since the description of the first DNA microarray in 1995 [38], microarrays have been intensively used to analyze transcripts, microRNAs, copy number, and promoters. Consequently, a great amount of microarray data is stored in freely accessible public repositories. The two major repositories, Gene Expression Omnibus and ArrayExpress, contain each more than 28,000 microarray datasets. Although these datasets have been originally generated as an output of protein-coding RNA expression profiles, they contain an un-exploited quantity of information.

With the development of high-throughput techniques such as next generation sequencing, an increasing number of RNA transcripts have been generated. The abundance of lncRNAs in the human genome was revealed in 2005 [17], and initially lncRNAs were believed to be non-functional. However, an increasing number of studies have shown that lncRNAs fulfil regulatory roles in many different ways, such as chromatin modification [3, 40], transcription factor co-actors [41], and RNA interference [42]. Although the lncRNAdb database is a useful resource to study the biological function of lncRNAs, analytical tools for lncRNA research are still limited. Mapping probes of different microarray platforms on different public databases has shown promise to a) identify lncRNAs in the microarray probes that were inaccurately annotated; b) generate co-expression network with both non-coding and coding genes from the same microarrays; c) investigate lncRNA functions using the co-localization of non-coding and coding genes in the genome.

To summarize, mining gene expression microarrays is a promising strategy to investigate lncRNAs in different organisms. Since the amount of gene expression data obtained by microarrays is enormous and most often publicly-available, this approach may foster the elucidation of the pathophysiological roles of lncRNA in diseases.

References

[1] Kapranov P, Cheng J, Dike S, Nix DA, Duttagupta R, Willingham AT et al. RNA maps reveal new RNA classes and a possible function for pervasive transcription. *Science* 2007, 316(5830):1484-1488.

[2] Carrieri C, Cimatti L, Biagioli M, Beugnet A, Zucchelli S, Fedele S et al. Long non-coding antisense RNA controls Uchl1 translation through an embedded SINEB2 repeat. *Nature* 2012, 491(7424):454-457.

[3] Lai F, Orom UA, Cesaroni M, Beringer M, Taatjes DJ, Blobel GA et al. Activating RNAs associate with Mediator to enhance chromatin architecture and transcription. *Nature* 2013, 494(7438):497-501.

[4] Mercer TR, Mattick JS. Structure and function of long noncoding RNAs in epigenetic regulation. *Nat Struct Mol Biol* 2013, 20(3):300-307.

[5] Hung T, Wang Y, Lin MF, Koegel AK, Kotake Y, Grant GD et al. Extensive and coordinated transcription of noncoding RNAs within cell-cycle promoters. *Nat Genet* 2011, 43(7):621-629.

[6] Wapinski O, Chang HY: Long noncoding RNAs and human disease. *Trends Cell Biol* 2011, 21(6):354-361.

[7] Helgadottir A, Thorleifsson G, Manolescu A, Gretarsdottir S, Blondal T, Jonasdottir A et al. A common variant on chromosome 9p21 affects the risk of myocardial infarction. *Science* 2007, 316(5830):1491-1493.

[8] Ishii N, Ozaki K, Sato H, Mizuno H, Saito S, Takahashi A et al. Identification of a novel non-coding RNA, MIAT, that confers risk of myocardial infarction. *J Hum Genet* 2006, 51(12):1087-1099.

[9] Guttman M, Amit I, Garber M, French C, Lin MF, Feldser D et al. Chromatin signature reveals over a thousand highly conserved large non-coding RNAs in mammals. *Nature* 2009, 458(7235):223-227.

[10] Wilusz JE, Sunwoo H, Spector DL. Long noncoding RNAs. functional surprises from the RNA world. *Genes Dev* 2009, 23(13):1494-1504.

[11] Liao Q, Xiao H, Bu D, Xie C, Miao R, Luo H et al. ncFANs: a web server for functional annotation of long non-coding RNAs. *Nucleic Acids Res* 2011, 39:W118-124.

[12] Michelhaugh SK, Lipovich L, Blythe J, Jia H, Kapatos G, Bannon MJ. Mining Affymetrix microarray data for long non-coding RNAs. altered expression in the nucleus accumbens of heroin abusers. *J Neurochem* 2011, 116(3):459-466.

[13] Amaral PP, Clark MB, Gascoigne DK, Dinger ME, Mattick JS. lncRNAdb: a reference database for long noncoding RNAs. *Nucleic Acids Res* 2011, 39:D146-151.

[14] Pruitt KD, Tatusova T, Maglott DR. NCBI Reference Sequence (RefSeq): a curated non-redundant sequence database of genomes, transcripts and proteins. *Nucleic Acids Res* 2005, 33:D501-504.

[15] Hubbard T, Andrews D, Caccamo M, Cameron G, Chen Y, Clamp M et al. Ensembl 2005. *Nucleic Acids Res* 2005, 33:D447-453.

[16] Yamasaki C, Murakami K, Fujii Y, Sato Y, Harada E, Takeda J et al. The H-Invitational Database (H-InvDB), a comprehensive annotation resource for human genes and transcripts. *Nucleic Acids Res* 2008, 36:D793-799.

[17] Carninci P, Kasukawa T, Katayama S, Gough J, Frith MC, Maeda N et al. The transcriptional landscape of the mammalian genome. *Science* 2005, 309(5740):1559-1563.

[18] Lashkari DA, DeRisi JL, McCusker JH, Namath AF, Gentile C, Hwang SY et al. Yeast microarrays for genome wide parallel genetic and gene expression analysis. *Proc Natl Acad Sci USA* 1997, 94(24):13057-13062.

[19] Dai M, Wang P, Boyd AD, Kostov G, Athey B, Jones EG et al. Evolving gene/transcript definitions significantly alter the interpretation of GeneChip data. *Nucleic Acids Res* 2005, 33(20):e175.

[20] Lein ES, Hawrylycz MJ, Ao N, Ayres M, Bensinger A, Bernard A et al. Genome-wide atlas of gene expression in the adult mouse brain. *Nature* 2007, 445(7124):168-176.

[21] Mercer TR, Dinger ME, Sunkin SM, Mehler MF, Mattick JS. Specific expression of long noncoding RNAs in the mouse brain. *Proc Natl Acad Sci USA* 2008, 105(2):716-721.

[22] Pang KC, Dinger ME, Mercer TR, Malquori L, Grimmond SM, Chen W et al. Genome-wide identification of long noncoding RNAs in CD8+ T cells. *J Immunol* 2009, 182(12):7738-7748.

[23] Zhang J, Finney RP, Clifford RJ, Derr LK, Buetow KH. Detecting false expression signals in high-density oligonucleotide arrays by an in silico approach. *Genomics* 2005, 85(3):297-308.

[24] Barrett T, Wilhite SE, Ledoux P, Evangelista C, Kim IF, Tomashevsky M et al. NCBI GEO: archive for functional genomics data sets--update. *Nucleic Acids Res* 2013, 41:D991-995.

[25] Rustici G, Kolesnikov N, Brandizi M, Burdett T, Dylag M, Emam I et al. ArrayExpress update--trends in database growth and links to data analysis tools. *Nucleic Acids Res* 2013, 41:D987-990.

[26] Altschul SF, Gish W, Miller W, Myers EW, Lipman DJ. Basic local alignment search tool. *J Mol Biol* 1990, 215(3):403-410.

[27] Altschul SF, Madden TL, Schaffer AA, Zhang J, Zhang Z, Miller W et al. Gapped BLAST and PSI-BLAST: a new generation of protein database search programs. *Nucleic Acids Res* 1997, 25(17):3389-3402.

[28] Camacho C, Coulouris G, Avagyan V, Ma N, Papadopoulos J, Bealer K et al. BLAST+: architecture and applications. *BMC Bioinformatics* 2009, 10:421.

[29] Mattick JS, Makunin IV. Non-coding RNA. *Hum Mol Genet* 2006, 15:R17-29.

[30] Liao Q, Liu C, Yuan X, Kang S, Miao R, Xiao H et al. Large-scale prediction of long non-coding RNA functions in a coding-non-coding gene co-expression network. *Nucleic Acids Res* 2011, 39(9):3864-3878.

[31] Wahlestedt C. Targeting long non-coding RNA to therapeutically upregulate gene expression. *Nat Rev Drug Discov* 2013, 12(6):433-446.

[32] Pu S, Vlasblom J, Emili A, Greenblatt J, Wodak SJ. Identifying functional modules in the physical interactome of Saccharomyces cerevisiae. *Proteomics* 2007, 7(6):944-960.

[33] Mercer TR, Dinger ME, Mattick JS. Long non-coding RNAs. insights into functions. *Nat Rev Genet* 2009, 10(3):155-159.

[34] Ponjavic J, Oliver PL, Lunter G, Ponting CP. Genomic and transcriptional co-localization of protein-coding and long non-coding RNA pairs in the developing brain. *PLoS Genet* 2009, 5(8):e1000617.

[35] Mondal T, Rasmussen M, Pandey GK, Isaksson A, Kanduri C. Characterization of the RNA content of chromatin. *Genome Res* 2010, 20(7):899-907.

[36] Katayama S, Tomaru Y, Kasukawa T, Waki K, Nakanishi M, Nakamura M et al. Antisense transcription in the mammalian transcriptome. *Science* 2005, 309(5740):1564-1566.

[37] Ohhata T, Hoki Y, Sasaki H, Sado T. Crucial role of antisense transcription across the Xist promoter in Tsix-mediated Xist chromatin modification. *Development* 2008, 135(2):227-235.

[38] Schena M, Shalon D, Davis RW, Brown PO. Quantitative monitoring of gene expression patterns with a complementary DNA microarray. *Science* 1995, 270(5235):467-470.

[39] Roberts KG, Morin RD, Zhang J, Hirst M, Zhao Y, Su X et al. Genetic alterations activating kinase and cytokine receptor signaling in high-risk acute lymphoblastic leukemia. *Cancer Cell* 2012, 22(2):153-166.

[40] Khalil AM, Guttman M, Huarte M, Garber M, Raj A, Rivea Morales D et al. Many human large intergenic noncoding RNAs associate with chromatin-modifying complexes and affect gene expression. *Proc Natl Acad Sci USA* 2009, 106(28):11667-11672.

[41] Feng J, Bi C, Clark BS, Mady R, Shah P, Kohtz JD. The Evf-2 noncoding RNA is transcribed from the Dlx-5/6 ultraconserved region and functions as a Dlx-2 transcriptional coactivator. *Genes Dev* 2006, 20(11):1470-1484.

[42] Ogawa Y, Sun BK, Lee JT. Intersection of the RNA interference and X-inactivation pathways. *Science* 2008, 320(5881):1336-1341.

In: Microarrays: Principles, Applications and Technologies
Editor: James V. Rogers
ISBN: 978-1-62948-669-7

Chapter 4

An *In Situ* Electroporation and Flow Device for Mechanotransduction Studies

Zoltán Kis[1], Leila Towhidi[1], Henry Ip[1, 2], Emmanuel Drakakis[1], Anil Bharath[1] and Rob Krams[1,*]

[1] Department of Bioengineering, Faculty of Engineering, Imperial College London, London, UK

[2] Department of Computing, Faculty of Engineering, Imperial College London, London, UK

Abstract

The ability of cells to receive and respond to chemical and physical signals from beyond the plasma membrane is fundamental to life. Sensing and processing of stimuli (such as hormones, mechanical factors, light, heat, neurotransmitters, nutrients, odorants, etc.) is mediated by signal-transduction pathways, molecular circuits that detect, amplify, and integrate diverse external signals to govern functions such as cellular survival, proliferation, differentiation, and apoptosis.

These information-processing circuits that control intra- and inter- cellular activities at a molecular level are extremely widespread and complex. For example, approximately half of the 25 largest protein families encoded by the human genome, deal primarily with information processing. Characterizing signalling pathways would not only offer a better understanding of life at molecular level, but would also allow the treatment of diseases.

To characterize these extremely complex signalling pathways, high throughput systems are required that can add functions (gain-of-function) and delete functions (loss-of-function) from primary cells by inserting cDNA and/or siRNA, respectively. Furthermore, in order to mimic physiologically relevant conditions, such as blood flow, cells need to be exposed to biomechanical stimulus.

Here, we describe the design and experimental verification of a device for applying biomechanical stimulus in the form of fluid shear stress and electroporation transfection

[*] Correspondence (r.krams@imperial.ac.uk).

of primary adherent cells. To that end, primary cells are seeded in an incubation chamber on a coated conductive glass slide allowing in situ electroporation to be taken place. Electroporation with our device on hard-to-transfect endothelial cells yields an siRNA transfection efficiency of 70% while keeping cell viability above 95%. Next, the microfluidics functionality of the platform was proven by deleting the inherent flow alignment function of endothelial cells. For this, primary endothelial cells were electroporated with siRNA molecules (siKLF-2) that down-regulate the KLF-2 gene (indispensable for flow alignment) and cells were exposed to physiological levels (15-20 dyne/cm^2) of shear stress for 24 hours with our device. A quantitative image analysis of microscopic images showed the loss of cell alignment in endothelial cells. Furthermore, quantification of mRNA by qPCR reveals a 52% KLF-2 down-regulation in siKLF-2 electroporated sheared cells compared to non-electroporated sheared cells.

In conclusion, our novel electroporator yielded good transfection efficiencies for hard-to-transfect seeded primary cardiac cells, being able to deliver siRNAs that deleted the flow alignment function from primary endothelial cells. The electroporator was shown to be flow-compatible, opening avenues for signalling pathways studies in cardiac cells under physiologically relevant mechanical and flow conditions as well as under complex chemical environment.

Introduction

Signal Transduction

The ability of cells to receive and respond to external chemical and physical signals is fundamental to all three domains of life (Archaea, Bacteria and Eukaryota). Chemical signals consist of molecules or atoms, both with electrical charge (ions) and without electrical charge, which either bind to transmembrane proteins embedded in the plasma membrane or enter the intracellular space by directly traversing the plasma membrane. In the first case, the binding of extracellular molecules or atoms (extracellular ligands) triggers a conformational change of the transmembrane proteins, which can subsequently activate a cascade of protein interactions within the cell. This intracellular signal transduction couples to gene expression regulators and eventually gene activity is modified, by up- or down-regulating gene expression. Upon ligand binding, some transmembrane proteins can also fold into a channel conformation, allowing atoms and molecules to pass through this channel, in addition to other molecules that diffuse through the lipid bilayer. In the second case, when signaling molecules enter the intracellular space by traversing the plasma membrane (either by passing the lipid bilayer or transmembrane protein channels), chemical signals are directly transduced to intracellular proteins that directly or indirectly (through other intermediates) modify gene activity. Thus, chemical signals are mediated by signal-transduction pathways which modify gene expression to govern cellular functions.

Similarly, physical signals can also couple to molecular circuits that detect, amplify and integrate diverse external signals and modify gene expression patterns. The physical signals that cells sense and respond to include temperature, electromagnetic radiation (e.g., visible light), electrical potential, magnetism, and mechanical factors (pressure, force, strain and shear stress). The important role of mechanical factors on cellular and molecular processes is increasingly being recognized, and has led to the emerging field of mechanobiology. Mechanical stimulus is mediated by signal-transduction pathways to control functions such as cellular cell division, gene expression, cell migration, morphogenesis, cell adhesion, fluid homeostasis, ion channel gating, vesicular transport [1-3], and a variety of diseases, including hypertension and atherosclerosis [4].

Bio-mechanical Factors and Atherosclerosis

The crucial importance of identifying (mechanical) factors in the development and progression of atherosclerosis is of particular interest due to the extremely high human mortality rates caused by this condition. Atherosclerosis is, by far, the most important contributor to cardiovascular diseases, and cardiovascular diseases are the leading cause of death globally [5-7]. Atherosclerosis occurs in blood vessels (arteries) that transport oxygenated blood and nutrients (including lipids) from the heart towards the tissues and organs of the body, and causes sclerosis (or hardening) of the arterial wall, hence its name. Atherosclerosis is a lipid driven, chronic inflammatory disease leading to plaque formation and through plaque rupture, can cause heart attack, stroke, gangrene and loss of function in the extremities [8]. The progression of atherosclerosis has been associated with certain risk factors, including: hypercholesterolemia, hypertension, diabetes, obesity, cigarette smoking, inflammatory biomarkers, and others [9]; these risk factors also guide the diagnosis and treatment of the disease. In spite of the fact that the above mentioned risk factors are omnipresent in the entire arterial system, atherosclerosis evolves at well-defined sites such as curved vessel segments, bifurcation and side branching points [10-12].

It has been postulated that disturbances of the blood flow pattern at or near these atherosclerosis predilection sites play a crucial role in the initiation and progression of atherosclerosis [13-15]. While early studies [16, 17] focused on the role of blood flow in the initiation of the disease, recent studies also suggests an important role of blood flow in advanced stages of atherosclerosis [18], the formation of vulnerable plaques [18-21], in-stent restenosis [22, 23], and inflammation and plaque formation [21, 24, 25]. As a consequence, the theory relating bloody flow patterns with disease formation is of great interest to clinical diagnosis with a number of clinical validation studies being carried out [18, 24]. However, despite compelling evidence of blood flow playing a key role in all stages of atherosclerotic diseases, the exact interaction between the above mentioned risk factors and blood flow patterns is largely unknown.

The field of mechanobiology studies blood flow quantitatively, while bringing together engineering knowledge with cellular and molecular biology. The interior of blood vessels (including arteries) are lined with a layer of endothelial cells, which are in direct contact with blood. The flow of blood exerts a friction force on the endothelial surface, acting in the plane of the endothelium, and its magnitude scales with the velocity of the blood. This force acting per unit area is called shear stress (also referred to as hemodynamic shear stress), and it is measured in dyne/cm^2 or Newton/cm^2 (N/cm^2) or Pascal (Pa). Endothelial shear stress values in healthy arteries range from 10-70 dyne/cm^2 [26]; corresponding to 1-7 Pa.

The Complexity of Cardiovascular Mechanotransduction

At a molecular level, numerous molecules and cellular microdomains from the basal (at the interface of the endothelium with the rest of the arterial wall), junctional (between neighboring endothelial cells) and luminal (at the blood-endothelium interface) surfaces of endothelial cells have been found to detect and respond to shear stress. These include: ion channels [27], receptor tyrosine kinases [28], the apical glycocalyx [29-31], primary cilia [32], heterotrimeric G proteins and G protein coupled receptors [33], xanthine oxidase [34], growth factor receptors [35], nicotinamide adenine dinucleotide phosphate (NADPH) oxidase [36], caveolae [27, 37], cell membrane phospholipid bilayer [38], platelet EC-adhesion molecule (PECAM)-1 [39], vascular endothelial (VE)-cadherin [40], focal adhesion sites [26, 41], and the cytoskeleton [13, 43-47]. The understanding of the dynamics of individual mechanosensors is currently emerging, but their integration, and consequently, the mechanism by which endothelial cells sense hemodynamic shear stress is not fully understood. After stimulation by shear stress an (unknown) combination of these receptors is activated, which then activate 6-8 downstream signaling pathways, eventually leading to changes in gene expression patterns. These 6-8 mechano-transduction pathways regulate 8 acknowledged transcription factors, which leads to the activation of a total of ~2,000 genes in response to shear stress [14, 15, 24, 48, 49].

The immense complexity that results from the interactions of these high number of gene products (proteins, small interfering RNAs, microRNA) prohibits an intuitive, and coherent understanding of the processes involved. Mathematical modeling is often used to study these large networks, and computational modeling has indeed been able to predict new signaling pathway dynamics [50, 51]. Also, new emerging properties have been identified for several shear-sensitive pathways, these new properties include oscillatory behavior (IP3-pathway, MAPK-pathway, NF-κB-pathway), bi-stability (MAPK-pathway) and memory (MAPK-pathway). However, despite these early successes, the failure of computational modeling methods to accurately represent reality became apparent. One of the most important reasons for this failure in computational modeling is the high dimensionality, complexity and inter-connectedness of the signaling pathways which are not fully mapped out. In addition, there is a lack of sufficient experimental data for modeling and model validation. For instance, the MAPK-pathway has been modeled with >100 coupled ordinary differential equations including more than hundreds of kinetic parameters. The actual values of these parameters are often derived from several unrelated studies based on different cell types and/or different organisms making the prediction of these models rather unrealistic. Other theoretical methods, such as parameter estimation techniques, suffer from finding non-unique solutions in this high-dimensional parameter space and cannot be solved due to the lack of quantitative information of kinetic constants. The understanding, modeling and prediction of behaviors of mechano-transduction signaling pathways under mechanical stimulation is proven to be even more difficult due to relatively scarce and qualitative data available in the literature. Without a systematic way of generating a large amount of experimental data for characterizing mechanotransduction signaling pathways, there is little hope for unraveling the cause of atherosclerosis risk factor advancement at molecular level, and using this information to aid clinical decision that will eventually lead to disease prevention and improved patient recovery.

Device for Cardiovascular Mechanotransduction Studies

RNA Interference and Currently-available Electroporators

The solution that we propose to fully characterize these complex cardiovascular mechano-transduction pathways, typically containing thousands of interacting elements, is the development of high-throughput systems that can add functions (gain-of-function) or inhibit functions (loss-of-function) in cardiac cells by inserting cDNA or siRNA, respectively. At the same time, our platform should also be able to mimic physiologically relevant conditions, such as the exposure to shear stress in our *in vitro* cell culture. Thus, our platform consists of a device capable of transfecting primary cardiac cells and a microfluidics flow channel for exposing cells to controlled levels of shear stress. For high-throughput, mammalian gene expression loss-of-function data collection, RNA interference could be employed. RNA interference is a powerful and reliable tool for studies into these signaling pathways, and involves the insertion (also referred to as transfection) of short (20-25 bp) double stranded small interfering RNA (siRNA) molecules into cultured mammalian cells. Once inside the cells, these siRNA molecules take advantage of the cell's natural machinery to effectively knock down expression of the target genes (at the mRNA level) to which that particular siRNA anneals to due to Watson-Crick base pairs complementarity. Our platform would also allow monitoring gene interactions using a reversed engineering approach. For this, gene encoding cDNA will be inserted into endothelial cells, and these cDNAs will be silenced by inserting siRNAs into the same cells.

Transfection of primary cells is notoriously difficult, but recent reports indicate that this can be reliably achieved with membrane electroporation [52]. Electroporation may be defined as the application of electric pulses causing temporary pores on the cell membrane allowing membrane impermeable substances (e.g., DNA, RNA, proteins) to enter the cell. Electroporation was chosen because it is simple, high-throughput, adaptable, and gives good transfection yields for a variety of primary cell types. In order to achieve spatial and temporal control of the transfection process, primary endothelial cells were electroporated in their natural surface adherent state, also minimizing the disruption of their cell division cycle during detachment while exposed to trypsin [53-55]. To functionally characterize gene networks in cardiac cells, physiologically complex mechanical and chemical environment should be mimicked. Hence, the incorporation of an electroporator in a flow channel seems an efficient way to carry out mechano-sensitive pathway studies effectively. Out of the existing adherent mammalian cell electroporation techniques (summarized in Table 1), none is readily compatible with systems that expose cardiac cells to physiologically-relevant mechanical and chemical conditions. Thus, we have developed a platform that allows electroporation of hard-to-transfect adherent primary cardiac cells as well as the incorporation of a microfluidics device that recreates physiologically relevant fluid flow and chemical conditions. The construction and operation of the device relies on commercially-available components; the protocols presented here are thus readily reproducible by interested researchers. The operation of the device is validated by flow experiments following the electroporation of a siRNA knock down of a flow-response target gene (Krüppel-like factor 2, KLF-2) [56] in primary

Porcine Aortic Endothelial Cells (PAEC). Phenotypes of the cell culture as well as qPCR results confirm the successful knock down of the target gene.

Table 1. Summary of electroporation techniques for adherent mammalian cells

No.	Cell type	Flow compatibility	High-throughput	Transfection efficiency+	Survival rate+	Reference
1	Cell lines	No	No	-	-	[57]
2	Cell lines	No	No	-	-	[58]
3	Cell lines	No	No	-	-	[59]
4	Cell lines	No	No	-	-	[60]
5	Cell lines and primary aortic smooth muscle cells	No	No	43%	60%	[61]
6	Cell lines	No	No	-	-	[62]
7	Cell lines and progenitor hippocampus cells	No	No	Not specified	Not specified	[63]
8	Cell lines and hippocampal cultures	No	No	92%	92%	[64]
9	Primary neurons and glia	No	No	92%	Not specified	[65]
10	Cell lines and HUVEC	No	No	13-36%	Greater than 31%	[66]
11	Cell lines and HUVEC	No	No	Not specified	Not specified	[67]
12	Cell lines and primary neurons and embryonic fibroblasts	No	No	25%,40%	90%	[68]
13	Cell lines	No	No	-	-	[69]
14	Cell lines	No	No	-	-	[70]
15	Cell lines	No	No	-	-	[71]
16	Cell lines	No	No	-	-	[72]
17	Cell lines	No	No	-	-	[73]
18	Cell lines	No	No	-	-	[74]
19	Cell lines	No	No	-	-	[75]
20	Cell lines	No	Yes	-	-	[76]
21	Cell lines	No	No	-	-	[77]
22	Cell lines and primary neurons	No	No	Not specified	Not specified	[78]
23	Cell lines and primary macrophages	No	No	99% (PI)&	93%	[79]
24	Primary chondrocytes	No	No	30%, nearly 100%	Not specified	[80]
25	Cell lines and HUVEC	No	Yes	Not specified	Not specified	[81]
26	Cell lines and primary macrophages	No	Yes	60%	Not specified	[82]

+ Transfection efficiencies and survival rates are shown for primary cells only.
& Primary mouse macrophageswere electroporated with propidium iodide (PI).

Design and Operation of the Electroporation-flow Device

An overview of the platform is shown in Figure 1. A glass slide coated with indium tin oxide (ITO) was fitted inside a 90 mm diameter circular Petri dish, to aid the sterilization/incubation process.

The ITO coated glass slide was electrically conductive and optically transparent, allowing electroporation and subsequent visualization by visible light and fluorescent microscopy. Furthermore, ITO is very durable, inert and non-toxic to cells, promoting excellent cell adhesion and growth [57]. The corners of the glass slide were soldered to electrically conductive wires that connect to the negative outlet of the electrical pulse generator.

Subsequently, walls of small culturing chambers were created, consisting of a double layer of electronically isolating PCR tape that was stacked on top of the glass slide. The size of the electroporation/flow channel that was cut out in the PCR tape, was 0.5 cm X 5 cm with a height of 500 μm. For electroporation, a gold plated solid aluminum block was connected to the positive outlet of the electric pulse generator, and was placed on top of the cell monolayer creating a chamber with a height of 500 μm. This way, the positive electrode was in direct contact with the electrically isolating plate through electroporation solution that covered the cells, (Figure 1).

Electrical pulses for electroporation were generated using an Agilent 33220A 20MHz Function/Arbitrary Waveform Generator programmed with Visual Basic scripts. Delivered pulses were monitored with an Agilent DSO3062A Digital Storage Oscilloscope. To facilitate convenient optimization for electroporation parameters, multiple smaller wells with a size of 0.5 cm X 0.5 cm and a height of 500 μm, were used. After the electrical pulse was delivered, the top metal electrode and wires were removed, and cells were examined under the microscope.

Next, a flow channel was created by placing a bottomless self-adhesive flow chamber from Ibidi on top of the electroporation/flow chamber (Figure 1C). Cells can be subjected to shear stress at any time between 4 and 24 hours after electroporation of the confluent adherent cells.

Flow channels were connected to each other, and to a Cobe precision peristaltic blood pump using platinum-cured silicone tubing to carry out multiple experiments simultaneously. The Cobe precision blood pump was used to suck liquid from the flow channels, through a medium container, where the medium is recycled (Figure 1C). The pump generated pulsatile flow, and the flow rate was gradually increased during the initial two hours of the experiment up to a flow rate of 45-50 mL/min, which yielded a shear stress of 15-20 dyne/cm^2 in the 50 x 5 x 0.75 mm flow channel. This physiological shear stress was maintained for 24 hours at 37°C in humidified air that contained 5% CO_2. After the experiment, cell phenotypes were observed under the Leica DM IL LED–DFC290 phase contrast microscope using the 10X objective.

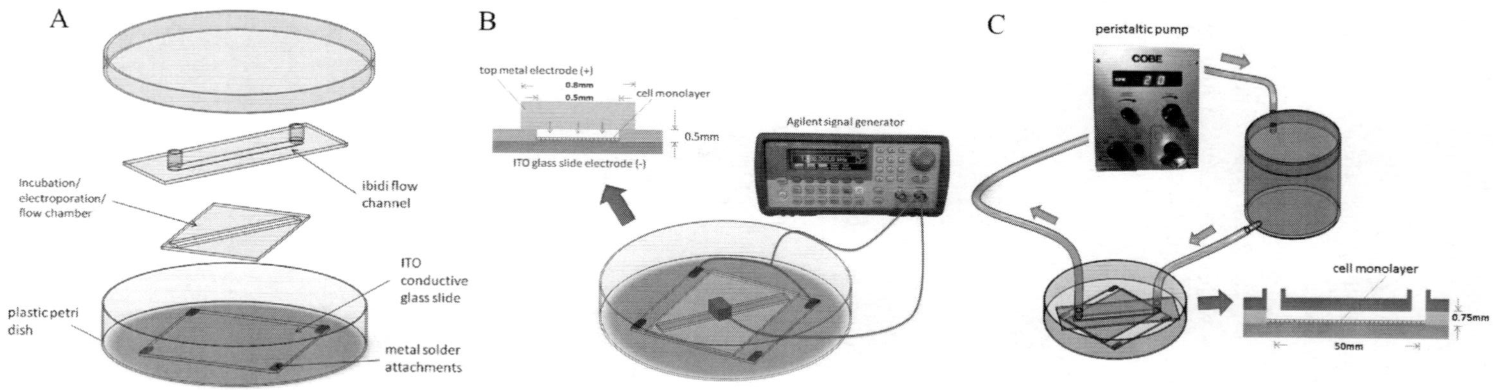

Figure 1. Scheme of the electroporation-flow device. A. Construction of the device. First an indium tin oxide (ITO) coated transparent and electrically conductive glass slide was placed in a Petri dish, and metal wires were soldered to metal connectors glued to corners of the glass slide, using electrically conductive glue. Next, the cell accommodating chamber was created by placing adhesive PCR tape double layer on top of the glass slide. After sterilization with UV light and coating with fibronectin, cells were seeded, left to reach confluence and electroporated in the chamber. Flow chambers were created by placing a bottomless flow channel on top of the cell containing chamber. B. Electroporation procedure. The glass slide on which cells adhered was connected to the negative outlet of the electrical pulse generator by wires. A gold plated solid aluminum block was wired to the positive outlet of the same pulse generator. The metal block came on top of the adherent cell monolayer in contact with the electroporation solution, and the pulse was delivered. C. Flow experiment procedure. Following electroporation the wires and the top electrode were removed, and a bottomless self-adhesive ibidi flow channel was placed on top of the cell containing well. The obtained flow chamber was connected by tubes to the culture medium reservoir and the peristaltic pump, facilitating the suction of liquid from the reservoir through the flow chamber into the pump, and pushing the liquid back into the reservoir.

In situ Electroporation of Adherent Primary Cells with Fluorescently-labeled siRNAs

Before knock-down effects on mechanotransduction pathways can be investigated, the electroporation parameters must be optimized for the target cell type. For the optimization of the electroporation protocol, electroporation wells (contained in a Petri dish) were sterilized with 254 nm UV light for 4 hours, and each of the 0.5 X 0.5 cm wells were coated with 15 μL of fibronectin solution (1 mg/mL) by incubation at 37°C for one hour. Cells were seeded in each well at an initial density of 38,000 cells/cm^2, and 12-24 hours after seeding the confluent adhering cell monolayer in each of these wells were washed with PBS to remove debris and dead cells. 15 μL fluorescent siRNA (1-20 μM) dissolved in PBS is pipetted into each electroporation well after which an electrical pulse train was applied between the top gold plated electrode (+) and conductive glass substrate (-). The following fluorescent siRNAs were used: AllStars Negative Control siRNA Alexa Fluor 488 conjugate from Qiagen; Silencer® Cy™3 labeled Negative Control No. 1 siRNA and Silencer® FAM labeled Negative Control No. 1 siRNA from Invitrogen. After electroporation with fluorescently-labeled siRNA, cells were visualized with either a Leica TCS SP5 DMI6000 confocal microscope or a **Zeiss Axiovert 200 inverted fluorescent microscope.**

A series of electroporation parameters and conditions were systematically changed, aiming to increase transfection efficiencies while keeping cell death to a minimum for primary PAECs. We have used a fluorescently-labeled siRNA (Alexa Fluor 488) that does not target any known DNA or RNA sequence in mammalian cells for this purpose. We determined that the optimal concentration of the fluorescently-labeled siRNA was 5 μM for transfection. We also observed that transfection efficiencies appeared to be significantly higher when the bottom glass slide acted as the negative electrode and the top metal electrode was positive (configuration I) compared to the reverse case when the glass slide was positive and the top metal was the negative electrode (configuration II). Using configuration I, a variety of square electrical pulses were systematically tested (i.e., single, double, triple, and multiple pulses), modifying the pulse duration from 1 to 1000 milliseconds (ms), modifying the electrical potential difference (or voltage) from 1 to 12 Volts (V), and also varying the number of pulses from 1 to 100. Pulses used during this optimization phase are represented in Figure 2. Out of these pulses, we found that a 3V pulse of 5 ms duration yielded maximal cell viability and transfection efficiency.

Transfection results produced using the 3V pulse of 5 ms are shown in Figure 3. Approximately 70% of cells electroporated with green fluorescently-labeled siRNA (Alexa Fluor 488) appeared green upon excitation with ~488 nm wavelength light. Viability of PAECs after electroporation was assessed using the trypan blue dye exclusion test as described before [83]. For this procedure, cells in each of the 0.5 X 0.5 cm wells were washed twice with PBS pre-warmed to 37°C. Subsequently, 15 μL of 0.4% (w/v) trypan blue solution was added to cells, and incubated 3 minutes at room temperature. Hereafter, cells in each well were washed twice with PBS pre-warmed to 37°C, and visualized under a Leica DM IL LED–DFC 290 phase contrast microscope. Non-viable cells were stained in blue. Post-electroporation cellular viability was also determined using propidium iodide (PI) [84]. Cells with 0.5 μg/mL of PI in culture media were incubated at 37°C for 10-30 minutes, and visualized under a Zeiss Axiovert 200 fluorescent microscope. Non-viable cells are red fluorescent due to propidium iodide uptake. Cell viability (100 minus number of red fluorescent cells divided by total number of cells present after electroporation) was >95%.

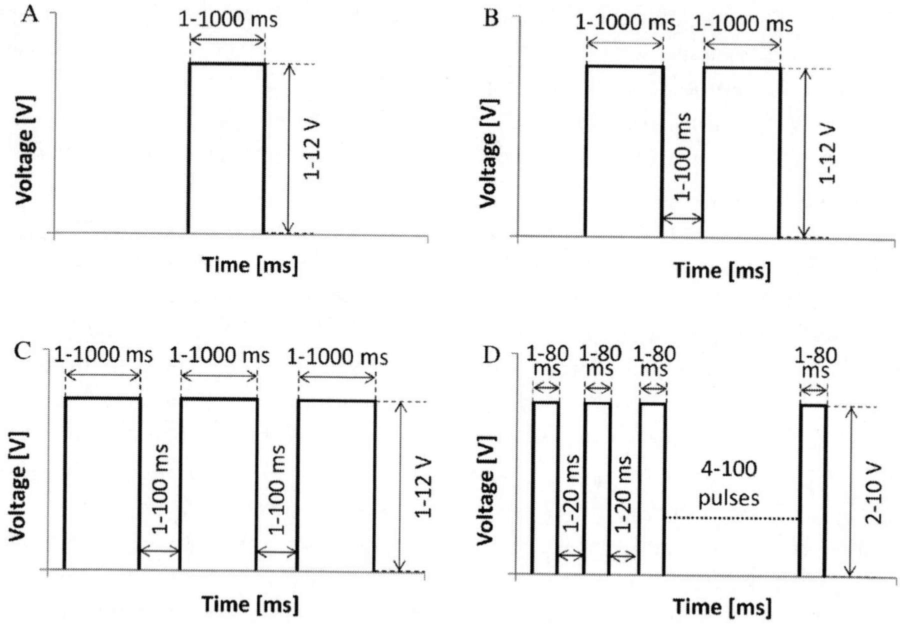

Figure 2. Electrical pulses applied for optimizing the electroporation parameters. A. Single pulses with magnitudes between 1 and 12 Volts (V), and with durations of 1 ms to 1 s. B. Double pulses of 1 to 12 V lasting between 1 ms and 1 s, with a gap of 1 to 100 ms in between pulses. C. Triple pulses with magnitudes between 1 and 12 V, with duration of 1 ms to 1 s and with pauses of 1 to 100 ms in between pulses. D. Multiple (4 to 100) pulses of 2 to 10 V with a duration of 1 to 80 ms and with gaps of 1 to 20 ms in between pulses. Out of these pulses the 3 V 5 ms pulses gave the highest transfection efficiencies with high cell viabilities.

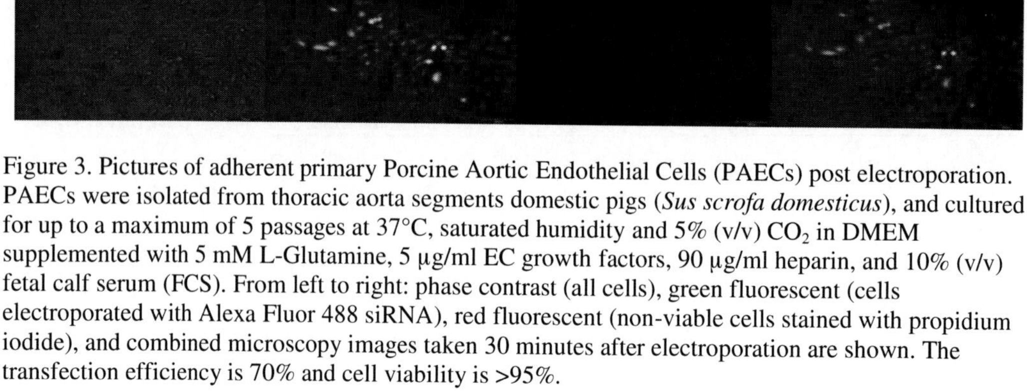

Figure 3. Pictures of adherent primary Porcine Aortic Endothelial Cells (PAECs) post electroporation. PAECs were isolated from thoracic aorta segments domestic pigs (*Sus scrofa domesticus*), and cultured for up to a maximum of 5 passages at 37°C, saturated humidity and 5% (v/v) CO_2 in DMEM supplemented with 5 mM L-Glutamine, 5 μg/ml EC growth factors, 90 μg/ml heparin, and 10% (v/v) fetal calf serum (FCS). From left to right: phase contrast (all cells), green fluorescent (cells electroporated with Alexa Fluor 488 siRNA), red fluorescent (non-viable cells stained with propidium iodide), and combined microscopy images taken 30 minutes after electroporation are shown. The transfection efficiency is 70% and cell viability is >95%.

Primary Cell Mechanotransduction Pathway Studies

The aim of our device is to facilitate the study of mechanotransduction pathways in primary cells. The study of mechanotransduction pathways is carried out in the framework of RNA interference. Therefore, it must provide reliable siRNA transfection through electroporation, facilitate the application of well controlled shear stress on seeded cells, and allow convenient readout, such as microscopy. For the electroporation/flow devices, the incubation chambers (0.5mm X 0.5cm X 5cm) were coated with 150 μL fibronectin solution (1 mg/mL) for cell adhesion. After reaching confluency, cells were washed with PBS immediately before the adherent monolayer was electroporated with 150 μL siRNA (33 μM) solution that silences the expression of Krüppel-like factor 2. For the application of shear stress, a bottomless self-adhesive microfluidics flow channel was placed on top of the electroporation well, and the resulting flow channel was connected to a peristaltic pump. Phase contrast light microscopy images of cells after flow experiments are shown in Figure 4.

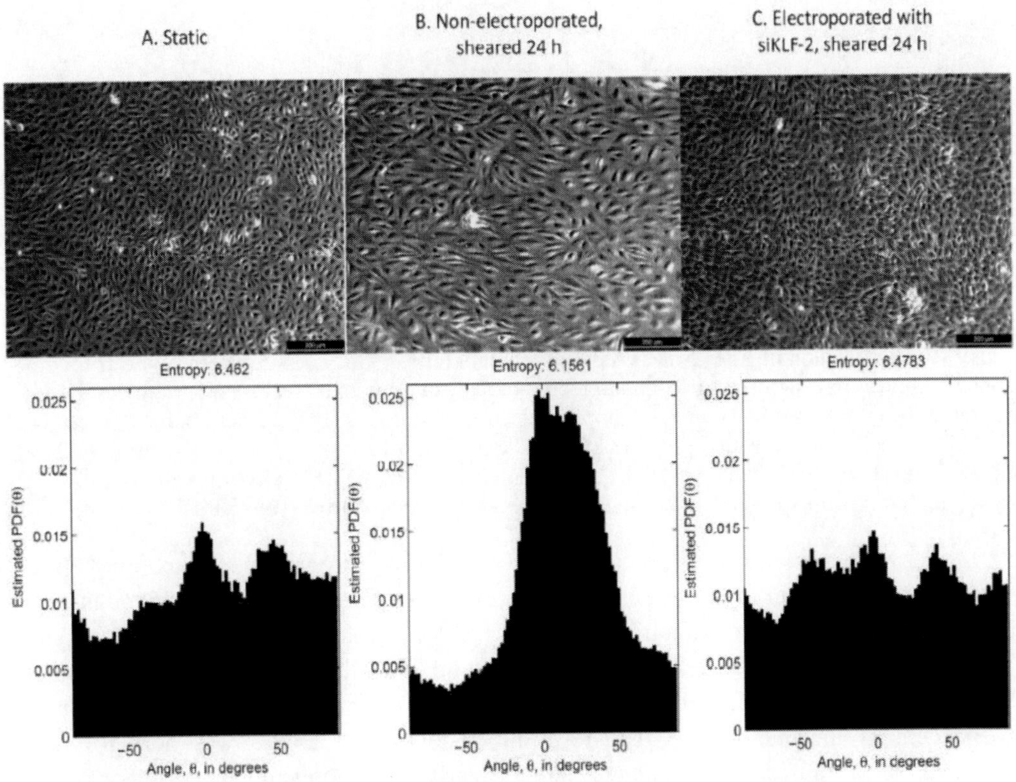

Figure 4. The novel electroporator is flow compatible and capable of delivering siRNA that modifies functions (alignment to flow) from primary endothelial cells. Phase contrast microscopy images of PAECs (top row) and orientation histograms (bottom row) showing the amount (fraction) of cells in function of orientation angle for the images above. The histograms are normalised, so that the vertical axis represents Probability Density Function (PDF) over angle, θ. A. Static (e.g., non-electroporated cells cultured under static conditions); B. No electroporation, sheared 24 h (cells not electroporated but sheared with 15-20 dyne/cm^2 for 24 hours); C. Electroporation with KLF-2 siRNA, sheared 24 h (cells electroporated with 33 μM siKLF-2 and exposed to 15-20 dyne/cm^2 of shear stress for of 24 hours).

Endothelial cells cultured under static conditions have a cobblestone like appearance. Whereas cells subjected to shear stress (20 dyne/cm^2 for 24 hours) stimulus appears elongated and aligns with the direction of the liquid flow. We have confirmed this with our setup, and obtained cobblestone and flow-elongated morphologies as shown in columns A and B of Figure 4, respectively. On the other hand, cells electroporated with KLF-2 targeting siRNA molecules (siKLF-2) and sheared with 20 dyne/cm^2 for 24 hours lost their ability to align to the direction of fluid flow, the aligment property being attributed to the transcription factor KLF-2 [85]. This is verified with our system where 24 hours of continuous flow is applied to cells transfected with siKLF-2 (see Figure. 4.C), providing evidence that our system is capable of delivering siRNAs to primary endothelial cells as well as inducing an appropriate amount of shear stress for mechanotransduction experiments.

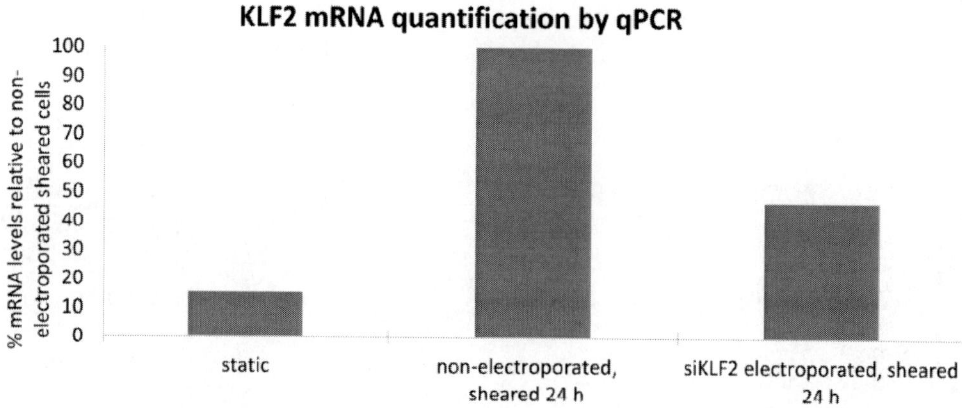

Figure 5. Quantification of KLF-2 mRNA down-regulation by qPCR. KLF-2 mRNA levels from "static" (PAECs from static culture); "non-electroporated, sheared 24 h" (cells exposed to 15-20 dyne/cm2 shear stress for 24 hours without prior electroporation); "siKLF-2 electroporated, sheared 24 h" (PAECs electroporated with 33 μM siKLF-2 and exposed to 15-20 dyne/cm2 shear stress for 24 hours). Considering "non-electroporated, sheared 24 h" as 100%, siKLF-2 electroporation yielded 52% KLF-2 mRNA down-regulation. mRNA levels were normalized to the enzyme HPRT1.

Images were analyzed in order to quantify cellular alignment to the direction of flow by determining the number of cells in each orientation angle. For this, angles of endothelial cells with respect to the flow direction were calculated from phase contrast images, applying a phase quadrature technique. Effectively, this method spatially convolves the phase-contrast image field with spatially-oriented complex filters to obtain the dominant direction, independent of image intensity. The resulting histogram of angles was normalized, and compared to a uniform or a single angle distribution applying the Kullback-Leibler divergence, or relative entropy. Obtained orientation histograms are shown below microscopy images in Figure 4. Histograms A and C show a relatively even distribution of cells by orientation angles whereas in case of histogram B a significant majority of cells are oriented at an angle of 15 degrees, ± 10 degrees. These histograms confirm that siKLF-2 electroporated, sheared cells (similarly to the static control) are less aligned than non-electroporated sheared cells. Entropy values show a similar trend, with low entropy (low degree of disorder) for shear exposed cells without siKLF-2 electroporation, and high entropy (high degree of disorder) for siKLF-2 electroporated and sheared cell as well as for the static control. In order to further quantify the knockdown of the gene that caused loss of flow alignment in endothelial cells, the mRNA levels of KLF-2 were quantified from 10 electroporation experiments using qPCR (see Figure 5 for details). Thus, immediately after completion of the flow experiment, cells were lysed and the total RNA was isolated using RNeasy Mini Kit from Qiagen. Next, the total RNA was converted into cDNA with the High Capacity RNA-to-cDNA Kit from Applied Biosystems. Levels of KLF-2 cDNA were quantified by real-time polymerase chain reaction (qPCR) using KLF-2 specific primers. mRNA levels in non-electroporated, sheared endothelial cells was considered as 100%, and electroporation with siKLF-2 yielded 52% KLF-2 mRNA down-regulation, a reduction comparable to those reported in literature [86].

Conclusion

Firstly, we have shown that our electroporation device is capable of transfecting primary PAECs while these cells remain adhered to surfaces, with transfection efficiencies of up to 70% and cell survival rate >95% (Figure 3). Secondly, we have demonstrated that we were able to suppress one of the most important mechanosensitive signaling pathways, and showed a consequent misalignment of endothelial cells. Electroporation with KLF-2 targeting siRNAs resulted in inhibition of the inherent flow induced alignment of primary endothelial cells, and in an approximate halving of the expected KLF-2 mRNA expression, as determined by qPCR.

In summary, our platform is capable of delivering cell membrane impermeable substances into hard-to-transfect primary endothelial cells allowing biomechanical studies under physiologically relevant complex flow and chemical conditions. As a future direction, the throughput of our system is currently being increased by adapting a "reversed" transfection of cells, for which siRNA or cDNA is deposited using a GeSIM robotic Nano-Plotter^{TM} system on a glass slide or deformable membrane before seeding cells for electroporation.

Acknowledgments

The authors thank Dr. Jennifer Frueh for providing siKLF-2, qPCR primers, PAECs, and Ms. Nataly Maimari and Dr. Jennifer Frueh for with helping KLF-2 mRNA quantification by qPCR. The Facility for Imaging by Light Microscopy (FILM) at Imperial College London is acknowledged for imaging support. Funding of M.Sc. studies for Z.K. by Dinu Patriciu Foundation (through an Open Horizons scholarship) is gratefully thanked. We acknowledge the British Hearth Foundation (RG 11/13/29055) for funding this work (to R.K.) and the Welcome Trust to H.I.

References

[1] Hamill OP, Martinac B. Molecular basis of mechanotransduction in living cells. *Physiol Rev* 2001 Apr;81(2):685-740.

[2] Kim DH, Wong PK, Park J, Levchenko A, Sun Y. Microengineered platforms for cell mechanobiology. *Annu Rev Biomed Eng* 2009;11:203-33.

[3] Eyckmans J, Boudou T, Yu X, Chen CS. A hitchhiker's guide to mechanobiology. *Dev Cell* 2011 Jul 19;21(1):35-47.

[4] Zou YP, Hu YH, Metzler B, Xu QB. Signal transduction in arteriosclerosis: Mechanical stress-activated MAP kinases in vascular smooth muscle cells (Review). *Int J Mol Med* 1998 May;1(5):827-34.

[5] World Health Organization. Cardiovascular diseases, Fact sheet N°317. 2013 [updated March 2013; cited 2013 08 May]; Available from: http://www.who.int/mediacentre /factsheets/fs317/en/index.html.

[6] Lloyd-Jones D, Adams R, Carnethon M, De Simone G, Ferguson TB, Flegal K, et al. Heart disease and stroke statistics--2009 update: a report from the American Heart Association Statistics Committee and Stroke Statistics Subcommittee. *Circulation* 2009 Jan 27;119(3):e21-181.

[7] World Health Organization. Global status report on noncommunicable diseases 2010. *World Health Organization,* editor. Geneva, 2011.

[8] Ross R. The pathogenesis of atherosclerosis: a perspective for the 1990s. *Nature* 1993 Apr 29;362(6423):801-9.

[9] Kannel WB, Sytkowski PA. Atherosclerosis Risk-Factors. *Pharmacol Therapeut* 1987;32(3):207-35.

[10] Gibson CM, Diaz L, Kandarpa K, Sacks FM, Pasternak RC, Sandor T, et al. Relation of Vessel Wall Shear-Stress to Atherosclerosis Progression in Human Coronary-Arteries. *Arterioscler Thromb* 1993 Feb;13(2):310-5.

[11] Friedman MH, Ehrlich LW. Effect of spatial variations in shear on diffusion at the wall of an arterial branch. *Circ Res* 1975;37(4):446-54.

[12] McDaniel MC, Galbraith EM, Jeroudi AM, Kashlan OR, Eshtehardi P, Suo J, et al. Localization of culprit lesions in coronary arteries of patients with ST-segment elevation myocardial infarctions: relation to bifurcations and curvatures. *Am Heart J* 2011 Mar;161(3):508-15.

[13] Chatzizisis YS, Coskun AU, Jonas M, Edelman ER, Feldman CL, Stone PH. Role of endothelial shear stress in the natural history of coronary atherosclerosis and vascular remodeling: molecular, cellular, and vascular behavior. *J Am Coll Cardiol.* 2007 Jun 26;49(25):2379-93.

[14] Tarbell JM, Pahakis MY. Mechanotransduction and the glycocalyx. *J Intern Med* 2006 Apr;259(4):339-50.

[15] Davies PF, Spaan JA, Krams R. Shear stress biology of the endothelium. *Ann Biomed Eng* 2005 Dec;33(12):1714-8.

[16] Caro CG, Fitz-Gerald JM, Schroter RC. Atheroma and arterial wall shear. Observation, correlation and proposal of a shear dependent mass transfer mechanism for atherogenesis. *Proc R Soc Lond B Biol Sci* 1971 Feb 16;177(46):109-59.

[17] Zarins CK, Giddens DP, Bharadvaj BK, Sottiurai VS, Mabon RF, Glagov S. Carotid Bifurcation Atherosclerosis Quantitative Correlation of Plaque Localization with Flow Velocity Profiles and Wall Shear-Stress. *Circ Res* 1983;53(4):502-14.

[18] Chatzizisis YS, Jonas M, Coskun AU, Beigel R, Stone BV, Maynard C, et al. Prediction of the Localization of High-Risk Coronary Atherosclerotic Plaques on the Basis of Low Endothelial Shear Stress. An Intravascular Ultrasound and Histopathology Natural History Study. *Circulation* 2008 Feb 4.

[19] Cheng C, Tempel D, van Haperen R, van Damme L, Algur M, Krams R, et al. Activation of MMP8 and MMP13 by angiotensin II correlates to severe intra-plaque hemorrhages and collagen breakdown in atherosclerotic lesions with a vulnerable phenotype. *Atherosclerosis* 2009 May;204(1):26-33.

[20] Cheng C, Tempel D, van Haperen R, de Boer HC, Segers D, Huisman M, et al. Shear stress-induced changes in atherosclerotic plaque composition are modulated by chemokines. *J Clin Invest* 2007 Mar;117(3):616-26.

[21] Cheng C, Tempel D, van Haperen R, van der Baan A, Grosveld F, Daemen MJ, et al. Atherosclerotic lesion size and vulnerability are determined by patterns of fluid shear stress. *Circulation* 2006 Jun 13;113(23):2744-53.

[22] Wentzel JJ, Kloet J, Andhyiswara I, Oomen JA, Schuurbiers JC, de Smet BJ, et al. Shear-stress and wall-stress regulation of vascular remodeling after balloon angioplasty: effect of matrix metalloproteinase inhibition. *Circulation* 2001;104(1):91-6.

[23] Wentzel JJ, Janssen E, Vos J, Schuurbiers JC, Krams R, Serruys PW, et al. Extension of increased atherosclerotic wall thickness into high shear stress regions is associated with loss of compensatory remodeling. *Circulation* 2003 Jul 8;108(1):17-23.

[24] Chatzizisis YS, Coskun AU, Jonas M, Edelman ER, Stone PH, Feldman CL. Risk stratification of individual coronary lesions using local endothelial shear stress: a new paradigm for managing coronary artery disease. *Curr Opin Cardiol* 2007 Nov;22(6):552-64.

[25] Chatzizisis YS, Jonas M, Beigel R, Coskun AU, Baker AB, Stone BV, et al. Attenuation of inflammation and expansive remodeling by Valsartan alone or in combination with Simvastatin in high-risk coronary atherosclerotic plaques. *Atherosclerosis* 2009 Apr;203(2):387-94.

[26] Malek AM, Alper SL, Izumo S. Hemodynamic shear stress and its role in atherosclerosis. *JAMA* 1999 Dec 1;282(21):2035-42.

[27] Oancea E, Wolfe JT, Clapham DE. Functional TRPM7 channels accumulate at the plasma membrane in response to fluid flow. *Circ Res* 2006 Feb 3;98(2):245-53.

[28] Chen KD, Li YS, Kim M, Li S, Yuan S, Chien S, et al. Mechanotransduction in response to shear stress. Roles of receptor tyrosine kinases, integrins, and Shc. *J Biol Chem* 1999 Jun 25;274(26):18393-400.

[29] Weinbaum S, Zhang X, Han Y, Vink H, Cowin SC. Mechanotransduction and flow across the endothelial glycocalyx. *Proc Natl Acad Sci USA* 2003 Jun 24;100(13):7988-95.

[30] Smith ML, Long DS, Damiano ER, Ley K. Near-wall micro-PIV reveals a hydrodynamically relevant endothelial surface layer in venules in vivo. *Biophys J* 2003 Jul;85(1):637-45.

[31] Weinbaum S, Tarbell JM, Damiano ER. The structure and function of the endothelial glycocalyx layer. *Annu Rev Biomed Eng* 2007;9:121-67.

[32] Hierck BP, Van der Heiden K, Alkemade FE, Van de Pas S, Van Thienen JV, Groenendijk BC, et al. Primary cilia sensitize endothelial cells for fluid shear stress. *Dev Dyn* 2008 Mar;237(3):725-35.

[33] Otte LA, Bell KS, Loufrani L, Yeh JC, Melchior B, Dao DN, et al. Rapid changes in shear stress induce dissociation of a G alpha(q/11)-platelet endothelial cell adhesion molecule-1 complex. *J Physiol* 2009 May 15;587(Pt 10):2365-73.

[34] McNally JS, Davis ME, Giddens DP, Saha A, Hwang J, Dikalov S, et al. Role of xanthine oxidoreductase and NAD(P)H oxidase in endothelial superoxide production in response to oscillatory shear stress. *Am J Physiol Heart Circ Physiol* 2003 Dec;285(6):H2290-7.

[35] Wang YX, Miao H, Li S, Chen KD, Li YS, Yuan SL, et al. Interplay between integrins and FLK-1 in shear stress-induced signaling. *Am J Physiol Cell Physiol* 2002 Nov;283(5):C1540-C7.

[36] Silacci P, Desgeorges A, Mazzolai L, Chambaz C, Hayoz D. Flow pulsatility is a critical determinant of oxidative stress in endothelial cells. *Hypertension* 2001 Nov;38(5):1162-6.

[37] Rizzo V, Morton C, DePaola N, Schnitzer JE, Davies PF. Recruitment of endothelial caveolae into mechanotransduction pathways by flow conditioning in vitro. *Am J Physiol Heart Circ Physiol* 2003 Oct;285(4):H1720-9.

[38] Tabouillot T, Muddana HS, Butler PJ. Endothelial Cell Membrane Sensitivity to Shear Stress is Lipid Domain Dependent. *Cell Mol Bioeng* 2011 Jun;4(2):169-81.

[39] Fleming I, Fisslthaler B, Dixit M, Busse R. Role of PECAM-1 in the shear-stress-induced activation of Akt and the endothelial nitric oxide synthase (eNOS) in endothelial cells. *J Cell Sci* 2005 Sep 15;118(18):4103-11.

[40] Tzima E, Irani-Tehrani M, Kiosses W, Dejana E, Schultz D, Engelhardt B, et al. A mechanosensory complex that mediates the endothelial cell response to fluid shear stress. *FASEB J* 2006 Mar 7;20(5):A1378-A.

[41] Tzima E, Irani-Tehrani M, Kiosses WB, Dejana E, Schultz DA, Engelhardt B, et al. A mechanosensory complex that mediates the endothelial cell response to fluid shear stress. *Nature* 2005 Sep 15;437(7057):426-31.

[42] Knudsen HL, Frangos JA. Role of cytoskeleton in shear stress-induced endothelial nitric oxide production. *Am J Physiol* 1997 Jul;273(1 Pt 2):H347-55.

[43] Berk BC, Traub O. Laminar shear stress - Mechanisms by which endothelial cells transduce an atheroprotective force. *Arterioscl Throm Vas* 1998 May;18(5):677-85.

[44] Li YSJ, Haga JH, Chien S. Molecular basis of the effects of shear stress on vascular endothelial cells. *J Biomech* 2005 Oct;38(10):1949-71.

[45] Tedgui A, Lehoux S, Castier Y. Molecular mechanisms of the vascular responses to haemodynamic forces. *J Intern Med* 2006 Apr;259(4):381-92.

[46] Hahn C, Schwartz MA. Mechanotransduction in vascular physiology and atherogenesis. *Nat Rev Mol Cell Biol* 2009 Jan;10(1):53-62.

[47] Yamamoto K, Ando J. New molecular mechanisms for cardiovascular disease:blood flow sensing mechanism in vascular endothelial cells. *J Pharmacol Sci* 2011 Aug 17;116(4):323-31.

[48] White CR, Frangos JA. The shear stress of it all: the cell membrane and mechanochemical transduction. *Philos Trans R Soc Lond B Biol Sci* 2007 Aug 29;362(1484):1459-67.

[49] Davies PF, Polacek DC, Shi C, Helmke BP. The convergence of haemodynamics, genomics, and endothelial structure in studies of the focal origin of atherosclerosis. *Biorheology* 2002;39(3-4):299-306.

[50] Endy D. Foundations for engineering biology. *Nature* 2005 Nov 24;438(7067):449-53.

[51] Kitano H. Computational systems biology. *Nature* 2002 Nov 14;420(6912):206-10.

[52] Kim JA, Cho KC, Shin MS, Lee WG, Jung NC, Chung CI, et al. A novel electroporation method using a capillary and wire-type electrode. *Biosensors Bioelectron* 2008 Apr;23(9):1353-60.

[53] Folkman J, Moscona A. Role of Cell-Shape in Growth-Control. *Nature* 1978;273(5661):345-9.

[54] Campisi J, Medrano EE. Cell-Cycle Perturbations in Normal and Transformed Fibroblasts Caused by Detachment from the Substratum. *J Cell Physiol* 1983;114(1):53-60.

[55] Vasiliev JM. Spreading of Non-Transformed and Transformed-Cells. *Biochim Biophys Acta* 1984;780(1):21-65.

[56] Boon RA, Leyen TA, Fontijn RD, Fledderus JO, Baggen JMC, Volger OL, et al. KLF2-induced actin shear fibers control both alignment to flow and JNK signaling in vascular endothelium. *Blood* 2010 Mar 25;115(12):2533-42.

[57] Raptis L, Firth KL. Electroporation of adherent cells in situ. *DNA Cell Biol* 1990 Oct;9(8):615-21.

[58] Zheng QA, Chang DC. High-efficiency gene transfection by in situ electroporation of cultured cells. *Biochim Biophys Acta* 1991 Jan 17;1088(1):104-10.

[59] Ghosh PM, Keese CR, Giaever I. Monitoring electropermeabilization in the plasma membrane of adherent mammalian cells. *Biophys J* 1993 May;64(5):1602-9.

[60] Raptis LH, Brownell HL, Firth KL, Mackenzie LW. A novel technique for the study of intercellular, junctional communication: electroporation of adherent cells on a partly conductive slide. *DNA Cell Biol* 1994 Sep;13(9):963-75.

[61] Bright GR, Kuo NT, Chow D, Burden S, Dowe C, Przybylski RJ. Delivery of macromolecules into adherent cells via electroporation for use in fluorescence spectroscopic imaging and metabolic studies. *Cytometry* 1996;24(3):226-33.

[62] Teruel MN, Meyer T. Electroporation-induced formation of individual calcium entry sites in the cell body and processes of adherent cells. *Biophys J* 1997 Oct;73(4):1785-96.

[63] Lundqvist JA, Sahlin F, Aberg MA, Stromberg A, Eriksson PS, Orwar O. Altering the biochemical state of individual cultured cells and organelles with ultramicroelectrodes. *Proc Natl Acad Sci USA* 1998 Sep 1;95(18):10356-60.

[64] Nolkrantz K, Farre C, Brederlau A, Karlsson RI, Brennan C, Eriksson PS, et al. Electroporation of single cells and tissues with an electrolyte-filled capillary. *Anal Chem* 2001 Sep 15;73(18):4469-77.

[65] Haas K, Sin WC, Javaherian A, Li Z, Cline HT. Single-cell electroporation for gene transfer in vivo. *Neuron* 2001 Mar;29(3):583-91.

[66] Lin YC, Li M, Fan CS, Wu LW. A microchip for electroporation of primary endothelial cells. *Sensor Actuat A Phys* 2003 Nov 15;108(1-3):12-9.

[67] Lin YC, Li M, Wu CC. Simulation and experimental demonstration of the electric field assisted electroporation microchip for in vitro gene delivery enhancement. *Lab Chip* 2004 Apr;4(2):104-8.

[68] Yamauchi F, Kato K, Iwata H. Spatially and temporally controlled gene transfer by electroporation into adherent cells on plasmid DNA-loaded electrodes. *Nucleic Acids Res* 2004;32(22).

[69] Jain T, Muthuswamy J. Microsystem for transfection of exogenous molecules with spatio-temporal control into adherent cells. *Biosens Bioelectron* 2007;22(6):863-70.

[70] Huang KS, Lin YC, Su CC, Fang CS. Enhancement of an electroporation system for gene delivery using electrophoresis with a planar electrode. *Lab Chip* 2007;7(1):86-92.

[71] Wang J, Stine MJ, Lu C. Microfluidic cell electroporation using a mechanical valve. *Anal Chem* 2007 Dec 15;79(24):9584-7.

[72] Jain T, Muthuswamy J. Bio-chip for spatially controlled transfection of nucleic acid payloads into cells in a culture. *Lab Chip* 2007;7(8):1004-11.

[73] Olofsson J, Levin M, Stromberg A, Weber SG, Ryttsen F, Orwar O. Scanning electroporation of selected areas of adherent cell cultures. *Anal Chem* 2007 Jun 15;79(12):4410-8.

[74] Ishibashi T, Takoh K, Kaji H, Abe T, Nishizawa M. A porous membrane-based culture substrate for localized in situ electroporation of adherent mammalian cells. *Sensor Actuat B Chem.* 2007 Dec 12;128(1):5-11.

[75] Koda S, Inoue Y, Iwata H. Gene Transfection into Adherent Cells Using Electroporation on a Dendrimer-Modified Gold Electrode. *Langmuir* 2008 Dec 2;24(23):13525-31.

[76] Njatawidjaja E, Iwata H. Gene delivery to cells on a miniaturized multiwell plate for high-throughput gene function analysis. *Anal Bioanal Chem* 2008 Oct;392(3):405-8.

[77] Vassanelli S, Bandiera L, Borgo M, Cellere G, Santoni L, Bersani C, et al. Space and time-resolved gene expression experiments on cultured mammalian cells by a single-cell electroporation microarray. *New Biotechnol* 2008 Jun;25(1):55-67.

[78] Chang WC, Sretavan DW. Single cell and neural process experimentation using laterally applied electrical fields between pairs of closely apposed microelectrodes with vertical sidewalls. *Biosens Bioelectron* 2009 Aug 15;24(12):3600-7.

[79] Jain T, McBride R, Head S, Saez E. Highly parallel introduction of nucleic acids into mammalian cells grown in microwell arrays. *Lab Chip* 2009;9(24):3557-66.

[80] Schonenberger C, Schutz A, Franco-Obregon A, Zenobi-Wong M. Efficient electroporation of peptides into adherent cells: investigation of the role of mechano-growth factor in chondrocyte culture. *Biotechnol Lett 2011* May;33(5):883-8.

[81] Huang HA, Wei ZW, Huang YY, Zhao DY, Zheng LH, Cai TJ, et al. An efficient and high-throughput electroporation microchip applicable for siRNA delivery. *Lab Chip* 2011;11(1):163-72.

[82] Jain T, Papas A, Jadhav A, McBride R, Saez E. In situ electroporation of surface-bound siRNAs in microwell arrays. *Lab Chip* 2012 Jan 16.

[83] Vassanelli S, Bandiera L, Borgo M, Cellere G, Santoni L, Bersani C, et al. Space and time-resolved gene expression experiments on cultured mammalian cells by a single-cell electroporation microarray. *N Biotechnol* 2008 Jun;25(1):55-67.

[84] Arndt-Jovin DJ, Jovin TM. Chapter 16 Fluorescence Labeling and Microscopy of DNA. In: Taylor DL, Yu-Li W, editors. *Methods in Cell Biology: Academic Press;* 1989. p. 417-48.

[85] Boon RA, Leyen TA, Fontijn RD, Fledderus JO, Baggen JMC, Volger OL, et al. KLF2-induced actin shear fibers control both alignment to flow and JNK signaling in vascular endothelium. *Blood* 2010;115(12):2533-42.
[86] Fledderus JO, Boon RA, Volger OL, Hurttila H, Yla-Herttuala S, Pannekoek H, et al. KLF2 primes the antioxidant transcription factor Nrf2 for activation in endothelial cells. *Arteriosc1 Throm Vas* 2008 Jul;28(7):1339-46.

In: Microarrays: Principles, Applications and Technologies
Editor: James V. Rogers
ISBN: 978-1-62948-669-7

Chapter 5

A New Application for Microarrays: Analysis of Global MicroRNA Expression Profiles in the Extracellular Microvesicles of Human Macrophage-Like Cells

Tadayoshi Kagiya[*1] *and Masayuki Taira*[2]
[1] Division of Functional Morphology, Department of Anatomy, Iwate Medical University, Iwate, Japan
[2] Department of Biomedical Engineering, Iwate Medical University, Iwate, Japan

Abstract

A microarray is one of the most powerful tools available for simultaneous determination of the expression levels of multiple genes. Microarray technology has recently been utilized to investigate intracellular microRNA expression. In this chapter, we propose a new microarray application: analysis of global microRNA expression profiles in extracellular microvesicles. MicroRNAs are small, noncoding RNAs that are involved in various biological processes, including cellular differentiation, proliferation, apoptosis, organ development and tumorigenesis. Recently, we and other researchers reported the presence of microRNAs in extracellular microvesicles. Although human macrophages secrete extracellular microvesicles containing miR-223, it is not known whether they contain other microRNAs. Here, we investigated the expression profiles of microRNAs in extracellular microvesicles in conditioned medium of cultured human macrophage-like cells using a microarray.

[*] Correspondence should be addressed: Tadayoshi Kagiya; Division of Functional Morphology, Department of Anatomy, Iwate Medical University, 2-1-1 Nishitokuta, Yahaba-cho, Iwate, 028-3694, Japan; Tel: +81 19 651 5111; Fax: +81 19 908 8010; E-mail: tkagiya@iwate-med.ac.jp.

THP-1, a human monocytic cell line, was cultured in the presence of 200 nM phorbol 12-myristate 13-acetate (PMA) for 72 h for differentiation into macrophage-like cells. Then the medium was changed and the cells were cultured without PMA for an additional 72 h. The conditioned medium cultured for 72 to 144 h was harvested and treated with ExoQuick-TC™ Exosome Isolation Reagent, according to the manufacturer's protocol, to collect the microvesicle pellets. The RNA was then isolated from the pellets. The samples were labeled and hybridized to an Agilent SurePrint G3 Human miRNA microarray. Of the 1,222 microRNAs investigated, 53 were detected by the microarray. miR-1246 was the most highly expressed, followed by miR-574-5p and miR-4281. The expression level of miR-223 was the 43rd highest. These results suggest that macrophages secrete extracellular microvesicles containing specific microRNAs. Thus, microarrays are useful for expression profiling of microRNAs in extracellular microvesicles.

Introduction

Microarray technology has developed since Schena et al. first reported that DNA microarray technology could monitor multiple gene expression profiles [1, 2]. Microarrays are one of the most powerful tools currently available for simultaneous determination of the expression levels of multiple genes. Microarray technology has recently been utilized to investigate intracellular microRNA (miRNA) expression [2, 3]. miRNAs are small, non-coding RNAs of approximately 22 nucleotides [3, 4] that are involved in various biological processes, including cellular differentiation, proliferation, apoptosis, organ development and tumorigenesis [3, 4].

miRNAs are incorporated into an RNA-induced silencing complex that contains argonaute-family proteins. The argonaute proteins recruit miRNAs specific for target mRNAs [3, 4], and the RNA-induced silencing complex either inhibits the translation of, or degrades, the target mRNAs [3, 4]. Consequently, miRNAs are involved in the post-transcriptional regulation of mRNA function and participate in the regulation of cellular differentiation, proliferation, apoptosis and organ development [3, 4].

Recently, we and other researchers reported the presence of miRNAs in extracellular microvesicles [4-10]. Extracellular microvesicles are membranous extracellular organelles [11] and microvesicles originated from endosomal compartment (40–100-nm diameter) are defined as exosomes [4, 9, 10].

Following endocytosis into early endosomes, the cargo is packaged into intraluminal vesicles (ILVs) within multivesicular bodies (MVBs). MVBs can then fuse with lysosomes resulting in degradation of the cargo, or alternatively, the MVBs can fuse with the plasma membrane, resulting in release of ILVs as exosomes [9, 11, 12]. Exosomes contain many types of miRNAs, mRNAs and proteins [4-12] (Figure 1). Extracellular microvesicles are considered to be important in cell-to-cell communication via the transfer of miRNAs, mRNAs, proteins and bioactive lipids to target cells [4-12]. The secretion of microvesicles containing miRNAs depends on cell type, biological condition and the kinds of miRNAs the cells contain [4, 6, 8, 10, 12].

Although human macrophages secrete extracellular microvesicles containing miR-223 [13], it is not known whether they contain other miRNAs. We investigated the expression profiles of miRNAs in extracellular microvesicles in conditioned medium of cultured human macrophage-like cells using a microarray.

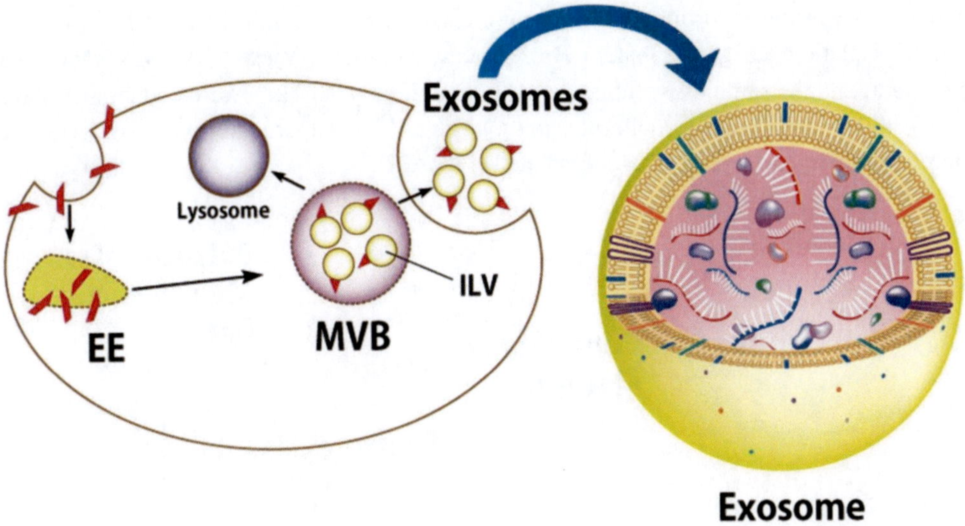

Figure 1. Exosome biogenesis and contents. Following endocytosis into early endosomes (EE), the cargo is packaged into intraluminal vesicles (ILVs) within multivesicular bodies (MVBs). MVBs can then fuse with lysosomes, resulting in degradation of the cargo, or alternatively, the MVBs can fuse with the plasma membrane, resulting in release of ILVs as exosomes. Exosomes contain many types of microRNAs, mRNAs and proteins.

Methods

Cell Culture

THP-1, a human monocytic cell line obtained from the RIKEN BioResource Center (Tsukuba, Japan) was seeded on six 35-mm-diameter culture dishes (Falcon, Lincoln Park, NJ) at a density of 20,000 cells/cm^2 and cultured in RPMI 1640 medium (Invitrogen, Fredrick, MD) supplemented with 10% heat-inactivated fetal bovine serum (FBS; Lot 18300101, Moregate Biotech, Bulimba, QLD, Australia) in the presence of 200 nM phorbol 12-myristate 13-acetate (PMA) (Sigma–Aldrich, St. Louis, MO) for 72 h to differentiate into macrophage-like cells. The medium was then changed and the cells were cultured without PMA for an additional 72 h at 37°C in a humidified atmosphere containing 5% CO_2. The conditioned medium cultured for 72 to 144 h was harvested.

Extracellular Microvesicle Precipitation

Extracellular microvesicles exist not only in conditioned medium, but also in FBS [4]. To measure the background expression of miRNAs, we isolated microvesicles from FBS. After the culture period, 12 mL each of conditioned medium and RPMI 1640 containing 10% FBS without culturing were collected, to measure background levels. To pellet dead cells and cell debris, the samples were centrifuged at 1,400×*g* for 5 min and 10 mL of the supernatant was transferred to fresh tubes. To eliminate vesicles of more than 200-nm diameter, the supernatant was filtered through a 200-nm pore size filter. Two milliliters of ExoQuick-TC™ Exosome Isolation Reagent (System Biosciences, Mountain View, CA) was added to the supernatant, and the samples were incubated overnight at 4°C. They were centrifuged the next day at 1,400×*g* for 30 min to pellet the microvesicles. The supernatant was removed and the pellets containing microvesicles were isolated (Figure 2).

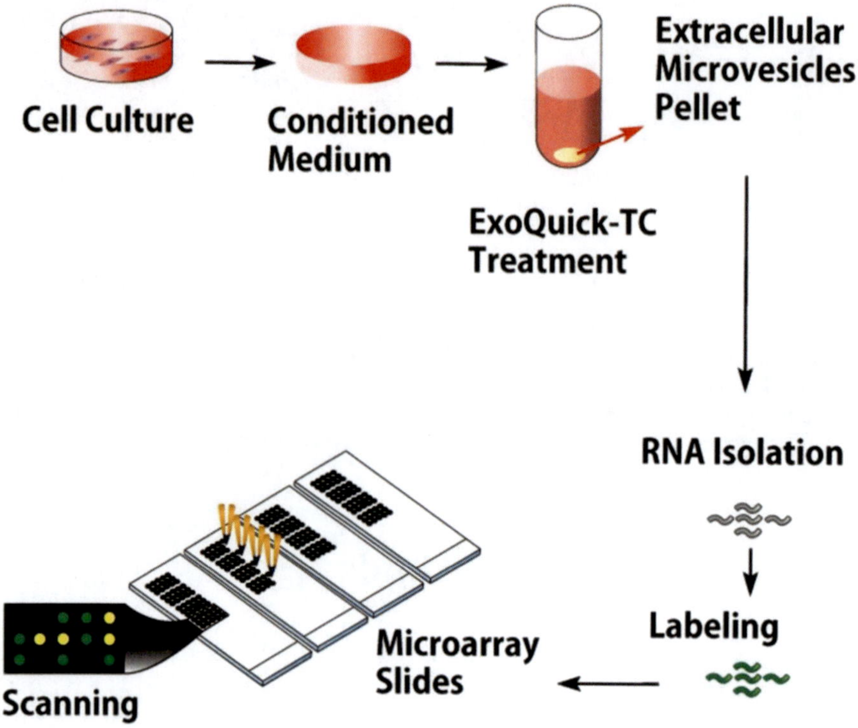

Figure 2. Experimental procedure. After the culturing period, the microvesicle-containing conditioned medium was collected. The medium was treated with ExoQuick-TC™ Exosome Isolation Reagent and then centrifuged to pellet the microvesicles. RNA was isolated from pellets containing microvesicles and labeled with Cy3. The samples were applied on microarray slides and hybridized with probes. After washing, the slides were scanned by Agilent Technologies Microarray Scanner. The data were analyzed using the Agilent Feature Extraction software.

Microarray Analysis

Using a mirVana™ miRNA isolation kit (Ambion, Austin, TX), RNA was harvested from the extracellular microvesicle pellets. The RNA samples were diluted in 70-μL nuclease-free water. RNA quality was assessed using an Agilent 2100 Bioanalyzer (Agilent Technologies, Santa Clara, CA) according to the manufacturer's protocol. After assessing RNA quality, the RNA samples (4 μL each) were labeled by Cy3 using a miRNA Complete Labeling and Hyb Kit (Agilent Technologies). The samples were then hybridized to an Agilent SurePrint G3 Human miRNA microarray (1,222 miRNAs based on miRBase database release 16.0). The data were analyzed using Agilent Feature Extraction software and GeneSpring GX software (Agilent Technologies) (Figure 2).

Quantitative RT-PCR (qRT-PCR) Analysis of miRNA Expression

The RNA samples (4 μL each) were reverse-transcribed using a TaqMan MicroRNA Reverse Transcription Kit (Applied Biosystems, Foster City, CA) with miRNA-specific primers (Applied Biosystems). The expression of mature miRNAs was analyzed using appropriate TaqMan miRNA assays (Applied Biosystems) and Premix Ex Taq probe qPCR (Takara Bio, Otsu, Japan) according to the manufacturer's protocol.

Results

Prior to microarray analysis, the existence of miRNAs in the extracellular microvesicles of PMA treated THP-1 cells was investigated using an Agilent 2100 Bioanalyzer. A quantity of RNA suitable for microarray analysis was observed in the microvesicles, peaking at around 500 nucleotides in length (Figure 3). This result suggests that the microvesicles contain miRNAs and mRNAs; we next subjected this RNA to miRNA microarray analysis.

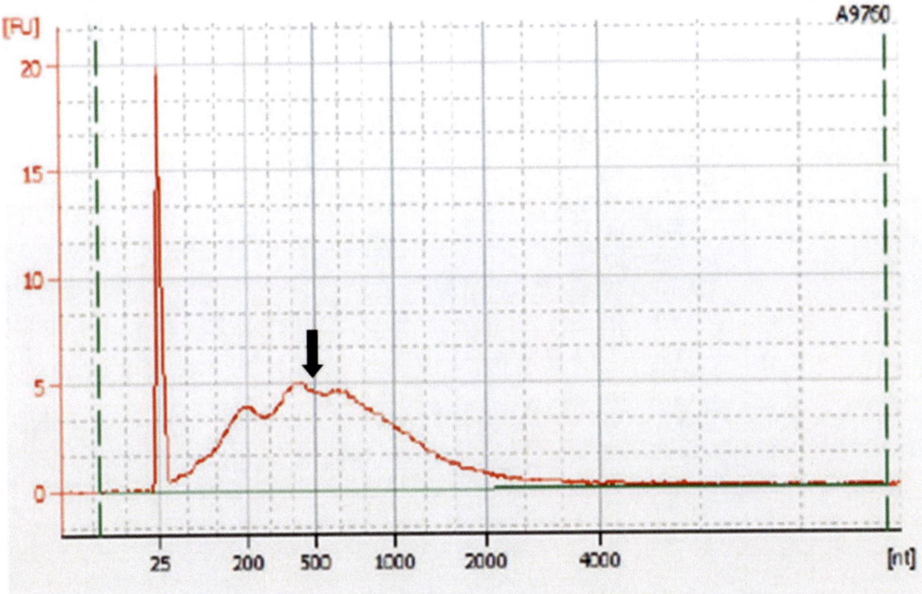

Figure 3. Analysis of RNA in the extracellular microvesicles of PMA treated THP-1 cells. THP-1 cells were cultured in the presence of 200 nM phorbol 12-myristate 13-acetate (PMA) for 72 h to differentiate into macrophage-like cells. The medium was then changed and the cells were cultured without PMA for an additional 72 h. The conditioned medium was cultured for 72 to 144 h, harvested and then the microvesicle pellet was collected. The RNA was then isolated from the pellet. Samples were analyzed using an Agilent 2100 Bioanalyzer. The arrow indicates 500 nucleotides in length.

To screen the expression profiles of miRNAs in the extracellular microvesicles of PMA treated THP-1 cells, we performed a microarray analysis. As extracellular microvesicles are released to extracellular environments, they exist not only in conditioned medium, but also in FBS. The expression of miRNAs in the microvesicles of FBS served as background for this study. To measure the background expression of miRNAs, we isolated the microvesicles in FBS and analyzed miRNA expression using a microarray. Table 1 shows the expression profile of miRNAs from the extracellular microvesicles in FBS. Of the 1,222 miRNAs we investigated, 28 miRNAs were detected. miR-451 was the most highly expressed, followed by miR-940 and miR-1234. Table 2 shows a list of miRNAs detected in the extracellular microvesicles of PMA treated THP-1 cells by microarray analysis. Of the 1,222 miRNAs we investigated, 53 miRNAs were detected. miR-1246 was the most highly expressed, followed by miR-574-5p and miR-4281. The expression level of miR-223, which is reportedly expressed in the extracellular microvesicles of macrophages [13], was the 43rd highest. To confirm the results of our microarray experiments, representative miRNAs were analyzed by qRT-PCR. Three miRNAs were investigated: miR-1246, which was the most highly expressed, miR-223, which is reportedly expressed in the extracellular microvesicles of human macrophages, miR-33a, which was not detected in this microarray analysis. The expression of miRNA determined by qRT-PCR was similar to the results of the microarray analysis (Figure 4). These data suggest that microarrays are useful for expression profiling of miRNAs in extracellular microvesicles.

Table 1. MicroRNA expression profile in FBS. This microarray experiment was performed once. Signal intensities were normalized by 90th percentile.

miRNA	Signal intensity
hsa-miR-451	354.4
hsa-miR-940	271.8
hsa-miR-1234	250.6
hsa-miR-4313	231.9
hsa-miR-720	226.6
hsa-miR-1280	225.3
hsa-miR-1238	207.5
hsa-miR-1228	203.0
hsa-miR-191*	187.3
hsa-miR-1281	182.4
hsa-miR-2116*	147.5
hsa-let-7b*	144.7
hsa-miR-1825	141.8
hsa-miR-3676	139.2
hsa-miR-1225-3p	138.8
hsa-miR-425*	138.6
hsa-miR-149	138.5
hsa-miR-1539	131.9
hsa-miR-23c	128.7
hsa-miR-4310	127.4
hsa-miR-223	113.7
hsa-miR-4312	110.1
hsa-let-7f-1*	109.2
hsa-miR-766	67.9
hsa-miR-1246	62.3
hsa-miR-4323	53.2
hsa-miR-933	31.6
hsa-miR-33b*	28.9

Table 2. MicroRNA expression profile in the extracellular microvesicles of PMA treated THP-1 cells. This microarray experiment was performed once. Signal intensities were normalized by 90th percentile and evaluated by those of FBS.

miRNA	Signal intensity	miRNA	Signal intensity
hsa-miR-1246	3527.1	hsa-miR-16	58.0
hsa-miR-574-5p	1102.9	hsa-let-7b*	47.3
hsa-miR-4281	622.5	hsa-miR-940	47.0
hsa-miR-574-3p	519.1	hsa-let-7f-1*	45.7
hsa-miR-1290	474.4	hsa-miR-33b*	43.7
hsa-miR-720	354.6	hsa-miR-4312	39.8
hsa-miR-3149	345.6	hsa-miR-1539	38.5
hsa-miR-32*	257.9	hsa-miR-4313	37.9
hsa-miR-197	226.3	hsa-miR-425*	34.8
hsa-miR-1281	195.9	hsa-miR-191*	33.8
hsa-miR-4290	193.7	hsa-miR-149	29.8
hsa-miR-483-3p	168.4	hsa-miR-1234	27.3
hsa-miR-1268	156.8	hsa-miR-223	25.8
hsa-miR-1825	149.6	hsa-miR-1280	22.2
hsa-miR-595	131.0	hsa-miR-1225-3p	21.0
hsa-miR-3148	124.8	hsa-miR-23c	19.7
hsa-miR-877*	118.0	hsa-miR-4323	19.7
hsa-miR-21	106.8	hsa-miR-2116*	15.4
hsa-miR-1260	105.2	hsa-miR-4310	13.4
hsa-miR-762	91.5	hsa-miR-1228	13.2
hsa-miR-23a	79.0	hsa-miR-1238	12.5
hsa-miR-630	77.3	hsa-miR-3676	9.3
hsa-miR-2278	72.5	hsa-miR-933	4.6
hsa-miR-766	72.2		
hsa-miR-4286	68.2		
hsa-miR-328	68.1		
hsa-miR-20a	60.7		
hsa-miR-92a	60.4		
hsa-miR-4284	59.7		
hsa-miR-1237	59.0		

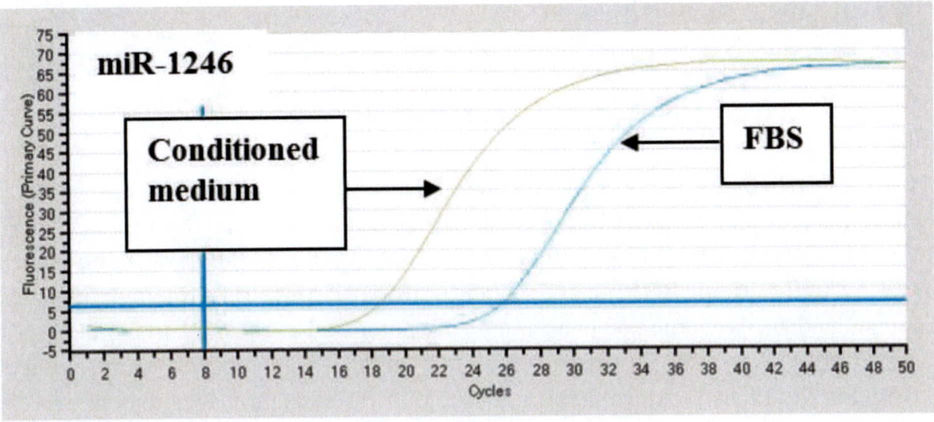

Figure 4. (Continued)

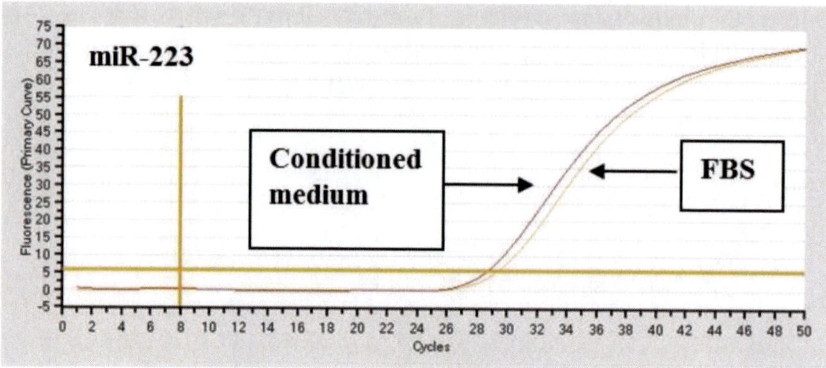

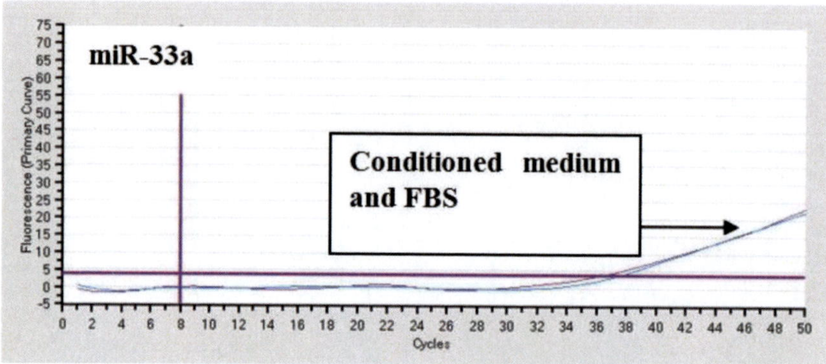

Figure 4. qRT-PCR validation of microarray data regarding microRNAs in the extracellular microvesicles of PMA treated THP-1 cells. THP-1 cells were cultured in the presence of 200 nM phorbol 12-myristate 13-acetate (PMA) for 72 h to differentiate into macrophage-like cells. The medium was then changed and the cells were cultured without PMA for an additional 72 h. The conditioned medium was cultured for 72 to 144 h, harvested, and the microvesicle pellet was collected. To measure background microRNA expression, we isolated the microvesicles present in FBS. RNA was then isolated from the pellets. MicroRNA expression in the samples was assessed by qRT-PCR. miR-1246 was highly expressed. On the other hand, miR-33a was not detected in the vesicles.

Conclusion

For the first time, we examined global miRNA expression in the extracellular microvesicles of PMA treated THP-1 cells using a microarray. Of the 1,222 miRNAs investigated, 53 were detected by the microarray. miR-1246 was the most highly expressed, followed by miR-574-5p. Both miR-1246 and miR-574-5p are serum circulating miRNAs and upregulated in the serum of cancer patients compared to healthy individuals [14-17]. Because at least a portion of serum circulating miRNAs are derived from extracellular microvesicles [10-12, 18], extracellular microvesicles from macrophages may be involved in cancer development. miR-150, which is the most highly enriched miRNA in the extracellular microvesicles of monocytic THP-1 cells [18], was not detected in this microarray analysis. Although human macrophages have been reported to secrete extracellular microvesicles containing miR-223 [13], the expression level of miR-223 was the 43rd highest. miR-223 is also present in extracellular microvesicles from murine osteoclasts, which are derived from macrophages [4]. miR-223 is highly expressed in myeloid and bone marrow cells [19].

In this chapter, we used a microarray to profile miRNA expression in extracellular microvesicles. There are some points to be considered. First, although we assessed miRNA expression using the Agilent SurePrint G3 Human miRNA microarray, this was not originally designed for investigation of extracellular microvesicles—it was designed for investigation of intracellular miRNA expression and optimized for total intracellular RNA, including ribosomal RNA, mRNA and miRNA. Accordingly, microarray results should be confirmed by qRT-PCR analysis, at least for representative miRNAs. Second, no internal control was included for comparison of miRNA expression levels in extracellular microvesicles [18]. When comparing more than two samples, several methods are proposed: (1) Start with the same quantity of sample (e.g., an identical volume of conditioned medium). To avoid differences in miRNA extraction efficiency, verify the samples by qRT-PCR normalized to an external control (e.g., spike with synthetic *Caenorhabditis elegans* miRNAs, cel-miR-39, cel-miR-54 or cel-miR-238, which are not found in humans, to the samples after denaturing RNase) [7, 17, 20]. (2) Start with the same quantity of extracellular microvesicle proteins [18]. Third, two principal methods of collecting extracellular microvesicles or exosomes may be used: with and without ultracentrifugation. The method used can affect the results. Although collected pellets contain both exosomes (40–100-nm diameter) and larger microvesicles (> 100 nm-diameter), the latter method is more reliable than the former for microarray analysis [21]. Thus, we used ExoQuick-TC™ Exosome Isolation Reagent in this study. Last, FBS contains extracellular microvesicles. Accordingly, the miRNA expression profile in FBS as a background must be evaluated (Tables 1 and 2) or culture cells in its absence [7].

Our data suggest that macrophages secrete extracellular microvesicles containing specific miRNAs. Thus, microarrays are useful for expression profiling of miRNAs in extracellular microvesicles. We anticipate that this new application for microarrays opens the door to investigation of the novel intercellular communication network mediated by extracellular microvesicles.

Acknowledgments

This study was supported, in part, by grants from the Scientific Research (C) (No. 23593042), and a Grant-in-Aid for Strategic Medical Science Research Center (2010–2014) from the Ministry of Education, Culture, Sports, Science and Technology, Japan.

References

[1] Schena M, Shalon D, Davis RW, Brown PO. Quantitative monitoring of gene expression patterns with a complementary DNA microarray. *Science* 1995; 270: 467-470.

[2] Sato F, Tsuchiya S, Terasawa K, Tsujimoto G. Intra-platform repeatability and inter-platform comparability of microRNA microarray technology. *PLoS One* 2009; 4: e5540.

[3] Kagiya T, Nakamura S. Expression profiling of microRNAs in RAW264.7 cells treated with a combination of tumor necrosis factor alpha and RANKL during osteoclast differentiation. *J Periodontal Res* 2013; 48: 373-385.

[4] Kagiya T, Taira M. Expression of MicroRNAs in the Extracellular Microvesicles of Murine Osteoclasts. *J Oral Tissue Engin* 2013; 10: 142-150.

[5] Valadi H, Ekstrom K, Bossios A, Sjostrand M, Lee JJ, Lotvall JO. Exosome-mediated transfer of mRNAs and microRNAs is a novel mechanism of genetic exchange between cells. *Nat Cell Biol* 2007; 9: 654-659.

[6] Skog J, Wurdinger T, van Rijn S, Meijer DH, Gainche L, Curry WT et al. Glioblastoma microvesicles transport RNA and proteins that promote tumour growth and provide diagnostic biomarkers. *Nat Cell Biol* 2008; 10: 1470-1476.

[7] Kosaka N, Iguchi H, Yoshioka Y, Takeshita F, Matsuki Y, Ochiya T. Secretory mechanisms and intercellular transfer of microRNAs in living cells. *J Biol Chem* 2010; 285: 17442-17452.

[8] Cantaluppi V, Gatti S, Medica D, Figliolini F, Bruno S, Deregibus MC et al. Microvesicles derived from endothelial progenitor cells protect the kidney from ischemia-reperfusion injury by microRNA-dependent reprogramming of resident renal cells. *Kidney Int* 2012; 82: 412-427.

[9] Bobrie A, Colombo M, Raposo G, Thery C. Exosome secretion: molecular mechanisms and roles in immune responses. *Traffic* 2011; 12: 1659-1668.

[10] Mathivanan S, Fahner CJ, Reid GE, Simpson RJ. ExoCarta 2012: database of exosomal proteins, RNA and lipids. *Nucleic Acids Res* 2012; 40: D1241-1244.

[11] Mathivanan S, Ji H, Simpson RJ. Exosomes: extracellular organelles important in intercellular communication. *J Proteomics* 2010; 73: 1907-1920.

[12] Bellingham SA, Guo BB, Coleman BM, Hill AF. Exosomes: vehicles for the transfer of toxic proteins associated with neurodegenerative diseases? *Front Physiol* 2012; 3: 124.

[13] Yang M, Chen J, Su F, Yu B, Su F, Lin L et al. Microvesicles secreted by macrophages shuttle invasion-potentiating microRNAs into breast cancer cells. *Mol Cancer* 2011; 10: 117.

[14] Jones CI, Zabolotskaya MV, King AJ, Stewart HJS, Horne GA, Chevassut TJ et al. Identification of circulating microRNAs as diagnostic biomarkers for use in multiple myeloma. *Br J Cancer* 2012; 107: 1987-1996.

[15] Takeshita N, Hoshino I, Mori M, Akutsu Y, Hanari N, Yoneyama Y et al. Serum microRNA expression profile: miR-1246 as a novel diagnostic and prognostic biomarker for oesophageal squamous cell carcinoma. *Br J Cancer* 2013; 108: 644-652.

[16] Foss KM, Sima C, Ugolini D, Neri M, Allen KE, Weiss GJ. miR-1254 and miR-574-5p: serum-based microRNA biomarkers for early-stage non-small cell lung cancer. *J Thorac Oncol* 2011; 6: 482-488.

[17] Wu L, Zhou H, Lin H, Qi J, Zhu C, Gao Z et al. Circulating microRNAs are elevated in plasma from severe preeclamptic pregnancies. *Reproduction* 2012; 143: 389-397.

[18] Zhang Y, Liu D, Chen X, Li J, Li LM, Bian Z et al. Secreted monocytic miR-150 enhances targeted endothelial cell migration. *Mol Cell* 2010; 39: 133-144.

[19] O'Connell RM, Zhao JL, Rao DS. MicroRNA function in myeloid biology. *Blood* 2011; 118: 2960-2969.

[20] Mitchell PS, Parkin RK, Kroh EM, Fritz BR, Wyman SK, Pogosova-Agadjanyan EL et al. Circulating microRNAs as stable blood-based markers for cancer detection. *Proc Natl Acad Sci U S A* 2008; 105: 10513-10518.

[21] Murakami Y, Toyoda H, Tanahashi T, Tanaka J, Kumada T, Yoshioka Y et al. Comprehensive miRNA expression analysis in peripheral blood can diagnose liver disease. *PLoS One* 2012; 7: e48366.

In: Microarrays: Principles, Applications and Technologies
Editor: James V. Rogers
ISBN: 978-1-62948-669-7

Chapter 6

Identification of Reference Genes by Meta-Microarray Analyses

***Issac H. K. Too*[1], *Sean S. J. Heng*[2], *Oliver Y. W. Chan*[2], *Bryan M. H. Keng*[2], *Ching-Yee Chia*[3], *Claudia W. X. Lim*[3], *Wan-Ting Leong*[3], *Qinghao Chu*[2], *Ernest J. G. Ang*[2], *Yongjie Lin*[2] *and Maurice H.T. Ling*[4,5*]**

[1]Department of Biological Sciences, National University of Singapore, Singapore
[2]Raffles Institution, Singapore
[3]Nanyang Girls' High School, Singapore
[4]Department of Zoology, The University of Melbourne, Australia
[5]School of Chemical and Biomedical Engineering, Nanyang Technological University, Singapore

Abstract

The expression levels of reference genes used in gene expression studies are assumed to not change under most circumstances. However, a number of studies have demonstrated that genes theoretically assumed to be stably expressed were found to vary under experimental conditions. In addition, previous studies have also reported that stably expressed genes in an organ, may not be stably expressed in other organs or in a different organism, suggesting the need to identify reference genes for each organ and each organism. Due to its ability to analyze the expression of thousands of genes in an experiment, microarrays present a suitable resource for the analysis and identification of reference genes. We present four cases on practical applications of microarrays whereby multiple published microarray data sets were examined to identify suitable reference genes using coefficient of variation (CV) and NormFinder. Our results suggest that microtubule affinity-regulating kinase 3 (MARK3) is a suitable reference gene for mouse liver, 40S ribosomal protein S29 (Rps29) is a suitable reference gene for mouse testes and pancreas, signal peptidase complex subunit 1 (SPCS1) and hydroxyacyl-CoA

* Corresponding author: mauriceling@acm.org.

dehydrogenase beta subunit (HADHB) are suitable reference genes for human lungs, and glucan biosynthesis protein G (mdoG) is a suitable reference gene for *Escherichia coli.* Further analysis suggests that the identified reference genes are involved in fundamental biochemical processes. This supports the theoretical basis and previous studies that housekeeping genes, on the whole, are generally stably expressed. However, our results also suggest that certain housekeeping genes that are stably expressed in one tissue or one organism may not be stably expressed in different tissues or organisms, supporting the need to identify reference genes for each tissue and organism.

Introduction

In the context of molecular cell biology, it is crucial to examine the variation in gene expression over the time in order to study the functions of individual genes. This provides valuable insight on the effects of differential gene expression on cell division, as well as biological responses in various conditions, such as developmental phases and effects of treatments. Profiling of gene expression by the mean of quantitative real-time polymerase chain reaction (qRT-PCR), Northern blot and DNA microarray analysis [1] have been well established. However, a number of variables, such as cell type, mRNA extraction and handling techniques, and analytical quantification approaches may result in different gene expression measurements [2]. This may affect analytical accuracy [3]. In order to address these variations, normalization usually involving a group of calibrating genes is employed in many gene expression quantification studies [3].

Housekeeping genes are generally used as calibrating genes in gene expression studies [3] due to their theoretical stable expression [4]. The ideal set of housekeeping genes should be expressed stably and consistently across all samples [5]. However, several studies have suggested that universal reference genes are relatively difficult to identify [6-9]. This corroborates several studies demonstrating that genes that are originally considered invariable in terms of expression may vary under different experimental conditions [10-12]. For an accurate comparison of DNA expression in different samples, it is necessary to use verified reference genes, such as GAPDH (glyceraldehyde-3-phosphate dehydrogenase) or UBQ (ubiquinone) [13], for normalization or determine new reference genes for each experimental system with varying external stimuli [6, 14]. However, it has been demonstrated that the expression of GAPDH [15] and UBQ [16] varied in some conditions. This indicates that GAPDH or UBQ may not be useful reference genes in different experimental platforms. Previous studies showed that ACTB (beta actin) and SDHA (succinate dehydrogenase complex, subunit A) were effectively used as reference genes in breast tumor studies [17], while IPO8 (importin 8) was stably expressed in human lung specimens. Another study also showed that EIF2A (eukaryotic translation initiation factor 2A, 65kDa) and PPIB (peptidylprolyl isomerase B, also known as cyclophilin B) were stably expressed in mouse liver and adrenal gland [18]. Other studies also demonstrated that EEF1A1 (eukaryotic translation elongation factor 1 alpha 1) expressed consistently in human cervical tissues [19], while B2M (beta-2-microglobulin) and RPL29 (ribosomal protein L29) were well used as reference genes in human stomach tissue studies [20]. High expression stability showed by RPL32 (ribosomal protein L32), GAPDH, POLR2A (DNA directed RNA polymerase II polypeptide A, 220kDa), TBP (TATA box binding protein), PGK1 (phosphoglycerate kinase

1) and RPL4 (ribosomal protein L4) are demonstrated as the most frequently observed reference genes in rodent and human heart gene expression studies [21], while RPS18 (ribosomal protein S18) was stably expressed in head and neck squamous cell carcinoma [22]. In addition, GUSB (glucuronidase, beta) was generally used as reference gene in human ovarian studies [23], while RPS4X (ribosomal protein S4, X-linked) and RPL13A (ribosomal protein L13a) were useful in quantifying the gene expression in larvae of flatfish [24] and osteoarthritic canine articular tissues [3] respectively. This suggests that expression of different housekeeping genes may be inconsistent among different species or organs.

To date, the selection of reference genes can be achieved by few mathematical methods, such as geNorm [25], BestKeeper© [26] and NormFinder [27] to evaluate gene expression stability in qRT-PCR [3]. BestKeeper© [26] and geNorm [25] are based on pairwise correlation has been shown to be sensitive to co-regulated genes [28]. On the other hand, NormFinder is less sensitive to co-regulated genes as it took account of the variations across subgroups [27]. However, these tools are usually applicable in a small sample size cDNA panel to determine the most stable reference genes [29]. Microarray is commonly used in relative quantification of gene expression because it affords speed and high-throughput [30, 31]. Due to its ability to analyze the expression of thousands of genes in a single experiment, the microarray presents a suitable resource for the analysis and identification of reference genes [13].

In this chapter, we present four cases of identifying reference genes by analyzing multiple microarray datasets using coefficient of variance (CV), followed by NormFinder analysis [27] whenever possible. In each dataset, the arithmetic mean and standard deviation were calculated for each gene and the CV was determined as a quotient of arithmetic mean and standard deviation. CV was used as a filter to select reference genes that showed low variation in expression. In the first and second cases, datasets from the different platform were used and direct comparison of probes by NormFinder [27] was not possible due to differences in probe sets. In the first case, we identified microtubule affinity-regulating kinase 3 (MARK3) as a suitable reference gene for mouse liver. In the second case, we identified 40S ribosomal protein S29 (Rps29) as a suitable reference gene for mouse testes and pancreas. In the third and fourth cases, datasets from the same platform were used, which enabled direct comparison. In the third case, we renormalized the entire set of microarray datasets and identified signal peptidase complex subunit 1 (SPCS1) and hydroxyacyl-CoA dehydrogenase beta subunit (HADHB) as suitable reference genes in human lungs. In the fourth case, we identified glucan biosynthesis protein G (mdoG) as suitable reference gene for 2 sub-strains of Escherichia coli K-12, MG1655 and W3110. Further analysis suggests that the identified reference genes are involved in fundamental biochemical processes.

Case 1: Finding Reference Genes for Mouse Liver Using Non-comparable Datasets [32]

Seven data sets of gene expression in the mouse liver under different conditions were obtained from Gene Expression Omnibus (GEO; www.ncbi.nlm.nih.gov/geo), National Center for Biotechnology Information (NCBI). These data sets used different platforms. Briefly, these data sets originated from the following studies:

1. LDL receptor-deficient mice fed on either a low fat diet or a high fat, western-style diet for 12 weeks with 12,548 genes located (GDS279)
2. Mice injected intraperitoneally with a cytokine and examined at 4 hours with 12,558 genes located (GDS280)
3. Analysis of animals fed a diet containing metformin, glipizide, rosiglitazone, or soy isoflavone extract and compared to hepatic gene expression profiles produced by long-term caloric restriction with 12,593 genes located (GDS1808)
4. Analysis of day 13.5 and 15.5 p38alpha-deficient hematopoietic mice with 45,101 genes located (GDS2693)
5. Comparison of embryonic day 13.5 and 15.5; analysis of livers from NMR-1 females fed human and chimpanzee diets for 2 weeks with 45,101 genes located (GDS3221)
6. Analysis of livers in C57BL/6 male mice maintained on a high-fat high-calorie diet supplemented with 0.04% resveratrol, a compound in red wine that extends lifespan of diverse species with 27,397 genes located (GDS2413)
7. Analysis of livers at 13 weeks following treatment with chemicals that test positive in two-year rodent cancer bioassays with 45,102 genes located (GDS2497).

A threshold of less than 0.1 CV was used to select stably expressed genes in each data set. The number of genes with CV ranging from <0.05 to <0.1 from each dataset were tabulated in Table 1. With GDS279 having the smallest number of gene probes compared to other data sets, it had limited the possible number of genes with CV <0.1 common in all data sets. Dataset GDS2497 has over 45102 probes, which resulted in the high number of genes having a CV lower than <0.1.

Despite the fact that there were close to 40,000 genes with CV <0.1 in datasets like GDS2497, GDS2693 and GDS3221, the expression level of many genes was not consistent under the different conditions, resulting in the small number of 39 genes with the CV of <0.1 present across the 7 datasets. Nephroblastoma Overexpressed gene (Nov), a gene that is essential for cell growth, was expressed with a low CV of 0.052 under the conditions of being p38alpha-deficient and hematopoietic (GDS2693), but under the conditions of being on different diets for 12 weeks (GDS279), it was expressed with a high CV of 0.561.

Analysis of the 7 datasets had identified 39 genes that were most stably expressed, with the criteria of having CV lesser than 0.1 across each of the conditions as shown in Table 2. Our results indicated that Microtubule Affinity-Regulating Kinase 3 (MARK3) was the most invariantly expressed gene with a CV of <0.08 - the lowest CV across all conditions, followed by 8 other genes with a CV of <0.09; namely Damage specific DNA Binding protein 1(Ddb1); D-fructose-1, 6-bisphosphate 1- phosphohydrolase 1 (Dnajc8); Fbp1, Ribosomal Protein L5 (Rpl5); Ribosomal Protein L10 (Rpl10), Ribosomal Protein L17 (Rpl7); Serine/arginine Repetitive Matrix 1(Srrm1) and Ubiquitin-Like 5 (Ubl5).

Of these 9 genes, at least 5 genes (MARK3, Ddb1, Rpl5, Rpl7, Srrm1 and Ubl5) are considered common housekeeping genes. Rpl5, Rpl10 and Rpl7 are ribosomal proteins, and ribosomal protein genes are constitutively expressed in all cell types and are essential for the biogenesis of new ribosomes for the synthesis of proteins, suggesting that ribosomal proteins are more likely to be stably expressed because any increase or decrease of the gene level will result in an abnormal amount of ribosomal level, which in turn may result in mutated or disabled cells as shown by Thorrez et al., [33]. In addition to coding for a ribosomal protein, Srrm1 is also involved in numerous pre-mRNA processing events [34]. Undoubtedly, varying availability of proteins in the translational process may result in variability in translational efficiency [35]. Taken together, it is likely that genes coding for proteins that are involved in translation of mRNA are more likely to be constitutively expressed at a constant level.

Although it has been shown that some housekeeping genes are stably expressed [36], other commonly used housekeeping genes like GAPDH was shown to have high fluctuations in the gene expression level. Our results show that CV of gene expression of GAPDH is 10% or more. Hence, GAPDH is not on our list of good housekeeping genes (Table 2). On the other hand, Mark3 appears to be stably expressed in mouse liver, suggesting that it is a suitable reference gene for mouse liver.

Table 1. Number of genes with coefficient of variation (CV) under values of 0.05, 0.06, 0.07, 0.08, 0.09 and 0.1 in their respective datasets

	CV Threshold					
	<0.05	**<0.06**	**<0.07**	**<0.08**	**<0.09**	**<0.1**
GDS279	1	10	37	86	169	310
GDS280	302	940	2060	3479	4928	6177
GDS1808	1	8	40	130	283	524
GDS2413	85	97	112	124	136	146
GDS2497	34075	39761	42263	43445	44018	44362
GDS2693	29494	35617	39607	41889	43020	43606

Table 2. Specific genes and their CV values

Genes	CV	Genes	CV
MARK3	<0.08	Eif2s3x	<0.1
4833439L19Rik	<0.09	Epn1	<0.1
9530068E07Rik	<0.09	H2-K1	<0.1
Ddb1	<0.09	Hdgf	<0.1
Dnajc8	<0.09	Mark2	<0.1
Fbp1	<0.09	Myo1b	<0.1
Rpl5	<0.09	Rpl3	<0.1
Rpl10	<0.09	Pcid2	<0.1
Rpl7	<0.09	Pex14	<0.1
Srrm1	<0.09	Pgm2	<0.1
Ubl5	<0.09	Pigs	<0.1
2410166I05Rik	<0.1	Prdx1	<0.1
Ahcyl1	<0.1	Rps3	<0.1
Ap3b1	<0.1	Rps3a	<0.1
Arpc5	<0.1	Stard3	<0.1
Bscl2	<0.1	Sucla2	<0.1
Cd151	<0.1	Trpc4ap	<0.1
Cox7c	<0.1	Zdhhc9	<0.1
Cstf1	<0.1	Zwint	<0.1
Cyb5	<0.1		

Case 2: Finding Reference Genes for Mouse Testes and Pancreas Using Non-Comparable Datasets [37]

Fifteen datasets were obtained from GEO, NCBI; of which, 8 were from mice testes and 7 were from mice pancreas. Briefly, the studies conducted with the datasets were as follows:

1. Analysis of mice testis from day 1 to adult to determine transcripts expressed specifically by male germ cells late in post-meiotic spermatogenesis (GDS410)
2. Spermatogenesis time course generated from mice testis collected from birth through adulthood to provide insight into genes implicated in maturation, maintenance, and function of testis and the integrated process of spermatogenesis (GDS607)
3. Analysis of CD4+ automimmune T and CD45+ hematopoeitic cells from non-obese diabetic (NOD) and transgenic mice with the BDC2.5 T cell receptor (TCR) from a diabetogenic T cell (GDS828)
4. Analysis of pancreatic islets from transgenic mice after treatment with 200 mg/kg cyclophosphamide (CY) at various time points up to 3 days (GDS1056)
5. Expression profiling of male germ cells lacking the telomerase RNA component, Terc, or the DNA repair/telomere binding protein, Ku86, or both to identify processes affected by dysfunctional telomeres (GDS1248)
6. Analysis of testes from Sertoli cell-selective androgen receptor knockouts (SCARKO) between postnatal day 8 and 20 to provide insight into molecular consequences of selective absence of androgen action in Sertoli cells at onset of spermatogenesis (GDS1367)
7. Analysis of the pancreas from 3-week-old transgenic mcie that express a diabetogenic T cell receptor on the NOD or NOD. scid genetic background (GDS1533)
8. Analysis of the pancreas from transcription factor Mist1 knockouts (Mist1KO) injected with caerulein to identify genes that may contribute to susceptibility and initiation of pancreatitis (GDS1731)
9. Expression profiling of testes from *Mus musculus* subspecies castaneus, domesticus, musculus, and a strain from Iran (GDS1884)
10. Analysis of transgenic pancreatic islet beta cells after activation of Myc for 24 hours or 21 days followed by deactivation for up to 6 days to elucidate the molecular basis of Myc-driven tumorigenesis in vivo (GDS2025)
11. Expression profiling of normal whole testes at various stages of development from gestation day 11 to postnatal day 2 (GDS2098)
12. Analysis of the epidididymis of animals at various ages from gestational day 12 to post-natal day 2 to provide insight into the molecular events involved in testicular development (GDS2202)
13. Analysis of ribonucleoprotein (RNP) and polysome fractions of prepuberal and adult testes to provide insight into the translation state of individual mRNAs as spermatogenesis proceeds (GDS2277)
14. Analysis of MIN6 pancreatic beta-cells treated for up to 24 hours with various concentrations of human islet amyloid polypeptide (GDS2945)

15. Analysis of pancreatic tissues from NMRI animals from embryonic day 12.5 to 16.5 to provide insight into the molecular mechanisms underlying the development of the pancreas (GDS2950).

One hundred probes with the lowest CV from each dataset identified and grouped based on the organ used, namely testes and pancreas. Non-unique probes in each organ list were isolated, indicating that they were the least variant genes for that platform and organ, supported by several datasets. Common non-unique probes, indicating low CV in both organs and supported by multiple data sets in each organ, were identified. The average CV and percentile rank were calculated to allow for comparison. A total of 5 genes were recorded as being shown to have a stable variation: Rps7 (40S ribosomal protein S7), Rps28 (40S ribosomal protein S28), Rps29 (40S ribosomal protein S29), Rpl31 (60S ribosomal protein L31), Rpl37a (60S ribosomal protein L37a) and MARK3. The gene with the lowest CV and percentile values was Rps29 (Table 3). All of the 5 genes identified to have a low variance were ribosomal proteins. In addition, the least variable gene from the commonly used housekeeping gene set was the ribosomal protein, Rpl13a. Thus, it can be concluded from the results that genes encoding for ribosomal proteins appear to be the most stably expressed. This coincides with results from previous findings also indicating that genes coding for ribosomal proteins have the least variance [33]. Ribosome has the vital role in cell maintenance, growth, survival and function. Cells with an abnormal level of ribosomes may result in cell mutation or death [33]. This suggests that most cells should possess comparable levels of ribosomes; hence, explaining the low expression variance of genes coding for ribosomal proteins.

Case 3: Finding Reference Genes for Human Lung Using Comparable Datasets [38]

Fourteen microarray data sets studying human lung epithelial from GEO were used. All of which employed Affymetrix Human Genome U133 Plus 2.0 Array (GPL570) containing 54,676 probes. This allows for comparisons across different data sets. The fourteen data sets were:

1. Bronchial epithelial cells at 4 and 24 hours following treatment with respiratory syncytial virus (GDS2023)
2. Airway epithelial cells of phenotypically normal smokers (GDS2486 and GDS2491)
3. Airway epithelia from healthy individuals 7 and 14 days following injury by epithelial denudation (GDS2495)
4. Epithelial and mesothelial lung cell lines at various time points up to 7 days after exposure to asbestos (GDS2604)
5. Twenty-eight day air-liquid cultured airway epithelial cells (GDS2615)
6. Analysis of resveratrol-treated lung carcinoma A549 cells (GDS2966).
7. Analysis of house dust mite (HDM) extract exposed H292 bronchial epithelial cells (GDS3003)
8. Bronchial epithelial cells exposed to cigarette smoke from a typical full flavor brand for up to 24 hours (GDS3493)

9. Bronchial epithelial cells exposed to cigarette smoke from a typical light flavor brand for up to 24 hours with (GDS3494)
10. Lung adenocarcinoma CL1-5 cells overexpressing Claudin-1 (CLDN1) (GDS3510)
11. Comparison of non-small cell lung cancer histological subtypes: adenocarcinomas (AC) and squamous cell carcinomas (SCC) (GDS3627)
12. Analysis of normal lung WI-38 fibroblasts exposed to various concentrations of the carcinogen benzo[a]pyrene-diol epoxide (BPDE) (GDS3706)
13. Analysis of A549 epithelial cells treated for up to 72 hours with TGF-beta (GDS3710).

The data sets were normalized by arithmetic mean transformation and Z transformation to construct parallel and comparable data sets based on the method described in [39]. In each dataset, the intensity value for each probe ($Probe_{Initial}$) was used to compute the average probe intensity for each dataset ($\mu_{InitialProbe}$). By assuming the standard arithmetic mean of all the probes as 1000 for the overall combined microarray data sets ($\mu_{Assumed}$), the correction factor for the average probe intensity for each data sets was constructed as a quotient of $\mu_{Assumed}$ and $\mu_{InitialProbe}$, i.e., Correction Factor = $\mu_{Assumed}/\mu_{InitialProbe}$. The transformed intensity of each original probe ($Probe_{Transformed}$) is then calculated using the equation, $Probe_{Transformed} = Probe_{Initial} \times$ Correction Factor. The $Probe_{Transformed}$ was then used to construct Z-score by the equation, $Z_{score} = (Probe_{Transformed} - \mu_{Assumed})/ SD_{Dataset}$, where $SD_{Dataset}$ is the standard deviation of the original dataset from which the probe ($Probe_{Initial}$) originates. The Z-scores for each probe across different original data sets would then be comparable. By constructing the CV values for all the probes of the chosen reference genes from the fourteen data sets, the mean CV values for each reference gene was generated. Using the Z-transformed dataset, 10% of the probes with the lowest CV were isolated and filtered. In this case, the CV would be the quotient of standard deviation of the Z-score to the arithmetic mean of the Z-score. For example, if geneA is represented in the microarray as 3 probes and all 3 probes are found in list of 10% lowest CVs, geneA is retained. On the other hand, if geneB is represented in the microarray as 3 probes and one of the probe is not in the list of 10% lowest CV, geneB is removed from the list. Only probes with gene symbols were included in the list. The filtered list was then analyzed with NormFinder [27] using the normalized microarray data as expression values. Due to the limitation that NormFinder has for analyzing zero and negative values, the normalized expression values of the fourteen data sets, which were lower than 0.0001 or close to zero value, were transformed into the absolute values or 0.0001, respectively.

By extracting the 10% probes with the lowest CV values from the transformed Z-score data, 5,458 probes were extracted from the original 54,676 probes. After removing gene probes without a specific gene name, the total probes number left was 2,213. Among these 2,213 probes, 743 genes fell within the lowest 10% CV, which were encoded by 932 probes. The analysis result showed that among the chosen 20 housekeeping genes from previous studies [3, 17, 19, 20, 22, 23, 32, 40-42], only EIF2A, PPIB, RPL4, 60S ribosomal protein L13a (RPL13A), 60S ribosomal protein L32 (RPL32) and 40S ribosomal protein S4 (RPS4X) were found in the lowest 10% CV subset. Expression stability analysis using NormFinder [27] showed that signal peptidase complex subunit 1 (SPCS1) and hydroxyacyl-CoA dehydrogenase/3-ketoacyl-CoA thiolase/enoyl-CoA hydratase, beta subunit (HADHB) had the highest stability of rank 1 and 2 with a stability index of 0.326 and 0.355 respectively, followed by 71 genes with a stability index of 0.360. EIF2A, PPIB, RPL4, RPL13A, RPL32 and RPS4X had the rank ranging from 305 to 730 (Table 4). The mean Z-score of the probes within the lowest 10% CV ranges from -0.4073 to 11.7692. The mean Z-scores of HADHB and SPCS1 were 1.8419 (61.8 percentile) and 2.8516 (75.8 percentile) respectively.

Table 3. The 5 identified invariant genes, and their CV and percentile values for both endocrine glands

Gene	Average CV	Average Percentile	Minimum CV	Maximum CV	Minimum Percentile	Maximum Percentile
Rps7	0.188608	0.218980	0.007615	1.075255	0.000619	0.985346
Rps29	0.136192	0.133532	0.006590	0.357252	0.000266	0.777627
Rps28	0.161726	0.230025	0.009225	0.603931	0.000088	0.916240
Rpl31	0.306197	0.295305	0.005310	1.181695	0.000220	0.965566
Rpl37a	0.192236	0.229087	0.009193	0.459498	0.000442	0.740711

Table 4. Stability of housekeeping genes generated by NormFinder [27]

Gene Symbols	NormFinder Stability Index	Rank
SPCS1	0.326	1
HADHB	0.355	2
EIF2A	0.691	371
PPIB	0.862	630
PPIB	0.867	634
RPL13A	0.651	305
RPL13A	0.811	548
RPL13A	0.832	587
RPL13A	0.911	689
RPL13A	0.950	730
RPL32	0.739	454
RPL4	0.723	428
RPL4	0.789	518
RPL4	0.805	541
RPS4X	0.777	499
RPS4X	0.838	592

By extracting the pool of Z-transformed probes intensities with the lowest 10% CV, the remaining 5,458 probes (out of 54,676 probes) were further selected to eliminate undefined genes. The selected 743 genes were input in NormFinder [27] to generate the stability value for each gene as a direct measure of the estimated expression variation; genes were then ranked accordingly. SPCS1 and HADHB were found to have the lowest CV values, demonstrated the lowest stability index, and were subsequently ranked as first and second, respectively.

The standard deviation of the CV values and NormFinder stability index were 0.077 and 0.265, respectively. Due to these values varying by more than 3 times the standard deviation, homoscedasticity (constant variance) was not assumed. Thus, Spearman's rank correlation coefficient was carried out to determine the correlation of stability index by NormFinder and CV values, which showed that the sum of d2 is 85109486 (Spearman's rank correlation coefficient = 0.369) and the p-value was 1.79×10-31. Since the p-value was less than 0.01, the null hypothesis is rejected, indicating that there is correlation between the NormFinder stability index and CV values.

Our results suggested that SPCS1 and HADHB were considerably stable among the datasets analyzed and can be used as suitable calibrating genes in human lung studies for both healthy and perturbed tissues. The mean Z-scores of HADHB and SPCS1 were at the 61.8 percentile and 75.8 percentile, respectively, suggesting that these genes can be used as suitable reference genes for genes with average to high expression. Further studies are needed to determine their validity for genes with low expression.

The SPCS1 gene is located on chromosome 3 at 3p21.1 and functions as signal peptidase complex in the signal sequences cleavage of most secretory and membrane proteins [43]. Most secretory proteins required to be translocated into endoplasmic reticulum (ER) membrane in order to allow the proteins to fold and assemble in a proper way before they are transported to the Golgi apparatus. Proteins that fail to assemble or fold into their native state will be translocated back across ER membrane to cytoplasm and undergo degradation [44]. The signal sequences on secretory proteins must be cleaved during protein synthesis and SPCS1 plays a role in cleavage of signal sequences of most secretory proteins to enable their translocation across the ER membrane [43].

HADHB is located on chromosome 2 at 2p23 and is involved in mitochondrial beta-oxidation of long chain fatty acids [45]. A previous study [46] had shown that different organs have different tendencies in fatty acid distribution, as liver is served as the largest reservoir for fatty acids followed by brain and lung. Since lung is one of the reservoirs for fatty acid, HADHB, a lipase, will be likelihood expressed more stably in the lung in order to degrade fatty acids and generate ATP. However, these two genes have not been used as reference genes in gene quantification studies.

Case 4: Finding Reference Gene for *Escherichia Coli* Using Comparable Datasets [47]

Four data sets of *E. coli* K-12 using Affymetrix *E. coli* Antisense Genome Array (GPL199) were obtained from Gene Expression Omnibus, National Centre for Biotechnology Information in which 3 were from *E. coli* K-12 sub-strain MG1655 and 1 from sub-strain W3110. Briefly, the studies conducted with the data sets are as follows:

1. MG1655 grown in either aerobic or anaerobic conditions, deleted for transcriptional regulators in oxygen response, used to validate a computational model of transcriptional and metabolic networks (GDS680)
2. MG1655 cells cultured aerobically in several media with varied carbon sources, including glucose, glycerol, succinate, L-alanine, acetate, and L-proline (GDS1099)

3. Analysis of derivatives of strain 1655: wild-type, fur mutant, and wild-type with added $FeSO_4$, induced to overexpress RyhB, a noncoding RNA regulated by the Fur repressor protein (GDS1494)
4. W3110 cells exposed to acidic, neutral, or alkaline pH in order to study acid and base responses (GDS1827).

For 4,631 genes, the top 10% with the lowest CV from each dataset were listed. The intersection between the 3 MG1655 datasets (GDS680, GDS1099, and GDS1494) was identified and analysed using NormFinder [27] to rank the stability of these genes. A weighted stability index for each gene was then calculated from the NormFinder stability index, and an average determined from the NormFinder stability indices multiplied by sample number.

A threshold of less than 10% CV was used to select the stably expressed genes across the three data sets GDS 680, 1099 and 1494 (MG1655). A total of 39 genes with consistently low variance were identified (Table 5) with the weighted CV values ranging from 0.099 to 0.138. mdoG was found to be most stable with both the lowest weighted CV value and weighted NormFinder Stability Index for MG1655. In GDS 1827 (W3110), mdoG was the most stable in the dataset, with a CV of 0.088 and NormFinder Stability Index of 0.078. The highest CV in GDS 1827 is 0.791 for hslV (peptidase component of the HslUV protease). Although it is not a well-established reference gene, our results suggest that mdoG may be a suitable reference gene across both *E. coli* strains W3110 and MG1655 and that mdoG may also be suitable for use as reference genes in other strains of *E. coli* K-12.

The mdoG gene has been shown to be involved in the formation of the β-1,6 glucose linkage [48] and in the periplasmic release of newly synthesized osmoregulated periplasmic glucans [49, 50] needed for the bacterial cell wall. Thus, it is plausible that the expression of mdoG is needed during binary fission. As *E. coli* divides rapidly, constant synthesis of the cell wall is needed, necessitating constitutive mdoG expression.

Table 5. Weighted mean CV and NormFinder stability index of 39 invariant genes across 3 data sets (MG1655)

Gene Symbol	Gene Name	Weighted CV values	Weighted NormFinder Stability Index
mdoG	glucan biosynthesis protein G	0.099	0.082
dapA	dihydrodipicolinate synthase	0.106	0.090
crp	DNA-binding transcriptional dual regulator	0.106	0.102
hslV	peptidase component of the HslUV protease	0.111	0.105
mrdB	cell wall shape-determining protein	0.101	0.114
fucU	L-fucose mutarotase	0.107	0.117
yjgP	LPS transport (lptF)	0.105	0.117
yigC	3-octaprenyl-4-hydroxybenzoate decarboxylase	0.126	0.117
sun	16S rRNA m(5)C967 methyltransferase, S-adenosyl-L-methionine-dependent	0.130	0.119
gor	glutathione oxidoreductase	0.117	0.126
hflB	ATP-dependent metalloprotease	0.127	0.130
yqiB	predicted dehydrogenase	0.125	0.134

murG	N-acetylglucosaminyl transferase	0.124	0.134
yrbG	predicted calcium/sodium:proton antiporter	0.122	0.134
yejK	nucleotide associated protein	0.120	0.141
yfgA	cytoskeletal protein required for MreB assembly	0.118	0.142
hflX	putative GTPase HflX	0.105	0.142
spoT	bifunctional (p)ppGpp synthetase II/ guanosine-3',5'-bis pyrophosphate 3'-pyrophosphohydrolase	0.117	0.143
holC	DNA polymerase III, chi subunit	0.134	0.144
xerD	site-specific tyrosine recombinase	0.114	0.146
tolB	periplasmic protein	0.115	0.146
yheS	fused predicted transporter subunits of ABC superfamily: ATP-binding components	0.110	0.146
ntpA	dihydroneopterin triphosphate pyrophosphatase	0.118	0.147
yabB	conserved protein, MraZ family	0.115	0.148
lolA	chaperone for lipoproteins	0.117	0.153
yggD	predicted DNA-binding transcriptional regulator	0.116	0.153
pnp	polynucleotide phosphorylase/polyadenylase	0.110	0.155
yrbB	ABC transporter maintaining OM lipid asymmetry, cytoplasmic STAS component	0.123	0.156
rnc	RNase III	0.117	0.157
xerC	site-specific tyrosine recombinase	0.138	0.160
rfaF	ADP-heptose:LPS heptosyltransferase II	0.120	0.161
yigP	conserved protein, SCP2 family	0.122	0.164
gyrB	DNA gyrase, subunit B	0.126	0.164
nagC	DNA-binding transcriptional dual regulator, repressor of N-acetylglucosamine	0.132	0.165
nrdR	Conserved protein	0.118	0.168
hemD	uroporphyrinogen III synthase	0.108	0.169
pheT	phenylalanine tRNA synthetase, beta subunit	0.124	0.171
frr	ribosome recycling factor	0.129	0.173
cls	cardiolipin synthase 1	0.129	0.181

Conclusion

The accuracy of most quantitative gene expression studies relied on the expressional stability of using the reference genes [5]. Previous studies [3, 22, 40, 51] had shown that not all commonly-used housekeeping genes such as GAPDH and beta-actin can be utilized as their gene expression may vary under different conditions [3] and the likelihood of suitable reference genes across many organisms and organs is low [7]. In addition, a recent study [52] examined the published microarray datasets from the livers of two closely related species of squirrels at comparable physiological states did not find suitable liver-specific reference genes. This suggests a need to identify reference genes for each organism and tissue. High-throughput transcriptome profiling technologies, such as microarray, presents a suitable resource for the analysis and identification of reference genes [13].

The advantage of CV over other more complex methods, such as NormFinder [27], is its capability to analyze as large a number of samples as required as the number of calculations increases proportionally to the sample size, resulting in linear complexity. NormFinder uses residual analysis between sample subgroup variation and the overall variation of the expression dataset to evaluate the variation contributed by each gene in the entire dataset [27]. Thus, the computational complexity of NormFinder increases exponentially as the number of samples increase; hence, NormFinder is only able to work with a small number of genes within reasonable time and computational resources. Given the advantageous ability of CV to process large amounts of data such as those derived from microarrays, it is plausible that CV can be used as a weaker filter for a broad category of genes with low expression variation, followed by stronger statistical analysis by NormFinder [27] to identify suitable reference genes.

CV does not take into consideration systematic variations such as those introduced by inaccuracy in sample preparation [53]. Thus, CV can only be used as a categorical filter of gene expression into a striation of classes. NormFinder is employed as the fine diagnostic tool to further determine the stability of gene expression within the class of genes with low expression variation to identify the most stably expressed gene. In the four case studies mentioned above, the identification of reference genes in multiple microarray datasets was achieved by analyzing the CV using either comparable or non-comparable microarray datasets on eukaryotic or prokaryotic biological samples. In the first case study, genes with the CV value of less than 0.1 were selected for further analysis while only the top 10% of genes that showed the lowest CV were chosen for downstream NormFinder [27] ranking. A threshold of 0.1 CV or 10% of lowest CV were set in these case studies. In Cases 3 and 4, the threshold functioned as a basic filter to limit the data for further processing by NormFinder [27]. By setting the threshold at 0.1 or extracting data only from the lowest 10% CV, many of the data were filtered out and left with the genes that showed low expression variation and able to be processed by NormFinder [27]. However, in the studies of comparable microarray datasets (Case 3 and 4) or different organs (Case 2), the threshold setting may exclude genes that showed low variation in "Dataset A" but not in "Dataset B" or in "Organ A" but not in "Organ B", or vice versa. The cross analysis of the data using the threshold in these scenarios will give rise to the identification of genes with the highest stability among the datasets. Nonetheless, in each dataset, there might be other genes that show lowest deviation in the gene expression. This could be one of the disadvantages that resulted from using CV to cross analyze the data from different datasets or different organs.

It is plausible that the results of CV and NormFinder [27] analysis may be correlated as both methods had been used to identify reference genes within a dataset [3]. Our results from Case 3 suggested that CV and NormFinder [27] stability were significantly correlated to each other (p-value = 1.79×10-31). However, the strength of this correlation is difficult to establish as the significance in p-value did not indicate the correlation strength. At the same time, the comparison between CV, NormFinder [27], geNorm [25] and BestKeeper© [26] requires a more comprehensive study using a datasets spanning across different tissues and organisms.

In all the cases presented, we utilized CV as a filter to screen all the genes of differential expression to reduce the number of genes/probes to be processed by NormFinder. However, it is possible that some stably-expressed genes may be excluded at this stage. Hence, using a 2-step process of CV and NormFinder may not be optimal. Due to this limitation of CV processing, a novel statistical tool with statistical stringency (measured by specificity and sensitivity) as NormFinder [27], geNorm [25] and BestKeeper© [26] but at the same time can extensively process the data as CV is necessary. This tool will allow the processing all the data within the microarray sets in a larger scale, thereby taking account of all the available data. This is likely to be statistically stronger than a 2-step approach of CV-NormFinder.

A possible way to achieve increased statistical stringency be expanding on the functions of NormFinder, which is usually used to process small sample size [27, 29, 54]. Hence, in this proposed extended version of NormFinder, the raw data in the microarray dataset are separated into to a few small manageable parts to run, for instance, "Run #1" and "Run #2" (Figure 1). In each run, NormFinder will generate a list of stability indices for the genes or probes. In the two lists generated, one gene or probe sample is overlapped across the two runs (Figure 1). By merging or normalizing the stability index for each particular probe in two runs, a new list of NormFinder stability indices that consist of all the genes or probe samples in both runs can be constructed. This process can be repeated to merge multiple NormFinder runs across large datasets using sets of overlapping samples, resulting in an overall comparison of expressional stability across entire genome.

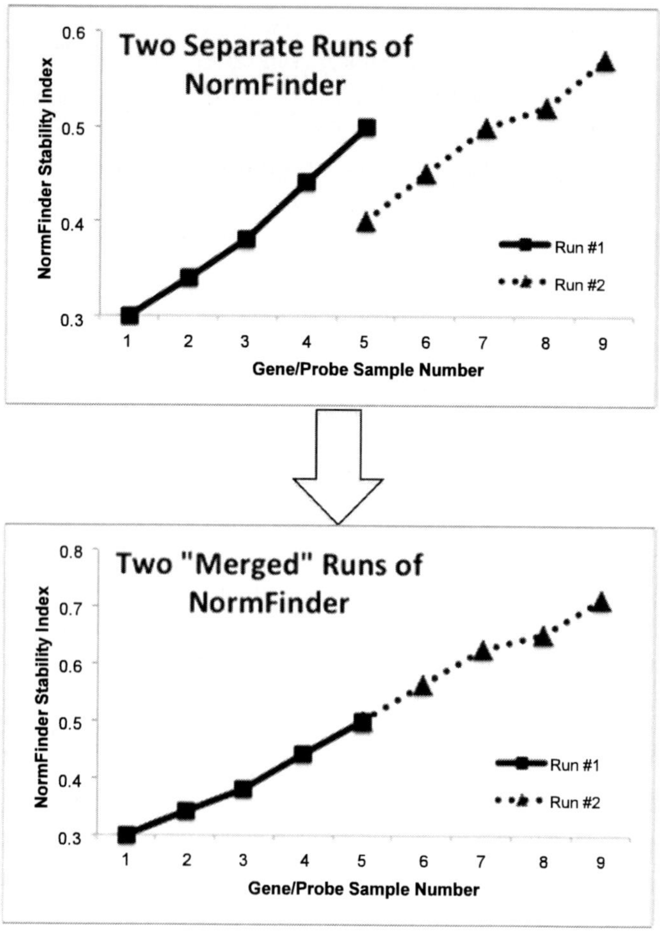

Figure 1. Two runs are performed on the same microarray dataset that has been fragmented into smaller manageable sample size to be analyzed by NormFinder. Each run generates a list of stability index for the gene/ probe sample. By merging or normalizing the stability index of the overlapped gene (Gene number "5"), a new list of stability index can be constructed that compromises all genes or probe samples within the microarray dataset.

Another possible method for microarray scale analysis might be the use of pairwise correlation of expression values in the entire dataset. Keng et al., [52] reported that CV/NormFinder-identified stably expressed genes demonstrated significantly less expressional correlation to other randomly selected genes. This agrees with a recent study by Ling et al., [55] suggesting that orthologous genes demonstrate higher degree of expressional correlation compared to random pairs of genes. Moreover, pairwise correlation of expression values may be the basis of a tool to identify reference genes from large datasets.

In this chapter, we are not suggesting that a standardized method should be employed in the process of analyzing and identifying a suitable reference gene. Instead, more comprehensive alternatives should be incorporated with a higher tolerance rate in sample size to allow an overall comparison of the gene expression in order to select the most appropriate housekeeping genes in each case.

Acknowledgments

The authors will like to thank the following colleagues for their comments on this manuscript: JSH Oon (University of Queensland, Australia), QMA Xie (University of Queensland, Australia), PCK Au (CSIRO, Australia), CC Goh (University of Virginia, USA), and YZ Koh (University of Portsmouth, UK). SSH, OYC, BMK, CC, CWL, WL, QC, EJA, and YL contributed to this work as part of the Science Mentorship Programme under Gifted Education Branch, Ministry of Education, Singapore.

References

[1] Chey, S; Claus, C; Liebert, UG. Validation and application of normalization factors for gene expression studies in rubella virus-infected cell lines with quantitative real-time PCR. *J Cell Biochem*, 2010, 110(1), 118-128.

[2] Bustin, SA; Nolan, T. Pitfalls of quantitative real-time reverse-transcription polymerase chain reaction. *J Biomolecular Tech*, 2004, 15(3), 155-166.

[3] Maccoux, LJ; Clements, DN; Salway, F; Day, PJ. Identification of new reference genes for the normalisation of canine osteoarthritic joint tissue transcripts from microarray data. *BMC Mol Biol*, 2007, 8, 62.

[4] Airoldi, EM; Huttenhower, C; Gresham, D; Lu, C; Caudy, AA; Dunham, MJ, et al. Predicting cellular growth from gene expression signatures. *PLoS Comput Biol*, 2009, 5(1), e1000257.

[5] Gubern, C; Hurtado, O; Rodriguez, R; Morales, JR; Romera, VG; Moro, MA; et al. Validation of housekeeping genes for quantitative real-time PCR in in-vivo and in-vitro models of cerebral ischaemia. *BMC Mol Biol*, 2009, 10:57.

[6] Czechowski, T; Stitt, M; Altmann, T; Udvardi, MK; Scheible, WR. Genome-wide identification and testing of superior reference genes for transcript normalization in Arabidopsis. *Plant Physiol*, 2005, 139(1), 5-17.

[7] Dundas, JB; Ling, MHT. Reference genes for measuring mRNA expression. *Theory in Biosci*, 2012, 131, 215-223.

[8] Jain, M; Nijhawan, A; Tyagi, AK. Validation of housekeeping genes as internal control for studying gene expression in rice by quantitative real-time PCR. *Biochem Biophys Res Commun*, 2006, 345, 646-651.

[9] Nicot, N; Hausman, JF; Hoffmann, L. Housekeeping gene selection for real-time RT-PCR normalization in potato during biotic and abiotic stress. *J Exp Botany*, 2005, 56, 2907-2914.

[10] Gibson, UE; Heid, CA; Williams; PM. A novel method for real time quantitative RT-PCR. *Genome Res*, 1996, 6, 995-1001.

[11] Sturzenbaum, SR; Kille, P. Control genes in quantitative molecular biological techniques: the variability of invariance. *Comp Biochem Physiol B Biochem Mol Biol* 2001, 130, 281-289.

[12] Takle, GW; Toth, IK; Brurberg, MB. Evaluation of reference genes for real-time RT-PCR expression studies in the plant pathogen Pectobacterium atrosepticum. *BMC Plant Biol*, 2007, 7, 50.

[13] Noriega, NC; Kohama, SG; Urbanski, HF. Microarray analysis of relative gene expression stability for selection of internal reference genes in the rhesus macaque brain. *BMC Mol Biol*, 2010, 11, 47.

[14] Remans, T; Smeets, K; Opdenakker, K; Mathijsen, D; Vangronsveld, J; Cuypers, A. Normalisation of real-time RT-PCR gene expression measurements in Arabidopsis thaliana exposed to increased metal concentrations. *Planta*, 2008, 227, 1343-1349.

[15] Glare, EM; Divjak, M; Bailey, MJ; Walters, EH. beta-Actin and GAPDH housekeeping gene expression in asthmatic airways is variable and not suitable for normalizing mRNA levels. *Thorax*, 2002, 57(9), 765-770.

[16] Gutierrez, L; Mauriat, M; Guénin, S; Pelloux, J; Lefebvre, JF; Louvet, R; et al. The lack of a systematic validation of reference genes: a serious pitfall undervalued in reverse transcription polymerase chain reaction (RT-PCR) analysis in plants. *Plant Biotechnol*, 2008, 6, 609-618.

[17] Gur-Dedeoglu, B; Konu, O; Bozkurt, B; Ergul, G; Seckin, S; Yulug, IG. Identification of endogenous reference genes for qRT-PCR analysis in normal matched breast tumor tissues. *Oncol Res*, 2009, 17(8), 353-365.

[18] Kosir, R; Acimovic, J; Golicnik, M; Perse, M; Majdic, G; Fink, M; et al. Determination of reference genes for circadian studies in different tissues and mouse strains. *BMC Mol Biol,* 2010, 11, 60.

[19] Shen, Y; Li, Y; Ye, F; Wang, F; Lu, W; Xie, X. Identification of suitable reference genes for measurement of gene expression in human cervical tissues. *Anal Biochem*, 2010, 405(2), 224-229.

[20] Rho, HW; Lee, BC; Choi, ES; Choi, IJ; Lee, YS; Goh, SH. Identification of valid reference genes for gene expression studies of human stomach cancer by reverse transcription-qPCR. *BMC Cancer*, 2010, 10, 240.

[21] Brattelid, T; Winer, LH; Levy, FO; Liestol, K; Sejersted, OM; Andersson, KB. Reference gene alternatives to Gapdh in rodent and human heart failure gene expression studies. *BMC Mol Biol*, 2010, 11, 22.

[22] Lallemant, B; Evrard, A; Combescure, C; Chapuis, H; Chambon, G; Raynal, C; et al. Reference gene selection for head and neck squamous cell carcinoma gene expression studies. *BMC Mol Biol*, 2009, 10, 78.

[23] Li, YL; Ye, F; Hu, Y; Lu, WG; Xie, X. Identification of suitable reference genes for gene expression studies of human serous ovarian cancer by real-time polymerase chain reaction. *Anal Biochem*, 2009, 394(1), 110-116.

[24] Infante, C; Matsuoka, MP; Asensio, E; Canavate, JP; Reith, M; Manchado, M. Selection of housekeeping genes for gene expression studies in larvae from flatfish using real-time PCR. *BMC Mol Biol*, 2008, 9, 28.

[25] Vandesompele, J; De Preter, K; Pattyn, F; Poppe, B; Van Roy, N; De Paepe, A; et al. Accurate normalization of real-time quantitative RT-PCR data by geometric averaging of multiple internal control genes. *Genome Biol*, 2002, 3(7), RESEARCH0034.

[26] Pfaffl, MW; Tichopad, A; Prgomet, C; Neuvians, TP. Determination of stable housekeeping genes, differentially regulated target genes and sample integrity: BestKeeper--Excel-based tool using pair-wise correlations. *Biotechnol Lett*, 2004, 26(6), 509-515.

[27] Andersen, CL; Jensen, JL; Orntoft, TF. Normalization of real-time quantitative reverse transcription-PCR data: a model-based variance estimation approach to identify genes suited for normalization, applied to bladder and colon cancer data sets. *Cancer Res*, 2004, 64(15), 5245-5250.

[28] He, JQ; Sandford, AJ; Wang, IM; Stepaniants, S; Knight, DA; Kicic, A; et al. Selection of housekeeping genes for real-time PCR in atopic human bronchial epithelial cells. *Eur Respir J*, 2008, 32(3), 755-762.

[29] Ren, S; Zhang, F; Li, C; Jia, C; Li, S; Xi, H; et al. Selection of housekeeping genes for use in quantitative reverse transcription PCR assays on the murine cornea. *Mol Vis*, 2010, 16, 1076-1086.

[30] De, RK; Ghosh, A. Interval based fuzzy systems for identification of important genes from microarray gene expression data: Application to carcinogenic development. *J Biomed Inform*, 2009, 42(6),1022-1028.

[31] Galiveti, CR; Rozhdestvensky, TS; Brosius, J; Lehrach, H; Konthur, Z. Application of housekeeping npcRNAs for quantitative expression analysis of human transcriptome by real-time PCR. *RNA*, 2010, 16(2), 450-461.

[32] Chia, CY; Lim, CW; Leong, WT; Ling, MH. High expression stability of microtubule affinity regulating kinase 3 (MARK3) makes it a reliable reference gene. *IUBMB Life*, 2010, 62(3), 200-203.

[33] Thorrez, L; Van Deun, K; Tranchevent, LC; Van Lommel, L; Engelen, K; Marchal, K; et al. Using ribosomal protein genes as reference: a tale of caution. *PLoS One*, 2008, 3(3), e1854.

[34] Le Hir, H; Maquat, LE; Moore, MJ. Pre-mRNA splicing alters mRNP composition: evidence for stable association of proteins at exon-exon junctions. *Genes Dev*, 2000, 14, 1098-1108.

[35] Meng, Z; Jackson, NL; Choi, H; King, PH; Emanuel, PD; Blume, SW. Alterations in RNA-binding activities of IRES-regulatory proteins as a mechanism for physiological variability and pathological dysregulation of IGF-IR translational control in human breast tumor cells. *J Cell Physiol*, 2008, 217, 172-183.

[36] Coulson, DTR; Brockbank, S; Quinn, JG; Murphy, S; Ravid, R; Irvine, GB; et al. Identification of valid reference genes for the normalization of RT qPCR gene expression data in human brain tissue. *BMC Mol Biol*, 2008, 9, 46.

[37] Chu, QH; Lin, YJ; Ang, EJG; Ling, MHT. Identification of transcriptional invariant genes in mouse endocrine glands from microarray data. *Proceedings of the 16th Youth Science Conference*, 2010, Singapore.

[38] Too, IH; Ling, MH. Signal peptidase complex subunit 1 (SPCS1) and hydroxyacyl-CoA dehydrogenase beta subunit (HADHB) are suitable reference genes in human lungs. *ISRN Bioinformatics*, 2012, 2012, Article ID 790452.

[39] Cheadle, C; Vawter, MP; Freed, WJ; Becker, KG. Analysis of microarray data using Z score transformation. *J Mol Diagn*, 2003, 5(2), 73-81.

[40] Nguewa, PA; Agorreta, J; Blanco, D; Lozano, MD; Gomez-Roman, J; Sanchez, BA; et al. Identification of importin 8 (IPO8) as the most accurate reference gene for the clinicopathological analysis of lung specimens. *BMC Mol Biol*, 2008, 9, 103.

[41] Liu, DW; Chen, ST; Liu, HP. Choice of endogenous control for gene expression in nonsmall cell lung cancer. *Eur Respir J*, 2005, 26, 1002-1008.

[42] Saviozzi, S; Cordero, F; Lo Iacono, M; Novello, S; Scagliotti, GV; Calogero, RA. Selection of suitable reference genes for accurate normalization of gene expression profile studies in non-small cell lung cancer. *BMC Cancer*, 2006, 6, 200.

[43] Kalies, KU; Hartmann, E. Membrane topology of the 12- and the 25-kDa subunits of the mammalian signal peptidase complex. *J Biol Chem*, 1996, 271(7), 3925-3929.

[44] Swanton, E; Bulleid, NJ. Protein folding and translocation across the endoplasmic reticulum membrane. *Mol Membr Biol*, 2003, 20(2), 99-104.

[45] Aoyama, T; Wakui, K; Orii, KE; Hashimoto, T; Fukushim, Y. Fluorescence in situ hybridization mapping of the alpha and beta subunits (HADHA and HADHB) of human mitochondrial fatty acid beta-oxidation multienzyme complex to 2p23 and their evolution. *Cytogenet Cell Genet*, 1997, 79(3-4), 221-224.

[46] Fu, Z; Attar-Bashi, NM; Sinclair, AJ. 1-14C-linoleic acid distribution in various tissue lipids of guinea pigs following an oral dose. *Lipids*, 2001, 36(3), 255-260.

[47] Heng, SSJ; Chan, OYW; Keng, BMH; Ling, MHT. Glucan biosynthesis protein G (mdoG) is a suitable reference gene in Escherichia coli K-12. *ISRN Microbiol*, 2011, 2011, Article ID 469053.

[48] Loubens, I; Debarbieux, L; Bohin, A; Lacroix, JM; Bohin, JP. Homology between a genetic locus (mdoA) involved in the osmoregulated biosynthesis of periplasmic glucans in Escherichia coli and a genetic locus (hrpM) controlling pathogenicity of Pseudomonas syringae. *Mol Microbiol*, 1993, 10(2), 329-340.

[49] Bohin, JP. Osmoregulated periplasmic glucans in Proteobacteria. *FEMS Microbiol Lett*, 2000, 186(1), 11-19.

[50] Page, F; Altabe, S; Hugouvieux-Cotte-Pattat, N; Lacriox, JM; Robert-Baudouy, J; Bohin, JP._Osmoregulated periplasmic glucan synthesis is required for Erwinia chrysanthemi pathogenicity. *J Bacteriol*, 2001, 183(10), 3134-3141.

[51] Meldgaard, M; Fenger, C; Lambertsen, KL; Pedersen, MD; Ladeby, R; Finsen, B. Validation of two reference genes for mRNA level studies of murine disease models in neurobiology. *J Neurosci Methods*, 2006, 156(1-2), 101-110.

[52] Keng, BMH; Chan, OYW; Hen, SSJ; Ling, MHT. Transcriptome analysis of Spermophilus lateralis and Spermophilus tridecemlineatus liver does not suggest the presence of Spermophilus-liver-specific reference genes. *ISRN Bioinformatics*, 2013, 2013, Article ID 361321.

[53] Sauer, U; Preininger, C; Hany-Schmatzberger, R. Quick and simple: quality control of microarray data. *Bioinformatics*, 2005, 21, 1572-1578.
[54] Silberberg, G; Baruch, K; Navon, R. Detection of stable reference genes for real-time PCR analysis in schizophrenia and bipolar disorder. *Anal Biochem*, 2009, 391, 91-97.
[55] Ling, MHT; Ban, Y; Wen, H; Wang, SM; Ge, SX. Conserved expression of natural antisense transcripts in mammals. *BMC Genomics*, 2013, 14(1), 243.

In: Microarrays: Principles, Applications and Technologies
Editor: James V. Rogers
ISBN: 978-1-62948-669-7

Chapter 7

Invertebrate DNA Chip: Opportunities and Challenges in the Development and Application of Microarrays for Marine Biodiversity Studies

Srujana Chitipothu[1,3], Alessia Cariani[2,4*], Fabio Bertasi[2,5], Marco Stagioni[2,4], Marc Kochzius[1,6], Dietmar Blohm[1], Fausto Tinti[2,4] and Monica Landi[2,7]

[1]Centre for Applied Gene Sensor Technology (CAG), FB2-UFT, University of Bremen, Bremen, Germany
[2]Department of Experimental Evolutionary Biology, University of Bologna, Bologna, Italy
[3]Current Affiliation: Vision Research Foundation, Sankara Nethralaya, Chennai, Tamilnadu, India
[4]Current affiliation: Department of Biological, Geological and Environmental Sciences, University of Bologna, Italy
[5]Current affiliation: Institute for Environmental Protection and Research, Rome, Italy
[6]Current affiliation: Marine Biology, Free University of Brussels (VUB), Brussels, Belgium
[7]Current affiliation: CBMA (Centre of Molecular and Environmental Biology), Department of Biology, University of Minho, Braga, Portugal

Abstract

DNA microarrays for marine biodiversity studies have been developed and tested in several groups, spanning different applications. They are employed to characterize ecological communities circumventing uncertainties and challenges associated with the

* Corresponding author: alessia.cariani@unibo.it.

conventional techniques of taxonomy employed to characterize these communities. Examples come from identifications of fishes and phytoplankton to monitoring of harmful algae. Although DNA microchips support a great diversity of applications and provide a wealth of findings in functional genomics and environmental studies, limits in ecological applications are known. So far, DNA microarrays used in biodiversity studies provide only qualitative data in terms of presence or absence of the species and its usage is greatly limited in providing the quantitative measurement of environmental samples. Moreover, DNA microarrays are commonly restricted to identify only those species that are targeted by the probes implemented on the chip. In turn, developing a broad spectrum array has technological challenges in terms of probe designing, experimental optimization, and statistical analysis. In this book chapter, we provide a critical assessment on promises and pitfalls of DNA microarrays as tool for marine invertebrate species identification. We implemented a DNA-chip prototype to identify 15 species of marine invertebrates from European Seas, including crustaceans, molluscs, and polychaetes, based on the two mitochondrial markers, cytochrome oxidase subunit I and 16S rRNA. Challenges involved in oligonucleotide probe design, in silico evaluation, and difficulties encountered through hybridization experiments are here explored. Specificity and sensitivity of the probes have also been critically evaluated to verify the suitability of the selected markers for microarray probe design. Since the ultimate application of DNA microarray to resolve complex environmental samples is a major challenge, we made an attempt to analyze gut contents of predator fishes. The problems encountered during this analysis, as the presence of target and not target species that could affect the specificity and the sensitivity of the DNA-chip to distinguish low and high abundant target species from a background of non-targets, were explored. We also reviewed advantages and disadvantages of DNA microarray technology compared to other molecular identification methods such as DNA barcoding and next generation sequencing. The outcome of the Invertebrate DNA Chip prototype served as a proof-of-concept for the identification of selected marine invertebrates and prey species of demersal fishes by DNA microarray. The potential of such microarrays can encompass several fields of scientific applications in marine biodiversity and ecosystem sciences, as marine ecosystem diversity and environmental monitoring, seafood quality control, and understanding food webs and ecosystem functioning.

Introduction

Challenges in Marine Biodiversity and Ecosystem Sciences

Compared to terrestrial ecosystems, very little is known about marine biodiversity, and its changes in species richness and ecosystem functioning, mostly due to sampling and taxonomic constraints [1]. Although the situation of marine biodiversity is alarming, expertise to assess biodiversity is currently being lost [2]. The necessary monitoring of marine ecosystem diversity is thus hampered, due to the lack of taxonomic expertise and tools for the identification of sampled organisms [3].

Monitoring marine biodiversity through the standard quali-quantitative analysis of fish gut content dates back in the early 80s [4, 5, 6]. Since then, such analysis has been used as a tool across several studies [7]. The stomach content analysis has been adopted as supportive tool in marine research, ranging from fisheries assessment [8], successful monitoring of marine ecology [9] and marine food web analysis [10]. The approach has been successful to detect abundant and well-known species in the gut contents. However, quantitative and qualitative constrains can be commonly identified [11, 12]. Because prey is often partially or fully digested, the diagnostic characters for species identifications are frequently lost, and the identification of organisms is then mainly based on their remaining parts or traces [11]. For such reason, unambiguous prey identification through taxonomic approaches undergoes a limiting bottleneck. In addition, the standard taxonomic identification of organisms by stereo-microscopic analysis is extraordinarily time consuming, prone to errors of inadequately trained operators and marginally useful for routine analysis, often yielding partial results.

Analyses of Fish Gut Contents - Traditional versus Molecular Approaches

Much of the knowledge on fish diet habits was initially based on analysis of stomach contents involving methods of occurrence, numerical, weight, and volumetric frequencies [6][13][14]. The advent of molecular tools employed in ecological and fisheries studies expanded and integrated the standard taxonomic approach [15][16][17][18]. Based on the nucleotide differences among species-specific DNA sequences, the concept of genetic markers for taxa identification has gained acceptance and widespread application [19][20][21][22][23]. In fact, molecular approaches to species identification are considered as the most efficient solution to inventory all life forms for biodiversity conservation and monitoring [15][24]. Successful reports were published about DNA-based identification of marine animals, such as eggs, larvae, and adults of fishes [20][25][26], zooplankton [27], and invertebrate larvae [28]. The sequences of partial mitochondrial Cytochrome *c* Oxidase subunit 1 (COI) gene along with those of the 16S rRNA (16S) and Cytochrome *b* (Cytb) genes were proposed as standard tools (e.g., barcodes) for a global bio-identification system of eukaryotes [15].

The need for simultaneous handling of large numbers of samples in mass scale studies has led to improved molecular techniques able to perform multiple sequences analyses, such as DNA microarray technologies [29]. Microarrays, or so-called microchips, are one of the most powerful innovations in molecular biology since their emergence nearly 30 years ago [30]. Allowing parallel hybridizations of hundreds to thousands of nucleic acids probes on a small surface area, DNA microchips transformed and accelerated the technical framework of several research disciplines. The first idea of identifying a DNA with radioactive-labeled short single-stranded DNA molecules started in 1975 [31], where the target DNA to be analyzed was immobilized on a membrane. After fourteen years, in 1989, Saiki et al., [30] reversed the design and started the microarray revolution, where many DNAs could be identified simultaneously by immobilizing many specific oligonucleotides on a solid support (i.e., the microarray slide). Since its first implementation [32], a rapid increase in novel technological solutions is being explored to date. In the last few decades, DNA microarrays have evolved as a promising tool for detecting thousands of targets from environmental samples in biodiversity studies [33][34][35]. Some research programs have highlighted the potential of the DNA-microarrays as high throughput molecular tools to detect marine organisms for effective monitoring of marine ecology [18].

DNA microarrays were used to identify different taxa in the marine realm: bacteria [36], pathogenic vibrios infecting shellfish and fishes [37][38], benthic fishes [29][39][40], phytoplankton [41][42] and harmful algae [43][44][45][46]. Invertebrates form a major part of marine ecosystems and largely contribute to the marine food webs [47]. Being a highly diverse ecological group of organisms, the identification of marine invertebrates at the species level is a real challenge. Until now, DNA-DNA hybridization techniques were developed to identify larvae of marine invertebrates [48]. Hence, one of the main objectives of this chapter was to develop a DNA microarray to detect marine invertebrate species in the gut content of benthic marine fish. At the methodological level, the potential for high-throughput and automation of microarray platforms is the critical factor driving interest towards the development of DNA microchip technology, even though sometimes the arrays are lacking consistency of results and involve complex process of data analysis [49].

DNA Microarrays – Challenges

Several types of challenges can be identified in the development and application of DNA microarrays for environmental studies [50], compared to the highly standardized functional genomics arrays, and to the most recently used approaches, such as the DNA barcoding methodology [15][16] [17].

1. *Biological challenges*: Due to the large number of species across different phyla of invertebrates, it is highly impossible to develop an array having probes specific for all the organisms. Such microarray can be limited only to a selected number of species with referenced DNA sequences available for probe designing. However, the use of probes to detect organisms at multiple hierarchical taxonomic levels can enhance the accuracy of the array. Developing a broad-spectrum microarray with probes suitable for the identification of organisms at species and higher taxonomic levels will be a task facing technological challenges in probe design and optimization of hybridization conditions [51][52].
2. *Selection of appropriate gene markers for probe design*: This is a critical step, as the candidate gene should have a nucleotide substitution rate able to discriminate even closely related species. For such reasons, the mitochondrial DNA genes 16S, COI, and Cytb are widely used markers in probe design for species identification. The COI has been widely found to be efficient in differentiating closely related species [15][24][40][53]. On the contrary, the 16S was more suitable for identifying inter-specific variation rather than intra-specific variation [54]. Some organisms, such as fishes, were successfully differentiated also using the Cytb [40].
3. *Technological challenges*: The selection of appropriate gene markers would require in-silico and in-vitro standardization experiments. Major technological challenges include 1) the design of species-specific oligonucleotide probes, 2) the amplification of the target genes from single- or multi-target DNA template, 3) the optimization of hybridization conditions for highest binding efficiency of target to probes, 4) the detection of the optimum signal intensity from the fluorescent labels, 5) the sensitivity of the array to detect target species in highly diverse environmental samples, and finally 6) the development of standard statistical methods for data analysis [55][56].

Invertebrate DNA Chip - Development

This chapter deals with the development and implementation of an Invertebrate Chip (hereafter INV-CHIP) for the identification of selected species of marine invertebrates and the critical assessment of its reliability in species identifications. The INV-CHIP targets important prey species of demersal fishes in the Mediterranean Sea and bioindicator organisms, such as polychaetes, that are difficult to be identified by morphological characters. The INV-CHIP aimed at enhancing ecosystem research on energy flow, food web structure, and environmental monitoring, for its potential to facilitate the identification of digested prey in gut contents analysis. By increasing the accuracy of multi-target identifications and the performance for large mass-scale biodiversity studies, the INV-CHIP could further strengthen the regular monitoring of bioindicator organisms in marine ecological research. The specificity and sensitivity of the INV-CHIP were tested using single and multiple targets as well as environmental samples such as gut contents of demersal fish, whose prey species composition was in parallel analysed by standard morphological and taxonomic analysis.

Materials and Methods

Sample Collection and DNA Extraction

A total of 267 specimens from 15 species of crustaceans (7 species), molluscs (4) and polychaetes (4) were collected from different areas of the Adriatic Sea and North Sea (Table 1) during scientific expeditions. Species were selected on the basis of their ecological traits and geographical range, focusing on prey of demersal fish and bioindicator organisms. Simultaneously, specimens of tub gurnard, *Chelidonichthys lucerna* (Linnaeus, 1758), were collected by trawling from the North Adriatic Sea in order to obtain suitable gut samples for testing the INV-CHIP. All the individual specimens collected were morphologically identified at species level and stored in ethanol 96% at 4°C. A collection of voucher specimens and tissue samples was created and genomic DNA was extracted from each sample as previously described [29]. After recording of biological data, the entire gut and stomach content of the tub gurnard individuals were excised and stored in ethanol 96% at -20°C. Genomic DNA extraction of individual gut content was performed by collecting biological materials (i.e., particles and prey remains) from the stomach. The taxonomic analysis of the prey species in the gut content was performed with standard morphological approach under a stereo-microscope. The selected biological materials were then homogenised with a pestle or a homogenizer to break down agglomerates. An aliquot of approximately 200μL was pelleted by centrifugation and ethanol was removed. The pellet was dried at 37°C. The total genomic DNA extraction was performed following the protocol from Deagle et al., [57].

Amplification and Sequencing of Genetic Markers

The COI and 16S markers were targeted for amplification and sequencing. The COI and 16S were PCR amplified using the universal primer pairs LCO 1490-HCO 2198 [58], and

16SarL-16SbrH [59], respectively. All PCR products were quali-quantitatively checked by 1.5% agarose gel electrophoresis. Exo-Sap purified PCR amplicons were sequenced using the Big Dye terminator cycle sequencing kit (Applied Biosystem) according to the manufacturer's protocol. Both DNA strands were sequenced. Electropherograms were visually inspected and edited for indels and substitutions, by using the software MEGA version 5 [60]. After editing, the identity of the target sequences was checked by nucleotide BLAST search against the NCBI GenBank [61].

Table 1. List of 15 invertebrate species and respective area of collection

Taxon	Family	Species Name	Species Code	Collection Site	Individuals collected
Crustacea	Alpheidae	***Alpheus glaber***	**Ag**	Adriatic Sea	10
	Crangonidae	***Crangon crangon***	**Cc**	North Sea	64
	Goneplacidae	***Goneplax rhomboides***	**Gr**	Adriatic Sea	10
	Portunidae	***Liocarcinus depurator***	**Ld**	Adriatic Sea	10
	Portunidae	***Liocarcinus vernalis***	**Lv**	Adriatic Sea	10
	Euphausiidae	***Meganyctiphanes norvegica***	**Mn**	Adriatic Sea	10
	Grapsidae	***Pachygrapsus marmoratus***	**Pm**	Adriatic Sea	10
Mollusca	Octopodidae	***Eledone cirrhosa***	**Ec**	Adriatic Sea	10
	Corbulidae	***Lentidium mediterraneum***	**Lm**	Adriatic Sea	10
	Veneridae	***Chamelea gallina***	**Cg**	Adriatic Sea	10
	Ommastrephidae	***Illex coindetii***	**Ic**	Adriatic Sea	10
Polychaeta	Nereididae	***Hediste diversicolor***	**Hd**	Mediterranean Sea	22
	Nephtyidae	***Nephtys hombergii***	**Nh**	Mediterranean Sea & North Sea	56
	Pectinariidae	***Pectinaria koreni***	**Pk**	North Sea	13
	Sigalionidae	***Sigalion mathildae***	**Sm**	North Sea	12

Sequence Analysis and Capture Probe Designing

Sequences of at least two individuals for each species and additional sequences from NCBI were used for capture probe design. COI and 16S sequences of non-target crustaceans, molluscs, and polychaetes from European Seas were obtained from NCBI (Table 2). The oligonucleotide probes were designed by the Chip Designing Internet Service (CDIS) provided by the Miconet service of the Techno Mathematics Department of the University of Bremen (Microarray Collaboration Network, web page www.miconet.uni-bremen.de) [62].

The final criteria used for the 16S probe design were: the length of the oligo in the range of 23 to 50 bp, GC content of 35-55%, specific binding with the target sequences and hybridization temperature in the range of 70-80°C, considering a total 1M salt concentration.

The COI capture probes were designed as for the 16S but with much narrower criteria (oligo length from 23-35 bp and GC content of 40-50%). The probes were checked for any secondary structure, self-complementarities, and absence of cross-reactions against targets and non-targets. Minimal free energy (mfe) structures were computed by using the software RNA fold [63]. A capture probe was included in the experimental tests only if it recognized the proper target species with 0 mismatches and showed at least 3 mismatches distributed in the center of the oligo with respect to other target and non-target species. At least two capture probes for each marker and species were selected (Table 3).

Table 2. For each target and non target species, 16S and COI sequences used for probe design. A: sequences newly obtained for the INV-CHIP and B: sequences retrieved from the NCBI database

Taxon	Species	16S			COI		
		Targets		Non-targets	Targets		Non-targets
		A	B	B	A	B	B
Crustacea	*A. glaber*	4	0	601	3	0	1085
	C. crangon	5	0		2	0	
	G. rhomboides	3	0		3	0	
	L. depurator	3	0		3	1	
	L. vernalis	3	0		2	0	
	M. norvegica	4	1		3	2	
	P. marmoratus	4	1		2	0	
Mollusca	*E.cirrhosa*	3	3	726	3	1	1624
	L. mediterraneum	2	0		2	0	
	C. gallina	2	3		3	2	
	I.coindetii	2	1		3	1	
Polychaeta	*H. diversicolor*	4	0	162	5	0	461
	N. hombergii	13	0		9	0	
	P. koreni	3	3		4	0	
	S. mathildae	3	0		4	0	
	Total	70		1489	58		3170

Preparation of DNA Microarrays and Hybridization Experiments

Preparation of DNA microarray slides and hybridization experiments were carried out as described in details in Kochzius et al. [29][40], using a set of 96 spots arranged in 4 rows and 24 columns on a glass slide. The array layout of the prototype INV-CHIP is displayed in Figure 1.

Table 3. List of species-specific capture probes (COI and 16S) developed and implemented in the INV-CHIP prototype.

Species Name	Probe Name	Sequence in 5'- to 3'- direction	Length (bp)	%GC	Tm (°C)
A. glaber	Ag-16S_1	AGTTAGGTCATTATGCTGGGGCG	23	52	82,56
	Ag-16S_2	GTTTAAGTTAGGTCATTATGCTGGGGCGG	29	48	83,7
	Ag-COI_1	TCCTACTCAGGCTCCCAGTTCTA	23	52	82,13
	Ag-COI_2	TTAGGGATCTTCTCGCTACACCTCG	25	52	82,49
C. crangon	Cc-16S_1	GAATGATCGGACAAGGGGCTAACTG	25	52	82,4
	Cc-16S_2	GGAATGAATGATCGGACAAGGGGCTAAC	28	50	83,79
	Cc-16S_3	GGAATGAATGATCGGACAAGGGGCTAACTG	30	50	84,85
	Cc-COI_1	CCTGCTTTAACTCTTCTTCTATCTAGAGGA	30	40	79,49
	Cc-COI_2	CCTCCTGCTTTAACTCTTCTTCTATCTAGAGGA	33	42	81,95
	Cc-COI_3	CTAGAGGATTAGTAGAAAGAGGAGTAGGAACTGGA	35	43	82,36
G. rhomboides	Gr-16S_1	GAACTTGTATGAATGGTTGGACAAAGGAAAAGCTGTCTCTATTATA	46	35	83,83
	Gr-COI_2	CGCCTCTGTTGATATAGGTATTTTTTCCCTTCATT	35	37	82,1
	Gr-COI_1	ACTTCCTCCCTCTTTAACTCTCTTACTAAT	30	37	79,22
L. depurator	Ld-16S_1	CTTAGAAAAATGTATTGGGTTGGGGCGAC	29	45	82,51
	Ld-16S_2	CTTAGAAAAATGTATTGGGTTGGGGCGACTAAGGTATA	38	39	83,84
	Ld-COI_2	TTACTCCCACCTTCGCTAACTCTTCTCCTC	30	50	84,73
	Ld-COI_1	TCTACCCTCCCCTATCGGCTGCTATT	26	54	85,64
L. vernalis	Lv-16S_1	ATTCAGAAAAGTGTATTGGGTTGGGGCG	28	46	84,23
	Lv-16S_2	AGGTTTATTCAGAAAAGTGTATTGGGTTGGGG	32	41	82,98
	Lv-COI_1	GCTTCTTCTCATGAGAGGTATAGTCGAGAG	30	47	81,36
	Lv-COI_2	TGACCCAGTTCTCTATCAGCACTTGTTTTG	30	43	82,96
M. norvegica	Mn-16S_1	ATGGTCGGACAAGAAACAGACTGTCTTCA	29	45	84,11
	Mn-16S_2	GGTGTAGCAGCTTTAATAGAAGGTCTGTTCGAC	33	45	83,63
	Mn-COI_2	CCACCTTCTTTAACTCTTTTATTAGGCAGAGGTC	34	41	82,07
	Mn-COI_1	GGCAGAGGTCTTGTAGAAAGAGGAGTC	27	52	82,75
P. marmoratus	Pm-16S_1	TCGAGGGCTATAAAGGCTTGGTG	23	52	82,23
	Pm-16S_2	AGTCTAATTCGAGGGCTATAAAGGCTTGGTG	31	45	83,75
	Pm-COI_1	ACCTCCCTCCTTATCTCTCTTACTTACAAGA	31	42	81,83
	Pm-COI_2	CAACATACGCTCTTATGGTATGACAATAGACCA	33	39	81,66
E. cirrhosa	Ec-16S_2	GATTGGGGTGATCAAGGAATAAAAAAGA	28	36	78,62
	Ec-16S_1	GATAAATAAACCGAGAGTTTTGC	23	35	73,4
	Ec-COI_1	CCTCCCCTATCAAGTAATTTAGCCCACATAG	31	45	82,11
	Ec-COI_2	ACCCTCTGTTGACCTAGCAATTTTCTCTTTACATC	35	40	83,27

Species Name	Probe Name	Sequence in 5'- to 3'- direction	Length (bp)	%GC	Tm (°C)
L. mediterraneum	Lm-16S_1	TGGGGAAAGGTATGAAGGGACTG	23	52	81,98
	Lm-16S_3	CTGGAGGGCTAATCGAATGGAAAGT	25	48	81,9
	Lm-16S_2	CAGCGTAATTTTCTACTGGAGGGCT	25	48	81,97
	Lm-COI_1	CCCCATTATCCGGTAATACAGCTCACTC	28	50	82,81
	Lm-COI_2	AGACTTTCTTATTTTATCGCTACACCTCGGT	31	39	81,72
C. gallina	Cg-16S_1	GGAGAATGGTATGAATGGTTTAACGTAGAATAACTGTCTTTGGAA	45	36	83,93
	Cg-COI_2	CTTATCTAGGGCTCTGTCTCATTCGGG	27	52	82,51
	Cg-COI_1	CTTAGGTTCTGATTGTTGCCGGTG	24	50	81,25
I. coindetii	Ic-16S_1	GCTTGAATTTTTTAAAGGGACGAGAAGACCCTAATGAGCTTATAA	45	36	84,23
	Ic-COI_2	CCACCATCTTTAACTATATTACTAGCCTCTTCAGC	35	40	81,35
	Ic-COI_1	CCCTTATCTAGAAATTTATCTCATGCTGGACCC	33	42	81,97
H. diversicolor	Hd-16S_1	GCTAATTAATCACACACACACCCAA	25	40	78,96
	Hd-16S_2	TGGGACAACCTAAAGACAAATAAACCTCTTAGCT	34	38	83,06
	Hd-COI_1	GTCCGTCAGTAGACCTTGCAATCTTC	26	50	81,91
	Hd-COI_3	ATGAACAGTATACCCGCCATTAGCCAGAAATATTG	35	40	83,44
	Hd-COI_2	GGTCTCATCTATTTTGGGAGCCCTAAAC	28	46	81,65
N. hombergii	Nh-16S_1	AAAAGGAACAAGTTTGGTTGGGGC	24	46	82,3
	Nh-16S_2	AAAAGGAACAAGTTTGGTTGGGGCGACAAAG	31	45	85,92
	Nh-COI_2	CCCTTCTTTAATTCTTCTTGTTATATCCGCAGCTG	35	40	82,4
	Nh-COI_1	TATCCGCAGCTGTAGAAAAAGGAGTC	26	46	81,45
P. koreni	Pk-16S_2	GGGGCGGCTGAGGAAAATTAAATC	24	50	81,83
	Pk-16S_1	GAGGGCCAAGCTGTCTCTTTAGT	23	52	82,19
	Pk-16S_3	GGATCAAAGAAAATAGCTACCTCGGGG	27	48	81,73
	Pk-COI_1	CTTTTCTCTCCACTTAGCCGGGATC	25	52	81,93
	Pk-COI_2	CCCATTAATACTTGCTGCTCCAGAC	25	48	80,67
	Pk_COI_3	CCCCTTATCAAGAAACCTTGCACATGCG	28	50	84,63
S. mathildae	Sm-16S_1	CGACCCAGGAGCATTTAAACCCT	23	52	82,79
	Sm-16S_2	GAAGGCTGGAATGAACGGATAAACGAG	27	48	82,23
	Sm-COI_3	CCTTCAGTTGATCTCGCTATCTTTTCTCTTCATATTGC	38	39	82,94
	Sm-COI_2	ACTCCCTCCTTCTCTAATTCTTCTTTTATCTTCA	34	35	80,73
	Sm-COI_1	TCTTCAAGAGCCGTTGAAAAAGGAGTT	27	41	81,92

PC	**NC**	**NC**	**PC**
Cc-COI_1	Cc-16S_1	Ag-COI_1	Ag-16S_1
Empty	Cc-16S_3	**PC**	Empty
Lv-16S_1	Ld-COI_2	Ld-16S_1	Cc-COI_3
Mn-16S_1	**PC**	Lv-COI_1	Empty
Hd-16S_1	Pm-COI_1	Pm-16S_1	Mn-COI_2
Empty	Hd-COI_2	**PC**	Hd-COI_1
Pk-COI_1	Pk-16S_2	Nh-COI_2	Nh-16S_2
Pk-COI_3	**PC**	Pk-16S_3	Empty
Ec-16S_2	Sm-COI_1	Sm-COI_3	Sm-16S_1
Empty	Gr-16S_1	**PC**	Ec-COI_1
Lm-16S_2	Lm-COI_1	Lm-16S_1	Gr-COI_2
Cg-COI_2	**PC**	Cg-16S_1	Empty
Ag-16S_2	Empty	Ic-COI_2	Ic-16S_1
Empty	Cc-16S_2	**PC**	Ag-COI_2
Lv-16S_2	Ld-COI_1	Ld-16S_2	Cc-COI_2
Mn-16S_2	**PC**	Lv-COI_2	Empty
Hd-16S_2	Pm-COI_2	Pm-16S_2	Mn-COI_1
Empty	Nh-16S_2	**PC**	Hd-COI_3
Sm-16S_2	Pk-COI_2	Pk-16S_2	Nh-COI_1
Ec-16S_1	**PC**	Sm-COI_2	Empty
Lm-COI_2	Lm-16S_3	Gr-COI_1	Ec-COI_2
Ic-COI_1	Empty	Empty	Cg-COI_1
PC	**NC**	**NC**	**PC**

Figure 1. Layout of the prototype INV-CHIP. Probes name as listed in Table 3; PC: positive control; NC: negative control; Empty: empty spots.

Target Amplification - Multiplex 16S and COI PCR

In order to reduce the number of PCR reactions needed to carry out all single target hybridizations, a multiplex PCR was optimized combining the single 16S and COI protocols. Multiplex PCR conditions were similar to those used in single PCR, except for the annealing temperature set to 45°C. PCR reactions were performed in 100μL with 0.3μM of each primer. Target DNAs for microarray hybridization experiments were amplified using Cy5-5-labelled forward primers as described in Kochzius et al. [29][40].

Experiments for Testing Probe Specificity

Hybridization settings were evaluated through a initial series of optimization experiments. Single target hybridizations were carried out at four different hybridization temperatures (50°C, 55°C, 60°C, and 65°C), over three different hybridization time-spans (2h, 3h, and 4h). The overall signal pattern was evaluated in terms of the relationship between the number of specific signals and any false positive and false negative signals produced by the probes in order to define the optimal hybridization temperature and time-span on this microarray.

After assessing the optimized hybridization conditions, a series of single target hybridizations experiments were performed for both 16S and COI markers to test the specificity of the probes selected. In order to check the behavior of each target species on the chip and also the signal patterns obtained by hybridizing different markers, the first series of hybridization experiments were performed amplifying and hybridizing the two markers separately. Single target hybridizations were realized with three individuals for each target species and each hybridization experiment was replicated three times. According to these results, oligos were selected for the final INV-CHIP, which included both 16S and COI probes. A total of 64 capture probes with satisfactory behaviour were chosen for the combined INV-CHIP (Table 3, Figure 1). Both 16S and COI markers from all the 15 target species were amplified in a multiplex PCR with two sets of primers as described above. The final concentration of amplicons in the hybridization solution was 20nM instead of 10nM (as in the case of single marker hybridizations) assuming that the two markers were amplified with a comparable yield in the multiplex PCR. As all the hybridization parameters had already been checked and the signal pattern of every single target was known, two non-interfering targets were hybridized together on to the microarray to reduce the number and time of experiments.

Experiments for Testing Probe Sensitivity: Binary and Ternary Mixtures

The sensitivity of the microarray was tested using a hybridization series with two targets on the same chip, to check the reciprocal influence in term of the signal intensities given by the probes when the two targets were present at different concentration ratios (10:1; 3:1; 1:1; 1:3; 1:10). In this test, three sets of targets were selected from each of the three taxonomic groups represented on the INV-CHIP: Crustacea, Mollusca, and Polychaeta. The two targets were selected according to the criterion that, although closely related, they did not show any cross hybridization in the single target experiments.

The sensitivity was further assessed with similar three target hybridization series to check the number of targets that the microarray can identify without any interference. As in the two target experiments, three species belonging to the same three taxonomic groups were simultaneously hybridized on the microarray. In the first experiments, all the targets were added in equal amounts, while further experiments were carried out keeping one target at the initial concentration and adding two other targets in higher concentrations (1:1:1; 1:10:10; 10:1:10; 10:10:1). The PCR products were hybridized on to the INV-CHIP to check if there was a correlation between the initial template concentration, the PCR amplification of different targets and the signal intensities obtained.

Experiments on Environmental Samples

The suitability of the INV-CHIP in ecological studies was assessed by analyzing the gut content of seven individuals of the demersal fish *Chelidonechtys lucerna*. Previous work had analyzed the diet of several demersal fishery resources, and the preliminary results showed that *C. lucerna* feed on some of the crustacean species targeted by the INV-CHIP [64]. Several gut content samples were collected, total DNA was extracted as described above, and hybridized onto the chip. Verification of the hybridization results was carried out by cloning and sequencing the 16S and the COI amplicons obtained from the gut content samples.

Measurement of Fluorescence Signals and Data Analysis

Hybridization signals were measured as described in Kochzius et al., [40]. The hybridization spots were replicated according to the configuration of the array (Figure 1) ($n = 1$ for probes, $n = 14$ positive controls - PC and $n = 4$ for negative controls - NC). The mean and standard deviation of the pixels fluorescence in the spots were measured by the software Genepix pro v4.1 (Axon, Union City, USA), along with the local background. Local background was the mean of the pixels measured in a ring around the spot twice its diameter. Data were pre-processed by calculating the coefficient of variation, CV (namely, the ratio between standard deviation (SD) and mean) for each spot and excluding the signals with $CV > 1$ from the analysis.

Hybridization signals for all tests were analysed using the same procedure. Negative control signals were compared with local background values by means of Welch's *t* test for unpaired samples with pooled SD [65]. When significantly different at $p < 0.05$, the highest among negative control- and background- signal was retained as the control (CTRL) signal. If not significantly different, the two measurements were averaged as the CTRL signal for subsequent analyses. Differences between hybridization signals were tested by one-way ANOVA. Variances were preliminarily tested for heteroscedasticity using Bartlett test [66]. When necessary, data transformations were done to stabilize the variances. If variances could not be stabilized using appropriate transformations, ANOVA analysis was run on the untransformed data [67]. On significant ANOVA results, differences among groups were tested by multiple pairwise *t* tests. The Benjamini–Hochberg procedure [68] was used to control the false discovery rate and correct α values. For ANOVA analyses and pairwise tests the significance threshold was set at $p < 0.01$. All statistical analyses were performed using R Project for Statistical Computing v.2.15 [69].

Results

Sequencing of the Targets

For each target species, an initial set of 10 individuals was processed for DNA extraction, gene amplification and sequencing. At least two 16S and COI sequences for each target species were obtained (Table 2).

In Silico Characterization of Capture Oligonucleotides

For the 16S probe design, a series of different calculations were necessary to obtain oligos with satisfactory features for all target species, broadening the allowed range of thermodynamic properties. The length and GC-content parameters of the selected 16S probes are 23–46 bp and 35–52%, and those of the COI probes are 23–35 bp and 35–54%. All the probes selected are shown in Table 3 along with the parameters considered for selection.

Optimization of the INV-CHIP Hybridization Conditions

Single target hybridizations were carried out with three different targets at four different hybridization temperatures (50°C, 55°C, 60°C, and 65°C), over three different hybridization time-spans (2h, 3h, and 4h). The optimal hybridization temperature and time-span on this microarray were 60°C and 3h respectively (data not shown). According to these results, a series of optimization experiments was performed with the remaining 12 target species, testing the same individual for each species at three different hybridization temperatures (50°C, 55°C, 60°C) for 3h (data not shown). The overall signal pattern was evaluated in terms of the relationship between the number of specific signals and any false positive and false negative signals produced by the probes. Even though the specific signal was lost from some probes the unspecific signals were significantly reduced at 60°C, and all further experiments were always carried out at 60°C for 3h.

Specificity of INV-CHIP

A total of 64 capture probes with satisfactory behavior were chosen for the INV-CHIP. As all the hybridization parameters had already been checked and the signal pattern of every single target was known, two non-interfering targets were hybridized together on to the microarray. The overall signal pattern along with significance of the tested values against the CTRL values are shown in Table 4. At least one probe of each marker per target species gave a significant signal, except the 16S probes of *C. crangon*, *G. rhomboides* and *L. mediterraneum* which were lost, probably due to an uneven amplification of the two markers in the multiplex PCR reaction. Most of the specific signals were significant considering the stringent threshold of $p < 0.01$, while two COI probes, Lv-COI_1 and Nh-COI_1, gave a specific significant signal only if $p < 0.05$ (Table 4).

Few probes showed false positive signals (binding with non specific targets). This occurred for the COI probes Lv-COI_2 and Nh-COI_1, and for the 16S probes Ld-16S_1 and Ld-16S_2 which cross-hybridized between the two congeneric target species *L. depurator* and *L. vernalis*. Several probes did not give the expected hybridization signal when the specific target was hybridized on the INV-CHIP, leading to false negative signals. This negative performance of the INV-CHIP was much more relevant for the 16S probes (with 11 out of 30 probes not producing a detectable signal when compared to the CTRL), than for the COI probes (where only five out of 34 probes provided false negative signals).

Table 4. Summary of the hybridization results of the 15 targets on the INV-CHIP. Probe names are listed in Table 3. For species codes refer to Table 1. **: specific signals with p < 0.01; *: specific signals with p < 0.05; FN: False negative signals; FP: False positive signals, with associated significance

Probes	Ag+Cc	Ec	Gr+Pk	Ld+Ic	Lm+Nh	Lv+Cg	Mn+Hd	Pm+Sm
Ag-16S_1	FN							
Ag-16S_2	**							
Ag-COI_1	**							
Ag-COI_2	**							
Cc-16S_1	FN							
Cc-16S_2	FN							
Cc-16S_3	FN							
Cc-COI_1	**							
Cc-COI_2	**							
Cc-COI_3	**							
Cg-16S_1						**		
Cg-COI_1						FN		
Cg-COI_2						**		
Ec-16S_1		**						
Ec-16S_2		**						
Ec-COI_1		**						
Ec-COI_2		**						
Gr-16S_1			FN					
Gr-COI_1			FN					
Gr-COI_2			**					
Hd-16S_1							**	
Hd-16S_2							**	
Hd-COI_1							**	
Hd-COI_2							**	
Hd-COI_3							**	
Ic-16S_1				**				
Ic-COI_1				**				
Ic-COI_2				**				
Ld-16S_1				**		FP *		
Ld-16S_2				**		FP **		
Ld-COI_1				**				
Ld-COI_2				**				
Lm-16S_1					FN			
Lm-16S_2					FN			
Lm-16S_3					FN			
Lm-COI_1					**			
Lm-COI_2					**			
Lv-16S_1						**		
Lv-16S_2						**		
Lv-COI_1						*		
Lv-COI_2			FP **			**		
Mn-16S_1							FN	
Mn-16S_2							**	
Mn-COI_1							FN	
Mn-COI_2							**	
Nh-16S_1					**			
Nh-16S_2					**			
Nh-COI_1	FP **				*			

Probes	Ag+Cc	Ec	Gr+Pk	Ld+Ic	Lm+Nh	Lv+Cg	Mn+Hd	Pm+Sm
Nh-COI_2					**			
Pk-16S_1			FN					
Pk-16S_2			**					
Pk-16S_3			**					
Pk-COI_1			**					
Pk-COI_2			FN					
Pk-COI_3			**					
Pm-16S_1								FN
Pm-16S_2								**
Pm-COI_1								FN
Pm-COI_2								**
Sm-16S_1								**
Sm-16S_2								**
Sm-COI_1								**
Sm-COI_2								**
Sm-COI_3								**

Sensibility of the INV-CHIP

Multiple target hybridizations were performed to 1) check the influence of closely related targets upon each other when hybridized together and 2) evaluate the effect on the signal intensities of the probes of one target when other targets are present at different concentrations, and 3) assess the detection limit of a target in complex mixtures. Binary mixtures were set up by selecting targets from the three main taxonomic groups considered in this study (Figure 2): *C. gallina, E. cirrhosa* (Mollusca), *G. rhomboides*, *P. marmoratus* (Crustacea), *N. hombergii*, and *S. mathildae* (Polychaeta). Figure 2a shows the signal pattern obtained with the molluscs *E. cirrhosa* and *C. gallina*. In the experiments performed, the presence of a second target did not influence significantly the signal of specific hybridization, except for the *C. gallina* capture probe Cg-COI_1, whose hybridization signal was totally lost when the *E. cirrhosa* target was added at higher concentration ratios. A gradual correspondence could be observed between the hybridization signal intensity from probes of one target, when the concentration of the second was increased or decreased; however, a clear correlation between target amount and hybridization signal intensity was not observed.

The second set of the two target hybridizations was carried out with the crustaceans *G. rhomboides* and *P. marmoratus* (Figure 2b). According to the results obtained in the specificity tests, the two captures Gr-16S_1 and Gr-COI_1 did not produce any detectable signal when *G. rhomboides* was hybridized at standard concentration, while increasing amounts of target yielded a significant signal only for probe Gr-COI-1. Overall, all the specific probes gave hybridization signals patterns according to the relative increase in the target concentration.

The third experiment was conducted with various mixtures of polychaetes *N. hombergii* and *S. mathildae* (Figure 2c). A correlation between the signal intensity and the target concentration was observed only to a certain extent. The rapid increase in the amount of the second target did not affect the significance of signals of the first target and vice versa, except for the capture probe Nh-COI_1. In the latter case, the signal was quite low compared to the other capture probes and the significant signal was lost when the *S. mathildae* target was added in highest concentrations.

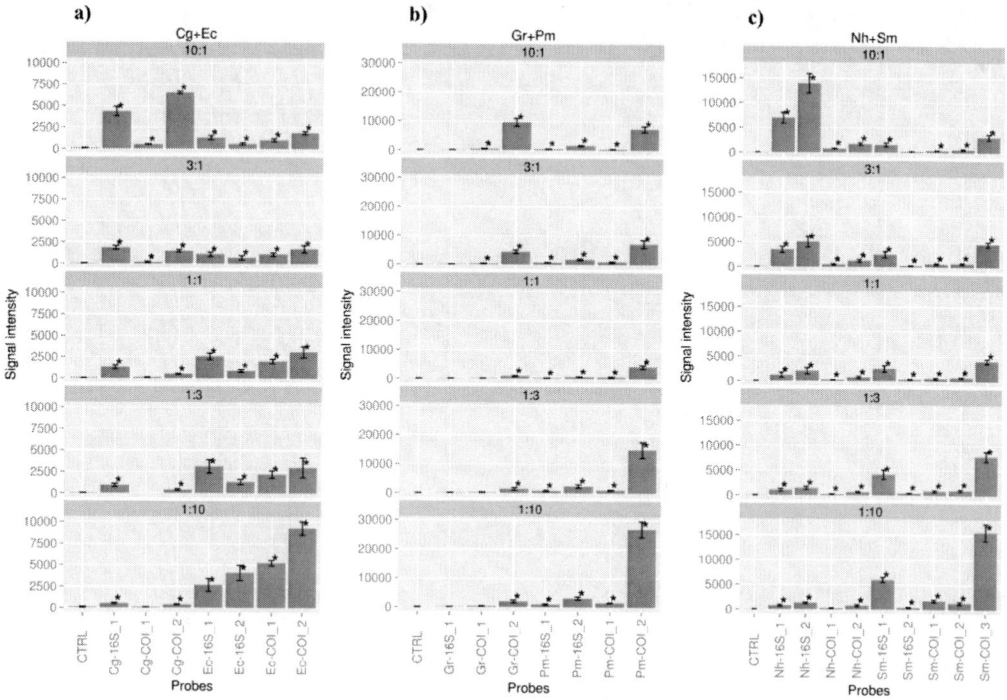

Figure 2. Signal pattern of binary mixture hybridizations; Probe names are listed in Table 3. For species codes refer to Table 1. a) *C. gallina, E. cirrhosa* (Mollusca); b) *G. rhomboides, P. marmoratus* (Crustacea); c) *N. hombergii, S. mathildae* (Polychaeta). * $p< 0.01$.

To further assess the detection limit of the INV-CHIP when analyzing complex mixtures, three target species of the same phyla were hybridized together. The molluscs *C. gallina, E. cirrhosa* and *Illex coindetii*, the crustaceans *G. rhomboides*, *P. marmoratus,* and *L. vernalis*, and the polychaetes *N. hombergii*, *S. mathildae,* and *P. koreni* were used in ternary mixtures hybridization experiments (Figure 3).

The results from the experiments with the ternary mixtures of molluscan targets are displayed in Figure 3a. Hybridization pattern showed that only the target *E. cirrhosa* was properly amplified in the multiplex reactions, while the other two targets were amplified with a minor yield. The observed signal intensity from these specific probes was very low compared to the single target experiments, thereby exhibiting decreased significance in many experiments. All the *E. cirrhosa* captures gave the expected signals when the target was equivalent to the other two templates as well as when the concentration of the target was lower respect to the other two. The overall signal from the *C. gallina* captures was very low, with no significant signal observed from both the *C. gallina* COI probes. This differed from the results of specificity tests, where the capture Cg-COI_2 gave the expected hybridization signal. The *C. gallina* 16S probe pattern reflected the concentration of the targets, providing a false negative result when the specific target was lower than the two other targets. Capture probes targeting *I. coindetii* gave significant hybridization signals in most of the experiments, with intensity values well correlated to the template concentrations in the multiplex PCRs. As observed for *C. gallina,* a false negative signal was obtained from two out of three *I. coindetii* probes when the specific target was at a lower concentration than the other two targets.

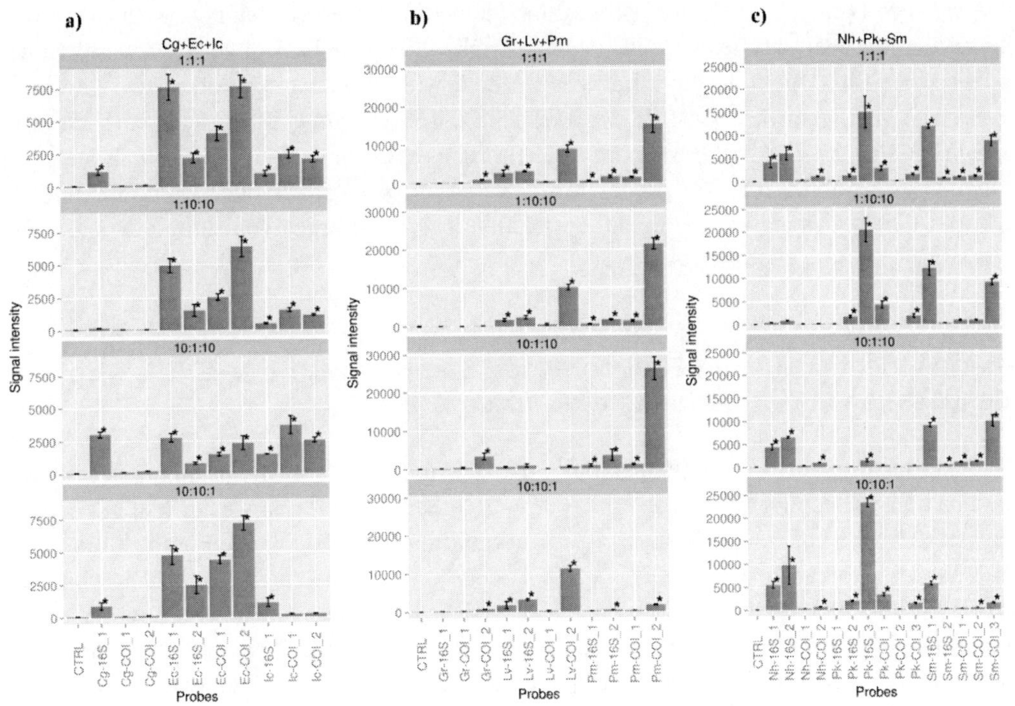

Figure 3. Signal pattern of ternary mixture hybridizations. Probe names are listed in Table 3. For species codes refer to Table 1. a) *C. gallina, E. cirrhosa* and *I. coindetii* (Mollusca); b) *G. rhomboides, P. marmoratus* and *L. vernalis* (Crustacea) and c) *N. hombergii, S. mathildae* and *P. koreni* (Polychaeta). * $p < 0.01$.

The second set of multi-targets consisted of three crustacean targets: *G. rhomboides, P. marmoratus* and *L. vernalis* (Figure 3b). All the three targets were amplified as expected since signals were obtained from at least one specific probe for each species. As expected, the two capture probes Gr-16S_1 and Gr-COI_1 did not produce any detectable signal either when *G. rhomboides* was hybridized at standard concentration or increasing its amounts in the ternary mixture. Moreover, the significant signal from probe Gr-COI_2 was lost when the concentration of the specific target was lowered with respect to the other two targets.

Hybridization patterns of *L. vernalis* were similar to those obtained from the other two crustacean species, losing all signals when the target was hybridized at the lowest concentration ratio. Three capture probes provided the expected specific signals, while the probe Lv-COI_1 did not produce any significant signal. The *P. marmoratus* capture probes showed a higher level of performance than that observed in the specificity test series, since all probes gave significant signals when the target was added at equal or higher concentration ratios. Only the capture probes Pm-16S_1 and Pm-COI_1 produced non-significant signals when *P. marmoratus* was analyzed at the lowest concentrations, confirming the expectation of false negative signals.

The third set of ternary mixture hybridization experiments included the polychaetes *N. hombergii, S. mathildae* and *P. koreni.* The signal pattern indicated that all three targets and both markers were successfully amplified in the multiplex PCR (Figure 3c). All capture probes produced at least one significant signal when the corresponding specific target was present in the mixture, with the exception of two *P. koreni* (Pk-16S_1 and Pk-COI_2) and one

N. hombergii (Nh-COI_1) capture probes that yielded false negative signals consistently with the results of specificity tests. False negative signals were indeed given by several capture probes when the target concentration was lowered with respect to the other two targets (*N. hombergii*: all capture probes; *S. mathildae*: Sm-16S_2, Sm-COI_1, and Sm-COI_2; *P. koreni*: Pk-16S_2). The higher number of COI probes giving false negative signals suggested a higher susceptibility of COI probes to the variation of target amount in the mixture. No hybridization series gave significant correlation between target concentration and signal intensity.

Species Identification in Environmental Samples with INV-CHIPs: Prey Identification in Gut Contents of *C. lucerna*

After testing the INV-CHIP specificity and sensibility with artificial mixtures of known composition, the applicability of the microarray to the taxonomic analysis of fish gut contents was tested by the gut content analysis of tub gurnard individuals collected in the Adriatic Sea [64]. In this geographic location, *C. lucerna* feeds on several crustacean species that were targeted by the INV-CHIP as shown in Table 5. This evidence represents on optimal test to assess the INV-CHIP specificity and sensitivity on field samples. The stomach contents of each individual was analyzed under the stereomicroscope and taxonomically recognized using identification keys, giving attention to assess the presence/absence of the species targeted by the INV-CHIP.

Table 5. Checklist of preys of *Chelidonichthys lucerna* and relative abundance indexes, modified from Stagioni et al., 2011. *, Species targeted by the INV-CHIP. %N, percentage in number; %W, percentage in weight; %F, frequency of occurrence; %IRI, percentage of index of relative importance of prey items; PSA, % prey taxon over all preys in all predators

PREY TAXA	%N	%W	%F	%IRI	PSA
CRUSTACEA	89.69	58.01	7.47	92.24	72.51
Philocheras spp.	44.38	1.53	14.52	26.51	23.98
Philocheras bispinosus	15.1	0.96	4.75	3.04	33.5
***Goneplax rhomboides**	14.16	29.52	28.95	50.3	60.41
***Liocarcinus* spp.**	7.09	4.47	10.95	5.04	59.1
Processa spp.	6.15	1.8	5.26	1.66	17.52
***Liocarcinus depurator**	5.96	20.32	10.02	10.47	73.7
***Alpheus glaber**	0.83	1.01	3.23	0.24	19.63
Solenocera membranacea	0.62	1.51	1.95	0.17	47.18
Squilla mantis	0.43	2.95	1.78	0.24	54.32
***Pachygrapsus marmoratus**	0.41	0.32	1.1	0.03	46.96
Macropodia spp.	0.28	0.16	0.34	0.01	26.02
PREY TAXA	**%N**	**%W**	**%F**	**%IRI**	**PSA**
Jaxea nocturna	0.21	0.51	0.93	0.03	16.05
Liocarcinus maculatus	0.23	0.43	0.68	0.02	33.92
Corystes cassivelaunus	0.11	1.79	0.51	0.04	81.96
Upogebia spp.	0.08	0.28	0.34	<0.01	16.83

PREY TAXA	%N	%W	%F	%IRI	PSA
Lophogaster typicus	0.08	0.02	0.34	<0.01	6.42
Liocarcinus vernalis	0.08	0.09	0.17	<0.01	41.35
Sicyonia carinata	0.06	0.09	0.17	<0.01	30.56
Pontophilus spinosus	0.06	0.09	0.17	<0.01	26.84
Parapenaeus longirostris	0.04	0.15	0.17	<0.01	17.35
Brachynotus spp.	0.04	0.05	0.17	<0.01	11.9
Chlorotocus crassicornis	0.04	0.05	0.08	<0.01	51.53
Liocarcinus pusillus	0.02	0.02	0.08	<0.01	16.68
Pisidia longimana	0.02	0.02	0.08	<0.01	10.67
Ebalia granulosa	0.02	<0.01	0.08	<0.01	52.46
Processa modica	0.02	<0.01	0.08	<0.01	1.29
Munida spp.	0.02	<0.01	0.08	<0.01	100
Melicertus kerathurus	0.02	<0.01	0.08	<0.01	2.99
TELEOSTEI	**6.72**	**39.74**	**1.92**	**7.45**	**74.65**
Engraulis encrasicolus	0.6	11.91	2.29	1.14	81.89
Gobius niger	0.6	8.77	2.04	0.76	86.93
Lesueurigobius friesii	0.49	1.57	1.53	0.13	52.88
Cepola macrophthalma	0.13	2.01	0.59	0.05	98.84
Lesueurigobius suerii	0.26	0.86	0.68	0.03	38.63
Trisopterus minutus	0.08	1.8	0.17	0.01	99.78
Arnoglossus laterna	0.08	1	0.25	0.01	73.34
Gobius spp.	0.08	0.52	0.34	0.01	27.4
Serranus hepatus	0.04	0.79	0.17	0.01	87.19
Callionymus maculatus	0.08	0.35	0.25	<0.01	56.29
Gaidropsarus biscayensis	0.08	0.18	0.34	<0.01	40.43
Merlangius merlangus	0.02	0.93	0.08	<0.01	100
Deltentosteus quadrimaculatus	0.04	0.15	0.17	<0.01	92.81
Callionymus spp.	0.04	0.13	0.17	<0.01	84.73
Callionymus risso	0.04	0.11	0.17	<0.01	100
Merluccius merluccius	0.02	0.24	0.08	<0.01	100
Microchirus variegatus	0.02	0.03	0.08	<0.01	79.17
Pomatoschistus minutus	0.02	0.01	0.08	<0.01	100
MOLLUSCA	**1.57**	**0.55**	**0.41**	**0.04**	**30.13**
Corbula gibba	0.34	0.19	1.44	0.03	4.69
Turritella communis	0.21	0.22	0.93	0.02	9.49
Anadara demiri	0.04	0.01	0.17	<0.01	2.94
Nassarius spp.	0.06	0.01	0.25	<0.01	0.48
Arca tetragona	0.04	0.02	0.17	<0.01	3.69
Tellina spp.	0.04	0.01	0.17	<0.01	3.01
Epitonium spp.	0.02	0.03	0.08	<0.01	10.65
ANELLIDA	**0.03**	**0.01**	**0.01**	**<0.01**	**4.92**
Aphrodita aculeata	0.02	0.01	0.08	<0.01	3.8
Sternaspis scutata	0.02	0.01	0.08	<0.01	9.67
NOT DETERMINED	**1.99**	**1.7**	**0.92**	**0.28**	**17.38**

The INV-CHIP hybridization experiments on gut contents were carried out according to the protocol optimized with artificial mixtures. The assay was replicated at least four times for each gut sample, due to high variability of the hybridization signals and the occurrence of various artifacts in the INV-CHIP hybridization results, probably due to the biological complexity of the environmental mixture.

Results were evaluated in terms of intensity and significance of hybridization signals. The inferred presence/absence of the INV-CHIP target species was compared to the results of the standard taxonomical analysis. The summarized INV-CHIP results were shown in Table 6 which displayed also the species identified by morphological keys, either at species level (SP) or at genus level (GE).

Table 6. Results of the INV-CHIP analysis of seven gut contents of *C. lucerna,* compared to those obtained by morphological identification. SP: morphological identification at species level. GE: morphological identification at genus level. Bold text: targets detected by morphological analysis and by the INV-CHIP. Shaded cells: target detected by the INV-CHIP but not by morphological analysis. St118-120: sampling station; TL1-4: individual code for each specimen collected in each sampling station

SAMPLES	Gut 1 St120 TL1	Gut 2 St118 TL2	Gut 3 St120 TL3	Gut 4 St118 TL3	Gut 5 St119 TL1	Gut 6 St118 TL4	Gut 7 St119 TL2
INV-CHIP Targets							
L. mediterraneum							
C. gallina							
I. coindetii							
E. cirrhosa							
M. norvegica							
C. crangon							
A. glaber						**X(GE)**	
G. rhomboids	**X(SP)**		**X(SP)**				
P. marmoratus							
L. depurator		**X(GE)**		**X(SP)**	**X(GE)**	**X(GE)**	**X(GE)**
L. vernalis		**X(GE)**			**X(GE)**	**X(GE)**	**X(GE)**
N. hombergii							
H. diversicolor							
P. koreni							
S. mathildae							
Non Target species							
Philocheras sp.	X	X	X	X	X	X	X
Gobidae (otolith remains)	X		X	X			
Processa sp.				X		X	
Callyonimus sp.		X			X	X	
Bony fish (trace)							X

Results for the gut content samples were also presented as box plot graphs in Figure 4, to point out the great variability of the signal patterns obtained. This suggested that additional replicates need to be performed for each samples in order to obtain more reliable results in terms of species identification.

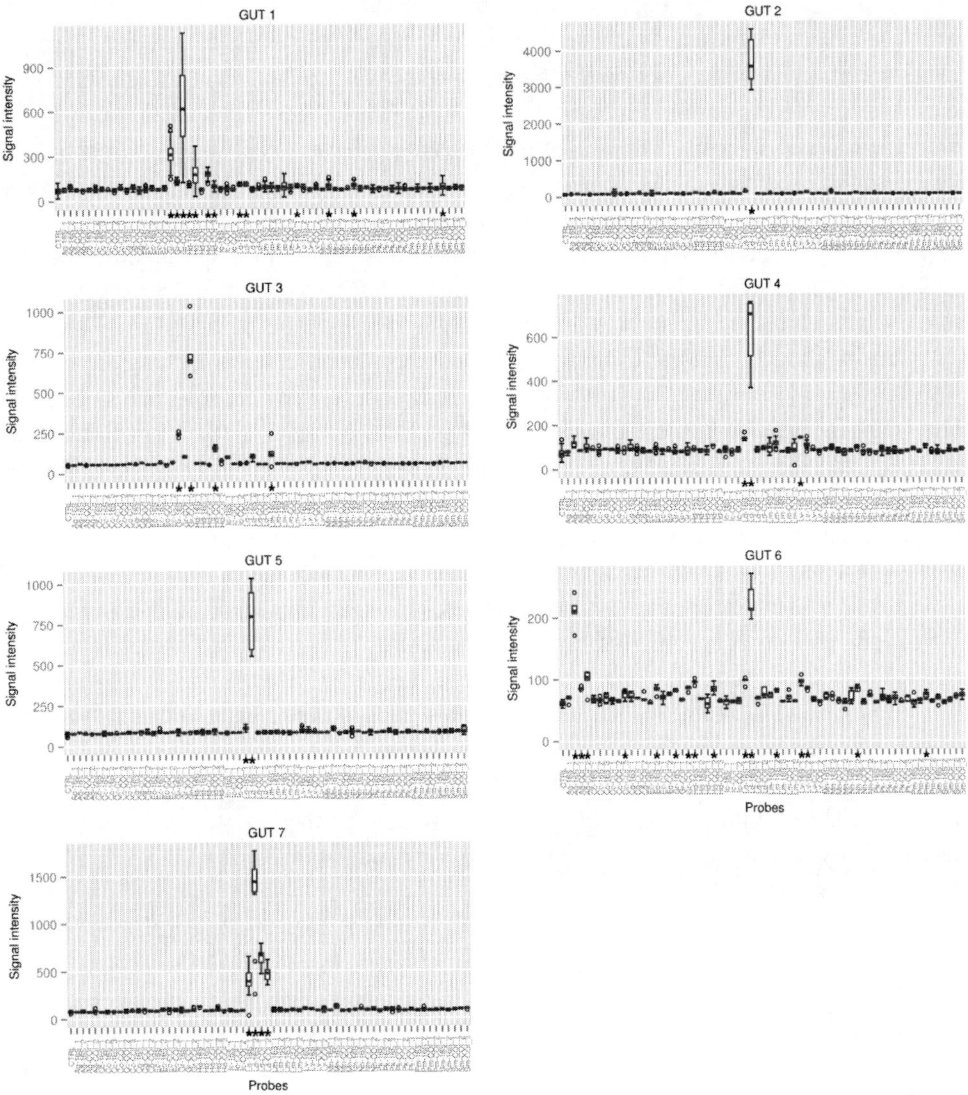

Figure 4. Boxplots graphs summarizing the gut contents hybridization experiments. Gut sample codes given as in Table 6.

All targets detected by the standard taxonomical analysis were identified in parallel by the INV-CHIP (Bold text in Table 6), but not all target detected by the INV-CHIP were detected by the expert taxonomist under the stereo microscope (Shaded cells in Table 6). The INV-CHIP results for the Gut3 sample were further confirmed by cloning and sequencing of both 16S and COI amplicons. Sequences produced were successfully blasted against the NCBI GenBank showing similarity to homologous sequences of taxa belonging to Crustacea, Decapoda. Once blasted against the target species sequences obtained in this study, cloned amplicon sequences gave the best alignment with sequences of the target species *G. rhomboides* (data not shown).

Conclusion

The specificity of the INV-CHIP target probes was satisfactory and the INV-CHIP appears to be an effective tool to identify target invertebrate species. Statistical analysis of the hybridization signal intensity gave effective measures of true positive identification, while ruling out both false positive and negative signals. The suitability of COI and 16S markers for species discrimination with a microarray platform was tested on vertebrates [29] and prompted us to test and compare their individual performance in discriminating the target invertebrate species. While 16S probes specifically identified 12 out of 15 targets, the COI capture probes gave specific signals in all 15 target species. The p-values associated to the signal intensity showed that the COI targets bound very specifically to the capture probes on the INV-CHIP and less false negative signals were observed with respect to 16S captures (Table 4). This suggests the mitochondrial COI gene as a useful marker for discriminating invertebrates at the species level on a microarray platform. The suitability of the COI gene as species identification marker has been proven in a great variety of invertebrate taxa, such as Copepoda [27][70], Lepidoptera [71][72], Culicidae [73], Araneae [74]. However, the combined COI-16S INV-CHIP microarray exhibited the advantage of a more robust pattern of hybridization signals for all 15 target species, providing internal validation of species identification. The target specific hybridization pattern can be used as reference for assessing the presence of the target in the analysis of unknown environmental samples (Figure 4). The PCR multiplexing of COI and 16S targets represents the additional value of the INV-CHIP by reducing cost and time of analysis.

The INV-CHIP specificity was also tested against non-target annelid worm species (class Polychaeta). The failure of hybridization corroborated the microarray specificity as demonstrated by a weak and non-significant cross-hybridization with a single *N. hombergii* probe (data not shown).

However, such preliminary validation of the INV-CHIP needs to be confirmed by further experiments using a higher number of non-target species closely related to all the 15 target species. A "PhyloChip" model, i.e., a chip carrying a hierarchical set of probes recognizing species at multiple taxonomic levels [52], could represent a suitable approach for enhancing microarray platform specificity. In the INV-CHIP, this approach was not accomplished because the design of capture probes suitable for the identification of the invertebrate taxa at higher taxonomic level was not feasible due to the unsatisfactory output generated by the CDIS analyses.

The reliable application of the INV-CHIP to biodiversity and ecosystem monitoring would require the simultaneous detection of multiple target species in a single experiment and efficient discrimination of the target species even in samples with high biological complexity. The results of the hybridization tests carried out with artificial binary and ternary mixtures of target species in different proportions demonstrated that the hybridization of one target was still possible in the presence of unbalanced amount of other targets. Even when a target species was underrepresented, the INV-CHIP was able to detect significant hybridization signals in the majority of experiments (Figure 2; Figure 3). Similar results were also obtained from the analysis of environmental samples (i.e., fish gut contents) whose quali-quantitative composition was actually unknown. However, the INV-CHIP cannot effectively be used for absolute, or relative target quantification as there was no deterministic correlation between the amount of the target in the mixtures and the intensity of the hybridization signals; however, increases in the signal intensity was observed at higher target concentrations.

Further sensitivity tests of the INV-CHIP were conducted to detect target species within artificial mixtures with large amount of non-target species (i.e., mimicking more complex environmental samples) by hybridizing the PCR-amplified targets along with a mix of genomic DNA from phytoplankton, polychaetes, and gut content of fishes. Even if target signals gradually decreased with increasing non-target concentrations, the target signal could still be detected in the presence of a 20-fold background concentration (data not shown).

Given the performance of the INV-CHIP in the specificity and sensitivity tests, results obtained from the analysis of the unknown environmental samples appeared satisfactory. The presence of target species in the gut contents was resolved by the INV-CHIP in the majority of samples analysed, and the comparative morphological and molecular analyses consistently identified most of the target invertebrate species. The gut contents analyses revealed a high frequency of crustacean Decapoda (Table 5), that are common prey in the diet of tub gurnard in the Adriatic Sea [64]. The positive hybridization signals given by the INV-CHIP, indicating the presence of the target species in the gut sample, were validated either by the standard morphological microscopic identification, or by cloning and sequencing PCR products of the gut content template DNA. COI and 16S sequences obtained from the preliminary molecular validation test showed high levels of BLASTn identity (>87%) with sequences of several crustacean Decapoda and a 100% identity with sequences of *Goneplax rhomboides* (data not shown). The consistency of results given by such independent approaches further validated the high power of INV-CHIP in identifying target invertebrate species.

Some discrepancies between morphological and molecular identification concerned species that were detected by the hybridization experiments, but not by the standard taxonomical stereo-microscopical analysis. Since these species are potential prey of tub gurnard, these targets might have been overlooked during the stereo-microscopical analysis. Although the advanced digestion of their remains in the fish stomach was apparent, allowing direct identification only to taxonomic levels higher than species, these species could still be detectable with the developed microarray assay.

As the feeding behaviour of fish might undergo spatio-temporal changes and the relative frequency between different prey species could be quite variable, quantification would be desirable; however, this is not possible at the current stage of INV-CHIP development. Methods already used to quantify microbial populations [75] can be adapted to marine organisms to quantitatively assess the diverse composition of natural populations. However, the INV-CHIP could be used for regular monitoring of the gut content for ecosystem functioning analysis and to investigate the common prey of commercially-relevant fishery resources.

Developing a microarray requires a particularly demanding initial phase of information gathering, in terms of sequence analysis of target and non-target species [76]. The critical threshold is to obtain the most comprehensive database as possible, on which the successive steps of probe design and in silico evaluation of oligos are based. Achieving such a sequence database for the target invertebrate species of the present research was not a trivial task, since only little information were available in terms of gene markers and molecular protocols. It was not possible to conduct a preliminary in silico assessment of the variation of selected genes at the between- and within-species levels due to few taxa and groups being well-represented in the public databases at the time this study was performed. The lack of a proper information background in terms of sequence variability at higher taxonomic levels (genus, family or higher) also influenced the strategy of downstream probe design. However, DNA barcoding is continuously increasing the inventory of animal biodiversity and, given the promising performance of COI marker in the INV-CHIP, exploiting the BOLD sequence information will allow further probes to be successfully designed for targeting single species or groups of species of high interest in future. [77][78]

Probe specificity is the first criterion to be considered during the in silico design and selection of capture probes. Since the sequences used in this chapter represented only a small subset of the gene diversity among individuals/populations of a given species, false negative signals were expected either from non-inventoried sequence variants of target species, or from unspecific binding with sequences of non-target species uncovered by the developed microarray. Even if the non-target sequences dataset included as much of the gene diversity at the species level of the taxa available in the public sequence databases, it cannot represent the real and comprehensive species diversity of the invertebrate taxa. Therefore, non-target DNAs from unknown and/or non-sequenced species can give false positive signals in the hybridization of environmental samples. Moreover, single nucleotide polymorphisms can occur within target species, due to natural biodiversity. In an attempt to reduce such problems, redundant probes were designed for each species and no probes were removed from the microarray even if their signal intensity was found to be very low.

A major concern in using DNA microarray for the detection and monitoring of environmental samples is the misidentification of species due to cross-hybridization and non-specific binding [50]. This might be caused by the high number of unknown organisms that can be present in environmental samples. On the contrary, capture probes on the microarray were designed only for a limited number of known and sequenced organisms; therefore, it is impossible to detect the organisms not included as targets on the array [79]. In environmental applications, DNA microarrays can also fail in identifying species if unknown haplotypes and deep phylogeographic structure variants are present [80]. In this case, the development of a "PhyloChip" microarray approach might help to overcome these empirical constraints [52].

The preparation of PCR amplicons as DNA targets required a considerable effort in terms of setting up best conditions of DNA extraction, PCR amplification and direct sequencing. The taxonomic differences among selected target organisms (belonging to three taxa Crustacea, Mollusca, and Polychaeta) must be considered in relation to the strategy to use universal primers for multi-target amplification. Finally, the use of PCR amplicons as targets in the microarray experiments was controversial, as the differential amplification of targets could produce incorrect quantification of target and consequently introduce bias in the hybridization results [78][81]. An accurate relationship between the amount of target and the intensity of hybridization signal is currently obtained using the isolated RNA for direct hybridization on the microarray, avoiding the PCR amplification [82]. However, the RNA is difficult to extract, subject to degradation and with a rather low concentration in environmental samples [82]. The RNA content per cell may vary under different environmental conditions and different users could possibly isolate different amounts of RNA. Therefore, the microarray-based quantification can be considered only as semi-quantitative.

The strategy used in this chapter was to PCR-amplify and label the target DNA sequences from the total genomic DNA template for subsequent hybridization. The suitability of using multiplexed 16S and COI amplicons for hybridization on the INV-CHIP was verified by assessing the yield of both gene markers on the agarose gel and on the assumption that both markers were amplified with similar ratio. The same approach was also used to amplify both genes from two-, three-, and mixed- templates, in the sensitivity experiments. The 16S amplicons from multiple target species could be discriminated on the gel, as they did vary between different species because of indels in their sequences (looped structure). Assuming that the amplification would occur in the desired ratios, further assessments were carried out at the level of the hybridization results, but the overall intensity of the signal was very low. The reasons of such behavior could be related to competition among the primer pairs and their differential efficiency to bind the template DNA. Another explanation could be due to the high number of amplification cycles (40) used for labeling, which could produce a plateau status in the PCR yield where amplification rate cannot be estimated. Nevertheless, all targets used in the multiplex PCR gave positive hybridization signals on the INV-CHIP, suggesting that all template DNA targets were amplified. For more quantitative results, a calibration method is needed to estimate PCR efficiency, since the advantage of using universal PCR primers for all targets should not be underestimated. Such a method would be very beneficial for field applications since an efficient multiplex PCR is essential to amplify the target DNA present in the environmental samples.

Data analysis and interpretation is one of the most critical methodological steps in the interpretation of the hybridization results of microarrays [83]. The different types of technology available pose new challenges to define standardized methods to process microarray information. Unlike gene expression, where numerous algorithms exist for statistical analysis, normalization, fold change analysis and biological interpretation, no statistical tools have been specifically developed for the analysis of DNA microarray data. Thus, the development of statistical algorithms for the analysis of quantitative data on the abundance of the species in a mixture of environmental samples represents a challenge. A typical analytical framework include data standardization, image acquisition and analysis, normalization, exploratory data analysis, statistical significance inference, as well as various considerations relevant to their implementation [56].

After the image acquisition, the data quality check can provide information on outlier spots and identification of problematic slides. Data quality check is an essential part of the data analysis, but it is still not easy to perform objectively or in an automated manner. Even though techniques for the removal of outlier spots and to improve data quality are still under development, some procedures seem to be more promising [84]. However, the usefulness of many quality control measures is unsubstantiated and no specific quality control method has been standardized yet [85].

The choice of the most appropriate statistical test that permits discrimination between low and high abundant species against a background of non-targets is a challenging issue. The detection of a meaningful response from the hybridization signals obtained in defined conditions is often accomplished from the comparison of signal intensity between reference and target spots, and by evaluating the significance of the deviation of signal intensity in relation to a fixed threshold. Methods of data analysis relying on the intensity threshold appear to be very limited from a statistical point of view. Therefore, a Student's t-test, with appropriate correction for unequal variances, was adopted to compare local background with negative controls. A one-way ANOVA analysis allowed testing the significance of hybridization signals versus controls. In most cases, fluorescence signals showed very large variance, which commonly cause problems in the application of parametric data analysis techniques. ANOVA relies on the assumption of heteroscedasticity [86]. Data transformations are commonly adopted to reduce the effects of heterogeneous variances in the comparative analysis of groups of samples. Though it is a common practice to perform statistical analysis on a log-transformed data, variances may remain heterogeneous even after transformation. In these cases some authors [67] suggest to perform analyses with untransformed data. When departures from assumptions of ANOVA occur, several authors prefer to switch to non-parametric methods, such as Kruskal-Wallis test, even if variance heterogeneity still can weaken consistency of the results [87]. In this chapter, a comparison of tests carried out both with transformed and untransformed data showed that in the former case a dramatic increase in false positives occurred (data not shown). Therefore, untransformed data were analysed.

The identification of target hybridization signals higher than those of the controls was performed using a one-way ANOVA with a customized pairwise one-tailed t-tests approach. Multiple post-hoc tests were performed following planned comparisons procedure between controls and probes (like the Dunnett's procedure), with p-values correction made by Benjamini-Hochberg's FDR (False Discovery Rate) [68].

The adoption of this pairwise procedure was decided after an accurate comparison of results of several tests most commonly recommended in literature for data with unequal variances and unequal sample sizes. The detection of false positive and negative signals has revealed to be one of the most tedious problems to solve. In this chapter, the t-test corrected with FDR performed very satisfactorily in terms of avoiding Type I errors, maintaining a good power of the test (low rate of false negatives) even in presence of an unbalanced design or small sample size.

A thoughtful experimental design is the most important step at the beginning of a microarray experiment. A well-planned experiment can maximize the efficiency of subsequent data analysis. Experimental design should include an adequate number of replicates at different levels to separate effects of different sources of variation, which may be technical (e.g., dye, batch) or biological (e.g., genotype, sex, time) [55]. Probe level variation carries important information useful to control technical sources of variation, outlier removal, and subsequent statistical analyses [88]. In some cases, the technological platform can permit increases in the technical replication and model the statistical properties of the data [89], thus improving the quality of the analysis. The control of biological variation is also challenging not only for analytical procedures, but also for costs associated with sampling of additional individuals [55].

The adoption of a balanced experimental design can help to avoid difficulties about the choice of an appropriate test, reducing the probability of Type I and Type II errors. Although there is no consensus about what is the best procedure to define the optimal sample size, there is consensus that power analyses should be performed applying the newer methods specifically implemented for microarray data, and that more replicates generally provide greater power [55].

Challenges in data analysis are often linked to the development of new technologies. Microarray analysis is evolving rapidly, as new and more complex analyses appear, making it easy for the researcher to get lost in endless new methods and software. Collaborations between biologists, statisticians, and mathematicians is advisable for proper planning and analyses [56]. Commercial and open-source software for performing data analyses are now widely available. The development of dedicated R packages [69] and increasing availability of libraries for the analysis of microarray can help researchers apply "canned" procedures as well as to customize data analysis tools according to the particular needs of the data structure.

Although DNA chip technology was applied for the first time nearly 30 years ago [30] to discriminate six DNA genotypes and nine common allelic mutations in one single experiment, this technique has helped to circumvent much of the problems associated with the conventional techniques used in taxonomy. The past decade experienced great advances and changes in molecular methods for species identification, spanning from approaches allowing the identification of one single specimen at once, to high-throughput techniques allowing the identification of hundreds or thousands of specimens at a time. The DNA barcoding approach, first introduced in 2003 [15], allows species identification by analysing single short sequences of the mitochondrial COI and compare it to all information available in public libraries [90][91]. The high-throughput next generation sequencing (NGS) technologies firstly implemented in 2005 [92][93] can perform massive, parallel analyses of a vast amount of specimens and species, even from fragmented DNA. NGS has revolutionized genomics research by providing large amounts of information quickly and efficiently [94]. NGS have recently been applied in species identification and ecological studies, since NGS platforms allow the direct analyses of DNA sequences from environmental samples [95][96].

Complete genome re-sequencing using NGS provides greater depth in terms of sample sequence coverage and helps in identifying unknown diversity in ecological communities [97][98]. Compared to NGS, the disadvantage of using microarrays in environmental and ecological studies is due to the identification of samples restricted to those species for which probes are captured on the array, and are limited in recognizing unknown, or novel species [99][100]. The development of microarray for new species identification and sample processing requires an intense procedure of benchmarking and optimization that outperform results.

Using microarray as a validation tool of NGS data is quite common when amplicon information from NGS is used for probe designing in microarray experiments for species identification [101]. NGS is the most preferred technology as data from one single experiment can provide in-depth analysis of the species diversity [102][103]. In some applications, DNA microarrays are sought after over NGS for routine identification of species in which sequence information is already known and specific probes are designed for the microarray. Despite drawbacks, microarrays have already been developed for species identification mainly in ecological and biodiversity assessment studies of microbial communities and algal blooms [34][43][36]. Overcoming the above mentioned problems would further increase the applicability of microarrays as an identification tool for identifying any organism down to species level [15][80].

Acknowledgments

The research leading to these results has received funding from the European Community's Sixth Framework Programme (http://ec.europa.eu/research/fp6/index_en.cfm), as Specific Targeted Research Project (STREP) funded by the European Commission under the contract no. 505491: "Fish & Chips": Towards DNA chip technology as a standard analytical tool for the identification of marine organisms in biodiversity and ecosystem research (www.fish-and-chips.uni-bremen.de).

The authors thank the whole Fish & Chips consortium for many insightful discussions. In particular: Sven Roll and Frank Meyerjürgens that produced the microarrays; Hannes Weber and Manfred Nölte who developed CDIS and designed oligonucleotide probes for the INV-CHIP; Antonios Magoulas and Christos Arvanitidis from HCMR Crete for helping with the taxonomical identification of Polychaete species and Venugopal Moleyur, College of Fisheries, Mangalore, India for his contribution in INV CHIP optimization experiments. A special thanks to Reinhard Zelm for his continuous support and help in both experimental and data analysis tasks.

References

[1] Vermeulen, N. From darwin to the census of marine life: marine biology as big science. *PloS One*, 2013, 8, e54284.

[2] Bianchi, CN; Morri, C. Marine biodiversity of the Mediterranean Sea: situation, problems and prospects for future research. *Mar Poll Bull*, 2000, 40, 367–376.

[3] Bortolus, A. Error cascades in the biological sciences: the unwanted consequences of using bad taxonomy in ecology. *J Hum Environ*, 2008, 37, 114–118.

[4] Daan, N. Data base report of the stomach sampling project 1981. ICES Cooperative Research Report No 164. *International Council for the Exploration of the Sea*, 1989.

[5] Hislop, J; Bromley, PJ; Daan, N; Gislason, H; Heessen, HJL; Robb, AP;et al. Database report of the Stomach Sampling Project, 1991. ICES Cooperative Research Report No. 219. *International Council for the Exploration of the Sea*, 1997.

[6] Hyslop, EJ. Stomach contents analysis-a review of methods and their application. *J Fish Biol*, 1980, 17, 411–429.

[7] Gorokhova, E; Fagerberg, T; Hansson, S. Predation by herring (*Clupea harengus*) and sprat (*Sprattus sprattus*) on *Cercopagis pengoi* in a western Baltic Sea bay. *ICES J Mar Sci*, 2004, 61, 959–965.

[8] Espinoza, M; Wehrtmann, IS. Stomach content analyses of the threadfin anglerfish *Lophiodes spilurus* (Lophiiformes: Lophiidae) associated with deepwater shrimp fisheries from the central Pacific of Costa Rica. *Revista de Biología Tropical*, 2008, 56, 1959–1970.

[9] Cocheret, de; la Morinière, E; Pollux, B; Nagelkerken, I; Hemminga, M; Huiskes, A; van der Velde, G. Ontogenetic dietary changes of coral reef fishes in the mangrove-seagrass-reef continuum: stable isotopes and gut-content analysis. *Mar Ecol Prog Ser*, 2003, 246, 279–289.

[10] Rosel, P; Kocher, T. DNA-based identification of larval cod in stomach contents of predatory fishes. *J Exp Mar Biol Ecol*, 2002, 267, 75–88.

[11] Baker, R,; Buckland, A; Sheaves, M. Fish gut content analysis: robust measures of diet composition. *Fish Fish*, 2013, doi, 10.1111/faf.12026.

[12] Rindorf, A; Lewy, P. Bias in estimating food consumption of fish from stomach-content analysis. *Can J Fish Aquat Sci*, 2004, 61, 2487–2498.

[13] Hynes, HBN. The food of fresh-water sticklebacks (*Gasterosteus aculeatus* and *Pygosteus pungitius*), with a review of methods used in studies of the food of fishes. *J Anim Ecol*, 1950, 19, 36–58.

[14] Bowen, S. Quantitative description of the diet. In: Nielsen LA, Johnson DL (eds). Fisheries techniques. *American Fisheries Society,* Bethesda, Maryland, 1983, 325–336.

[15] Hebert, PDN; Cywinska, A; Ball, SL; deWaard, JR. Biological identifications through DNA barcodes. *Proc Biol Sci R Soc*, 2003, 270, 313–321.

[16] Hebert, PDN; Stoeckle, MY; Zemlak, TS; Francis, CM. Identification of birds through DNA barcodes. *PLoS Biology*, 2004, 2, e312.

[17] Hebert, PDN; Gregory, TR. The promise of DNA barcoding for taxonomy. *Syst Biol*, 2005, 54, 852–859.

[18] Kochzius, M; Kappel, K; Dobitz, L; Silkenbeumer, N; Nolte, M; Weber, H;et al. The "Fish & Chips" project: microarrays as a tool for the identification of marine organisms in biodiversity and ecosystem research. OCEANS 2007 - Europe, *IEEE*,2007,1–4.

[19] Féral, J. How useful are the genetic markers in attempts to understand and manage marine biodiversity? *J Exp Mar Biol Ecol*, 2002, 268, 121–145.

[20] Ward, RD; Hanner, R; Hebert, PDN. The campaign to DNA barcode all fishes, FISH-BOL. *J Fish Biol*, 2009, 74, 329–356.

[21] Jérôme, M; Martinsohn, JT; Ortega, D; Carreau, P; Verrez-Bagnis, V; Mouchel, O. Toward fish and seafood traceability: anchovy species determination in fish products by molecular markers and support through a public domain database. *J Agr Food Chem*, 2008, 56, 3460–3469.

[22] Teletchea, F. Molecular identification methods of fish species: reassessment and possible applications. *Rev Fish Biol Fisheries*, 2009, 19, 265–293.

[23] Espiñeira, M; González-Lavín, N; Vieites, JM; Santaclara, FJ. Development of a method for the genetic identification of flatfish species on the basis of mitochondrial DNA sequences. *J Agr Food Chem*, 2008, 56, 8954–8961.

[24] Costa, FO,; Carvalho, GR. The Barcode of Life Initiative: synopsis and prospective societal impacts of DNA barcoding of Fish. *Genomics Soc Policy*, 2007, 3, 29-40.

[25] Rocha-Olivares, A. Multiplex haplotype-specific PCR: a new approach for species identification of the early life stages of rockfishes of the species-rich genus *Sebastes Cuvier*. *J Exp Mar Biol Ecol*, 1998, 231, 279–290.

[26] Sweijd, NA; Bowie, RCK; Evans, BS; Lopata, AL. Molecular genetics and the management and conservation of marine organisms. *Hydrobiologia*, 2000, 420, 153–164.

[27] Bucklin, A; Guarnieri, M; Hill, RS; Bentley, AM; Kaartvedt, S. Taxonomic and systematic assessment of planktonic copepods using mitochondrial COI sequence variation and competitive, species-specific PCR. *Hydrobiologia*, 1999, 401, 239–254.

[28] Bell, JL; Grassle, JP. A DNA probe for identification of larvae of the commercial surfclam (*Spisula solidissima*). *Mol Mar Biol Biotechnol*, 1998, 7, 127–137.

[29] Kochzius, M; Nölte, M; Weber, H; Silkenbeumer, N; Hjörleifsdottir, S; Hreggvidsson, GO;et al. DNA microarrays for identifying fishes. *Mar Biotechnol*, 2008, 10, 207–217.

[30] Saiki, RK. Genetic analysis of amplified DNA with immobilized sequence-specific oligonucleotide probes. *Proc Natl Acad Sci USA*, 1989, 86, 6230–6234.

[31] Southern, EM. Detection of specific sequences among DNA fragments separated by gel electrophoresis. *J Mol Biol*, 1975, 98, 503–517.

[32] Lashkari, DA; DeRisi, JL; McCusker, JH; Namath, AF; Gentile, C; Hwang, SY;et al. Yeast microarrays for genome wide parallel genetic and gene expression analysis. *Proc Natl Acad Sci*, USA, 1997, 94, 13057–13062.

[33] Call, DR; Borucki, MK; Loge, FJ. Detection of bacterial pathogens in environmental samples using DNA microarrays. *J Microbiol Methods*, 2003, 53, 235–243.

[34] Franke-Whittle, IH,; Klammer, SH; Insam, H. Design and application of an oligonucleotide microarray for the investigation of compost microbial communities. *J Microbiol Methods*, 2005, 62, 37–56.

[35] Loy, A; Bodrossy, L. Highly parallel microbial diagnostics using oligonucleotide microarrays. *Clinica Chimica Acta*, 2006, 363, 106–119.

[36] Peplies, J; Lau, SCK; Pernthaler, J; Amann, R; Glöckner, FO. Application and validation of DNA microarrays for the 16S rRNA-based analysis of marine bacterioplankton. *Environ Microbiol*, 2004, 6, 638–645.

[37] Cariani, A; Piano, A; Consolandi, C; Severgnini, M; Castiglioni, B; G Caredda, G; M Candela, M; P Serratore, P; G De Bellis, G; F Tinti, F. Detection and characterization of pathogenic vibrios in shellfish by a Ligation Detection Reaction-Universal Array approach. *International Journal of Food Microbiology*, 2012, 153, 474–482.

[38] González, SF; Krug, MJ; Nielsen, ME; Santos, Y; Call, DR. Simultaneous detection of marine fish pathogens by using multiplex PCR and a DNA microarray. *J Clin Microbiol*, 2004, 42, 1414–1419.

[39] Yoon, HK; Jeong, D; Chung, IH; Jung, JW; Oh, MJ; Kim, S. Rapid species identification of elasmobranch fish (skates and rays) using oligonucleotide microarray. *Biochip*, 2009, 3, 87–96.

[40] Kochzius, M; Seidel, C; Antoniou, A; Botla, SK; Campo, D; Cariani, A;et al.Identifying fishes through DNA barcodes and microarrays. *PloS One*, 2010, 5, e12620.

[41] Metfies, K; Medlin, LK. DNA microchips for phytoplankton: The fluorescent wave of the future. *Nova Hedwigia*, 2004, 79, 321–327.

[42] Metfies, K; Berzano, M; Mayer, C; Roosken, P; Gualerzi, C; Medlin, L;et al. An optimized protocol for the identification of diatoms, flagellated algae and pathogenic protozoa with phylochips. *Molecular Ecol Notes*, 2007, 7, 925–936.

[43] Gescher, C; Metfies, K; Medlin, LK. The ALEX CHIP—Development of a DNA chip for identification and monitoring of *Alexandrium*. *Harm Algae*, 2008, 7, 485–494.

[44] Galluzzi, L; Cegna, A; Casabianca, S; Penna, A; Saunders, N; Magnani, M. Development of an oligonucleotide microarray for the detection and monitoring of marine dinoflagellates. *J Microbiol Methods*, 2011, 84, 234–242.

[45] McCoy, GR; Touzet, N; Fleming, GT; Raine, R. An evaluation of the applicability of microarrays for monitoring toxic algae in Irish coastal waters. *Environ Sci Poll Res Int*, 2012, doi 10.1007/s11356-012-1294-1.

[46] Kegel, J; Del Amo, Y; Costes, L; Medlin, L. Testing a microarray to detect and monitor toxic microalgae in Arcachon Bay in France. *Microarrays*, 2013, 2, 1–23.

[47] Pauly, D. Fishing down marine food webs. *Science*, 1998, 279, 860–863.

[48] Goffredi, SK; Jones, WJ; Scholin, CA; Marin, R; Vrijenhoek, RC. Molecular detection of marine invertebrate larvae. *Mar Biotechnol*, 2006, 8, 149–160.

[49] Kothapalli, R; Yoder, S; Mane, S; Loughran, T. Microarray results: how accurate are they? *BMC Bioinformatics*, 2002, 3, 22.

[50] Zhou, J; Thompson, DK. Challenges in applying microarrays to environmental studies. *Curr Opin Biotechnol*, 2002, 13, 204–207.

[51] Gescher, C; Metfies, K; Frickenhaus, S; Knefelkamp, B; Wiltshire, KH; Medlin, LK. Feasibility of assessing the community composition of prasinophytes at the Helgoland Roads sampling site with a DNA microarray. *Appl Environ Microbiol*, 2008, 74, 5305–5316.

[52] Metfies, K; Medlin, LK. Feasibility of transferring fluorescent in situ hybridization probes to an 18S rRNA gene phylochip and mapping of signal intensities. *Appl Environ Microbiol*, 2008, 74, 2814–2821.

[53] Kim, S; Eo, H-S; Koo, H; Choi, J-K; Kim, W. DNA barcode-based molecular identification system for fish species. *Mol Cells*, 2010, 30, 507–512.

[54] Peplies, J; Glöckner, FO; Amann, R. Optimization strategies for DNA microarray-based detection of bacteria with 16S rRNA-targeting oligonucleotide probes. *Appl Environ Microbiol*, 2003, 69, 1397–1407.

[55] Kerr, MK. Design considerations for efficient and effective microarray studies. *Biometrics*, 2003, 59, 822–828.

[56] Leung, YF; Cavalieri, D. Fundamentals of cDNA microarray data analysis. *Trends Genet*, 2003, 19, 649–659.

[57] Deagle, BE; Tollit, DJ; Jarman, SN; Hindell, MA; Trites, AW; Gales, NJ. Molecular scatology as a tool to study diet: analysis of prey DNA in scats from captive Steller sea lions. *Mol Ecol*, 2005, 14, 1831–1842.

[58] Folmer, O; Black, M; Hoeh, W; Lutz, R; Vrijenhoek, R. DNA primers for amplification of mitochondrial cytochrome c oxidase subunit I from diverse metazoan invertebrates. *Mol Mar Biol Biotechnol*, 1994, 3, 294–299.

[59] Palumbi, S; Romano, S; Mcmillan, WO; Grabowski, G. *The simple fool's guide to PCR*, 2002, 1–45. http://palumbi.stanford.edu/SimpleFoolsMaster.pdf.

[60] Tamura, K; Peterson, D; Peterson, N; Stecher, G; Nei, M; Kumar, S. MEGA5: molecular evolutionary genetics analysis using maximum likelihood, evolutionary distance, and maximum parsimony methods. *Mol Biol Evol*, 2011, 28, 2731–2739.

[61] Altschul, SF; Gish, W; Miller, W; Myers, EW; Lipman, DJ. Basic local alignment search tool. *J Mol Biol*, 1990, 215, 403–410.

[62] Nolte, M. Optimization of oligonucleotide sets for DNA microarrays. PhD Thesis University of Bremen, Germany, 2002.

[63] Hofacker, IL; Fontana, W; Stadler, PF; Bonhoeffer, LS; Tacker, M; Schuster, P. Fast folding and comparison of RNA secondary structures. *Monatshefte Für Chemie/Chemical Monthly*, 1994, 125, 167–188.

[64] Stagioni, M; Montanini, S; Vallisneri, M. Feeding of tub gurnard *Chelidonichthys lucerna* (Scorpaeniformes: Triglidae) in the north-east Mediterranean. *J Mar Biol Assoc* UK, 2011, 92, 605–612.

[65] Welch, BL. On the comparison of several mean values: An alternative approach. *Biometrika*, 1951, 38, 330–336.

[66] Snedecor, GW;Cochran, WG. Statistical Methods. 8th Editio. Iowa State University Press; 1989.

[67] Underwood, A. Techniques of analysis of variance in experimental marine biology and ecology. *Oceanogr Mar Biol*, 1981, 16.

[68] Benjamini, Y; Hochberg, Y. Controlling the false discovery rate: a practical and powerful approach to multiple testing. *J R Stat Soc B*, 1995, 57, 289–300.

[69] R Development Core Team. {R: A language and environment for statistical computing}. Vienna, Austria: R Foundation for Statistical Computing; 2009.

[70] Bucklin, A; Frost, B; Bradford-Grieve, J; Allen, L; Copley, N. Molecular systematic and phylogenetic assessment of 34 calanoid copepod species of the *Calanidae* and *Clausocalanidae*. *Mar Biol*, 2003, 142, 333–343.

[71] Brown, B; Emberson, RM; Paterson, AM. Mitochondrial COI and II provide useful markers for Wiseana (Lepidoptera: Hepialidae) species identification. *Bull EntomolRes*, 1999, 89, 287–293.

[72] Janzen, DH; Hajibabaei, M; Burns, JM; Hallwachs, W; Remigio, E; Hebert, PDN. Wedding biodiversity inventory of a large and complex Lepidoptera fauna with DNA barcoding. *Philos Trans R Soc Lond Biol Sci*, 2005, 360, 1835–1845.

[73] Shouche, YS; Patole, MS. Sequence analysis of mitochondrial 16S ribosomal RNA gene fragment from seven mosquito species. *J Biosci*, 2000, 25, 361–366.

[74] Barrett, RD; Hebert, PD. Identifying spiders through DNA barcodes. *Can J Zool*, 2005, 83, 481–491.

[75] Palmer, C; Bik, EM; Eisen, MB; Eckburg, PB; Sana, TR; Wolber, PK;et al. Rapid quantitative profiling of complex microbial populations. *Nucl Acids Res*, 2006, 34, e5.

[76] Schena, M; Shalon, D; Davis, RW; Brown, PO. *Quantitative monitoring of gene expression patterns with a complementary DNA microarray. Science*, 1995, 270, 467–470.

[77] Ratnasingham, S; Hebert, PDN. bold: The Barcode of Life Data System (http://www.barcodinglife.org). *Mol Ecol Notes*, 2007, 7, 355–364.

[78] Kanagawa, T. Bias and artifacts in multitemplate polymerase chain reactions (PCR). *J Biosci Bioeng*, 2003, 96, 317–323.

[79] Gentry, TJ; Wickham, GS; Schadt, CW; He, Z; Zhou, J. Microarray applications in microbial ecology research. *Microb Ecol*, 2006, 52, 159–175.

[80] Hajibabaei, M; Singer, GAC; Hebert, PDN; Hickey, DA. DNA barcoding: how it complements taxonomy, molecular phylogenetics and population genetics. *Trends Genet*, 2007, 23, 167–172.

[81] Medlin, LK; Metfies, K; Mehl, H; Wiltshire, K; Valentin, K. Picoeukaryotic plankton diversity at the Helgoland time series site as assessed by three molecular methods. *Microb Ecol*, 2006, 52, 53–71.

[82] Peplies, J; Lachmund, C; Glöckner, FO; Manz, W. A DNA microarray platform based on direct detection of rRNA for characterization of freshwater sediment-related prokaryotic communities. *Appl Environ Microbiol*, 2006, 72, 4829–4838.

[83] Blohm, DH; Guiseppi-Elie, A. New developments in microarray technology. *Curr Opin Biotechnol*, 2001, 12, 41–47.

[84] Kauffmann, A; Huber, W. Microarray data quality control improves the detection of differentially expressed genes. *Genomics*, 2010, 95, 138–142.

[85] Allison, DB; Cui, X; Page, GP; Sabripour, M. Microarray data analysis: from disarray to consolidation and consensus. *Nat Rev Genet*, 2006, 7, 55–65.

[86] Sokal, RR;Rohlf, FJ. Biometry. 3rd Editio. *Newyork Freeman*, 1995.

[87] Hollander, M; Wolfe, DA. *Nonparametric statistical methods*. Wiley, 1973.

[88] Ayroles, JF; Gibson, G. Analysis of variance of microarray data. *Methods Enzymol*, 2006, 411, 214–233.

[89] Lin, SM; Du, P; Huber, W; Kibbe, W; a. Model-based variance-stabilizing transformation for Illumina microarray data. *Nucl Acids Res*, 2008, 36, e11.

[90] Costa, FO; Landi, M; Martins, R; Costa, MH; Costa, ME; Carneiro, M;et al.A ranking system for reference libraries of DNA barcodes: application to marine fish species from Portugal. *PloS One*, 2012, 7, e35858.

[91] Bergsten, J; Bilton, DT; Fujisawa, T; Elliott, M; Monaghan, MT; Balke, M;et al. The effect of geographical scale of sampling on DNA barcoding. *Syst Biol*, 2012, 61, 851–869.

[92] Margulies, M; Egholm, M; Altman, WE; Attiya, S; Bader, JS; Bemben, La;et al. Genome sequencing in microfabricated high-density picolitre reactors. *Nature*, 2005, 437, 376–380.

[93] Shendure, J; Porreca, GJ; Reppas, NB; Lin, X; McCutcheon, JP; Rosenbaum, AM;et al. Accurate multiplex polony sequencing of an evolved bacterial genome. *Science*, 2005, 309, 1728–1732.

[94] Morozova, O; Marra, MA. Applications of next-generation sequencing technologies in functional genomics. *Genomics*, 2008, 92, 255–264.

[95] Ekblom, R; Galindo, J. Applications of next generation sequencing in molecular ecology of non-model organisms. *Heredity*, 2011, 107, 1–15.

[96] Sogin, ML; Morrison, HG; Huber, JA; Mark Welch, D; Huse, SM; Neal, PR;et al. Microbial diversity in the deep sea and the underexplored "rare biosphere". *Proc Natl Acad Sci*, USA 2006, 103, 12115–12120.

[97] Kermarrec, L; Franc, A; Rimet, F; Chaumeil, P; Humbert, JF; Bouchez, A. Next-generation sequencing to inventory taxonomic diversity in eukaryotic communities: a test for freshwater diatoms. *Mol Ecol Resources*, 2013, 13, 607–619.

[98] Bokulich, NA; Joseph, CML; Allen, G; Benson, AK; Mills, DA. Next-generation sequencing reveals significant bacterial diversity of botrytized wine. *PloS One*, 2012, 7, e36357.

[99] Gilad, Y; Rifkin, SA; Bertone, P; Gerstein, M; White, KP. Multi-species microarrays reveal the effect of sequence divergence on gene expression profiles. *Genome Res*, 2005, 15, 674–680.

[100] Kammenga, JE; Herman, MA; Ouborg, NJ; Johnson, L; Breitling, R. Microarray challenges in ecology. *Trends Ecol Evol*, 2007, 22, 273–279.

[101] Roh, SW; Abell, GCJ; Kim, KH; Nam, YD; Bae, JW. Comparing microarrays and next-generation sequencing technologies for microbial ecology research. *Trends Biotechnol*, 2010, 28, 291–299.

[102] Marinković, M; de Leeuw, WC; de Jong, M; Kraak, MHS; Admiraal, W; Breit, TM;et al. Combining next-generation sequencing and microarray technology into a transcriptomics approach for the non-model organism *Chironomus riparius*. *PloS One*, 2012, 7, e48096.

[103] Park, JY; Lee, SY;, An, CM; Kang, JH; Kim, JH; Chai, JC;et al. Comparative study between Next Generation Sequencing Technique and identification of microarray for Species Identification within blended food products. *BioChip J*, 2012, 6, 354–361.

In: Microarrays: Principles, Applications and Technologies
Editor: James V. Rogers
ISBN: 978-1-62948-669-7

Chapter 8

Using Microarrays to Decipher the Virus-Host Interactions in HCV Infected Patients: Implications for Viral Resistance, Drug Development and Personalized Medicine

Jing Sun[1], *Jianguo Yuan*[1], *Bing Liu*[2], *Shilin Li*[2], *Yujia Li*[2], *Xiaoqiong Duan*[2], *Chunhui Yang*[2], *Peibing Zeng*[2] *and Limin Chen*[2,3*]

[1]Sichuan Mingri Pharmaceutical Research Institute of New Technology, Chengdu, Sichuan, China

[2]Institute of Blood Transfusion, Chinese Academy of Medical Sciences and Peking Union Medical College, Chengdu, Sichuan, China

[3]Toronto General Research Institute, University Health Network, University of Toronto, Toronto, ON, Canada

Abstract

Microarray is one of the high-throughput gene expression profiling tools that allow simultaneous examination of expression levels of tens of thousands genes at the transcript level. cDNA microarray technology was applied to decipher the virus-host interactions in patient liver tissues chronically-infected with hepatitis C virus (HCV). We identified an 18-gene signature that differentiates treatment responders and non-responders to interferon therapy. Based on the expression levels of these 18 host genes, we can accurately predict whether an HCV patient will respond to interferon therapy or not, with a positive prediction value (PPV) higher than 96%. A prognostic tool is being developed to facilitate personalized medicine for HCV patients. Furthermore, a novel ubiquitin-like ISG15/USP18 signaling pathway that has a close link with interferon resistance was identified. Potential anti-HCV drugs can be developed from these findings.

* Corresponding author: limin_chen_99@yahoo.com or limin.chen@utoronto.ca. Phone: +86-28-61648530.

Introduction

HCV Infection and Treatment

Hepatitis C virus (HCV) is the only member of the genus *Hepacivirus* in the family *Flaviviridae* [1]. Identified as the pathogen underlying "non A-non B" hepatitis in 1989, chronic HCV infection is one of the most common chronic liver diseases, with 170 million people chronically infected worldwide [2]. HCV infects only humans and chimpanzees. After exposure to HCV, 60-80% of humans develop chronic infections despite the induction of HCV-specific antibodies and a HCV-specific cellular immune response [3]. Chronic infection is defined as HCV RNA persistence in the serum for more than 6 months as detected by PCR. HCV chronic infection frequently results in progressive liver fibrosis, cirrhosis, and an increased risk of hepatocellular carcinoma [4]. As such, HCV is a significant health burden, and it is currently the leading indication for liver transplantation in the US and Western Europe [1].

The HCV genome consists of a 9.6 Kb positive strand RNA, encoding a single polyprotein that is processed by both host and viral proteases to generate 3 structural (Core, E1, E2) and 7 non-structural (P7, NS2/3, NS3, NS4A, NS4B, NS5A, and NS5B) proteins [5]. There are six major HCV genotypes; genotype 1 is the most prevalent genotype in North American, Asia, and Europe, while genotype 4 is the most common in Egypt [5].

HCV infects and replicates almost exclusively in liver parenchymal cells (hepatocytes). The HCV life cycle starts with receptor-mediated viral particle endocytosis. Several receptors or co-receptors have been identified, including CD81 [6], the scavenger receptor class B type I (SR-BI) [7], claudin-1 [8], occludin [9], DC-SIGN and L-SIGN [10]. Once inside the cell, the HCV virion is de-coated and the positive strand RNA is released into the cytoplasm to function as an mRNA template to direct internal ribosomal entry-site (IRES)-mediated translation. This translation generates a single polyprotein (3,011 amino acids) that is co- and post-translationally cleaved into the 3 structural and 7 non-structural proteins noted above [5]. These viral proteins assemble with the newly-replicated viral RNA to form new virions that are released from cells to infect other infection-naïve hepatocytes.

There is no vaccine available to prevent HCV infection. Until very recently, the standard of antiviral therapy was a combination therapy with pegylated IFNα and ribavirin. This treatment is often inadequate. Although there is a 60-80% response rate in patients infected with genotype 2 and 3, there is only a 30-50% rate of response for genotype 1. The fact that the different genotypes respond dramatically differently to combination therapy also indicates that viral genotype plays an important role in determining the outcome of infection, although detailed mechanisms remain poorly understood. In North America and Asia countries, the predominant genotype is genotype 1; thus, a large fraction of patients chronically infected with Hepatitis C do not respond to treatment with combination PegIFN/Ribavirin. The treatment is also expensive – estimated at more than $25,000 per patient - and is associated with significant side effects over a prolonged period of treatment (12-24 months) [11].

Interferon-alpha has been the backbone of anti-HCV treatment for well over a decade, and may remain a component of therapy for the foreseeable future. HCV is an extremely mutable virus, and both IFNα and ribavirin play important roles in stimulating multiple aspects of the host innate immune response, thereby attacking the multiple viral clones that are a feature of chronic infection [12]. Interferon-alpha elicits its antiviral effect by the induction of a myriad of interferon-stimulated genes (ISGs) [12]. However, it is unclear which ISGs are important for viral control or interfere with viral clearance, and whether interferon induces ISGs differently in certain individuals [13].

With the development of HCV in vitro cell culture system and a better understanding of the viral life cycle, direct acting antivirals (DAAs) specifically targeting the HCV NS5B polymerase and the HCV NS3 protease have been developed. The first of these to be widely applied were approved for clinical use in 2011. One of two HCV protease inhibitors (PIs), telaprevir and boceprevir are now used in combination with pegylated interferon (P) and ribavirin (R) to treat patients infected with HCV genotype 1 and a higher sustained virological response (SVR, 63%-75%) was achieved [14]. However, this data is obtained from treatment-naïve patients and it has been shown that the SVR rate is lower in treatment-experienced patients [15].

Due to the high mutation rate of HCV, DAAs are currently used in combination with interferon and ribavirin to increase efficacy. The HCV virus replicates quickly but the viral RNA polymerase complex has a poor proof-reading capacity. As a result, there is a large pool of mutated HCV viral clones in any chronically-infected patient before any treatment is initiated. It has been reported that preexisting mutations to non-nucleotide NS5B inhibitors (NNIs), PIs, and NS5A inhibitors are present in 22.5%, 7.7%, and 16.2% of patients, respectively [16, 17]. These preexisting mutated clones may be transformed into dominant viruses under selection pressure during the course of combination therapy, and eventually lead to treatment failure [18]. At present, combination IFNα and ribavirin are used to overcome the issue of resistance. In the future, the use of newer agents that also target host immune pathways but with fewer side effects is an attractive goal. In any event, even with combination DAAs the SVR rate is unlikely to be 100%: thus, there will remain a significant fraction (and worldwide, a large number) of patients for whom alternative treatments are needed. Antivirals targeting host genes or pathways may be a good alternative to this end, including cyclophilin inhibitors [19] and potentially other ubiquitin-like proteins (Ubls). One attractive aspect of targeting the Ubls is that the antiviral effects may be relevant to multiple viral infections beyond HCV.

Microarray Gene Expression Profiling

Microarray gene expression profiling is one of the most commonly used high-throughput detection of gene transcript (mRNA) levels for tens of thousands of genes at the same time. Combined with a set of well-developed data analysis software, the microarray technique (also called gene chip) has been successfully used to profile gene expression changes in many diseases, such as various tumors and viral infections. Unbiased pathways identified by gene expression profiling help us further understand the molecular mechanism(s) of pathogenesis and point out many intervention targets that can be further manipulated for potential drug development. Although microarray analysis allows us to gain a lot of information that was obviously impossible to be revealed by traditional methods, false positive hits are likely. Therefore, identified genes with differential expression levels should always be verified (confirmed) by traditional RT-PCR. Only verified genes with differential expression levels can be further used for pathway analysis to identify particular pathways that are involved in different disease states, revealing novel molecular mechanisms of pathogenesis and drug resistance.

Using Microarrays to Decipher Host Responses to HCV Infection

Hepatic Gene Response Signature

More than 85% of patients infected with HCV cannot clear the virus, and they will develop into chronic infections leading to higher incidence of liver fibrosis and hepatocellular carcinoma (HCC). Although the HCV genome was successfully cloned and decoded in 1989, an effective vaccine is still not available. Treatment options are limited until the DAAs were developed and administered in combination with the traditional pegylated-interferon alpha and ribavirin as “triple therapy”, which increases the HCV cure rate up to 90%. But still there is a considerable portion of patients who do not respond to this “triple therapy”.

In order to understand why some patients respond to treatment and some others do not, we used cDNA microarray gene expression profiling to compare gene expression levels in the pretreatment HCV infected liver tissues between treatment responders(Rs) and non-responders (NRs). A few hundred genes whose expression levels were changed in the HCV infected livers have been identified and 18 genes show differential expression between two groups [20]. A majority of these 18 genes are ISGs whose expression levels are more highly expressed in pretreatment NR liver tissue. Therefore, we proposed that pre-activation of the endogenous interferon signaling pathway leading to increased expression of a set of ISGs contributes to HCV resistance to interferon therapy. This hypothesis has been supported by data from other groups [21, 22].

Novel Ubiquitin-like ISG15/USP18 Signaling Pathway and Viral Infection

Three out of the 18 differentially-expressed hepatic genes (ISG15, USP18, and CEB1) between Rs and NRs are linked into the same ubiquitin-like signaling pathway (Figure 1). ISG15 was the first ubiquitin-like protein identified. Manipulation of ISG15 and ISGylation may be useful in countering viral infection, both because ISG15 plays an important role in the host innate immune response and because it can have discrete functions that impact the lifecycles of specific viruses. Several lines of evidence suggest that ISG15 conjugation plays an important role in the innate antiviral response or viral life cycles [23]. These include: 1) ISG15 is targeted by multiple viruses. For example, the non-structural protein (NS-1B) of Influenza B virus binds to the free form ISG15, preventing ISGylation [24]; 2) ISG15 inhibits HIV release (but not virus replication) [25]; 3) over-expression of ISG15 in IFN-α/β receptor knockout mice protects them from Sindbis virus (SNV)-induced lethality, and is associated with decreased SNV replication in multiple organs [26]; 4) Mice lacking the ISG15 deconjugation enzyme USP18/UBP43, resulting in increased ISGylation, are resistant to lymphocytic choriomeningitis virus (LCMV) and vesicular stomatitis virus (VSV) infection [23].

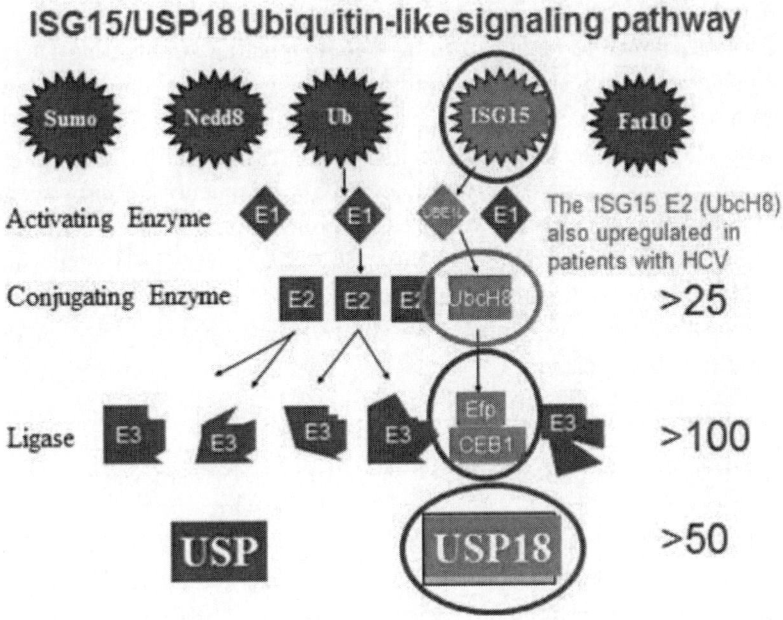

Figure 1. ISG15/USP18 ubiquitin-like signaling pathway. Like ubiquitin, ISG15 can be conjugated to target proteins through a sequential action of E1 activating enzyme (Ube1L), E2 conjugating enzyme (UbcH8), and E3 ligase (Efp, CEB1). USP18 is a cysteine protease that specifically cleaves ISG15 from its targets.

Two possible mechanisms for ISG15 antiviral activity have been proposed. First, ISG15 may conjugate to key cellular proteins or viral proteins and inhibit virus replication. For example, in HIV infection, the Gag protein is ubiquitinated in order to recruit the endosomal complex required for transport (ESCRT-I) to the plasma membrane for viral budding. ISG15 has been shown to conjugate with Gag and thus might prevent Gag from being ubiquitinated. As a consequence, HIV release is inhibited [27]. Conjugation of ISG15 to host interferon regulatory factor 3 (IRF3), a key signal-transducing factor in the activation of the antiviral innate immune response, protects IRF3 from ubiquitin-mediated degradation by the 26S proteasome [28]. In addition, ISG15 conjugation system broadly target newly-synthesized proteins, such as viral proteins, within cells. For HPV 16 pseudoviruses, co-translational modification of viral L1 capsid protein could sterically inhibit capsid assembly and HPV infectivity [29]. Second, ISG15 may also act alone, as a cytokine. The free form of ISG15 has been shown to activate natural killer (NK) and cytotoxic T-lymphocytes (CTL) and to stimulate IFN-γ production [30], leading to induced dendritic cell maturation and neutrophil recruitment.

ISGylation and ISG15 may play a role in modulating the host innate antiviral response. One of the key effectors of this response is IFN. Although it was initially suggested that ISGylation plays an important role in the regulation of the JAK-STAT pathway and IFN signaling [31], IFN signaling is intact in ISG15 knock-out mice [32]. In vivo analyses of ISG15 antiviral activity report contradictory findings. Although the replication of the Sindbis virus-expressing ISG15 was inhibited in IFNAR1 deficient mice [33], and ISG15 null mice have an increased susceptibility to Sindbis, Influenza, and HSV-1 virus infections [34], there was no difference in the replication of VSV and LCMV in ISG15 null and wild type mice [32,

34]. These data suggest that the antiviral activity of ISG15 might be virus-specific. More recently, ISG15 was shown to be able to inhibit IFN signaling by targeting RIG-I [35] or by boosting the innate antiviral response through decreased IRF3 degradation [28]. Surprisingly, we and others have found that ISG15 promotes HCV production in a cell culture model (see below for details) [36, 37]; thus, decreasing ISGylation by inhibition of ISG15 or its Ube1L E1 activating enzyme may represent a new anti-HCV treatment. We subsequently showed that increased ISGylation not only promotes HCV replication in an in vitro cell culture model but also inhibits interferon anti-HCV activity [36]. Silencing Ube1L, the sole activating enzyme for ISG15 conjugation pathway, inhibits ISG15 conjugation to target proteins. HCV replication, especially egress, is inhibited by Ube1L silencing, implying that ISGylation is essential for the HCV life cycle.

USP18 is a Negative Regulator of Type I Interferon Signaling

ISG15 conjugation is reversible and controlled by USP18 (UBP43), an IFN-inducible cysteine protease of the ubiquitin-specific protease (USP) family [38]. USP18 appears to counteract the effects of interferon in which the lack of USP18 results in enhanced and prolonged STAT1 phosphorylation, DNA binding, and increased induction of hundreds of ISGs [39]. Perhaps as a result of increased IFN signaling and effect, USP18 knockout mice show greater resistance to the cytopathic effects of a number of viruses, including LCMV, VSV, and SNV [40]. USP18-deficient cells exhibit high levels of ISG15 modified proteins (ISGylation). Furthermore, they are hypersensitive to type I IFN and undergo apoptosis upon IFN stimulation [41]. Thus, USP18 appears to be a negative regulator of IFN signaling.

Although ISG15 may play a role in the anti-HCV response, the ability of USP18 to regulate the anti-HCV interferon response may be independent of its ability to deconjugate ISG15. Ablation of ISG15 [42] or its E1 activating enzyme Ube1L in mice [43] did not reverse the phenotype of the USP18 knockout, nor affect IFN-induced JAK/STAT signaling, indicating that neither ISG15 nor ISGylation is essential in JAK/STAT signaling. It was recently reported that USP18 negatively regulates JAK–STAT signaling independently of its isopeptidase activity [44]. In that study, USP18 action was specific to type I IFN responses and achieved through a direct interaction between USP18 and the IFNAR2 subunit of the type 1 IFN receptor. Binding of exogenous and endogenous USP18 to IFNAR2 in vivo interfered with the JAK-receptor interaction and led to inhibition of the downstream phosphorylation cascade and other signaling events. Whether this is a cell- or species-specific mechanism remains to be determined.

Most recently, USP18 was found to modulate the expression levels of the EGF surface receptor at the protein level. Silencing USP18 by specific siRNA resulted in decreased epithelial growth factor receptor (EGFR) expression by 50-80%, while overexpression of USP18 stimulated EGFR protein translation in an USP18 protease activity-dependent manner [45].

Host Factors to Predict Treatment Response to Interferon Therapy

18-gene Signature

Host genes respond differently to HCV infection. Based on expression levels of a set of host genes, response to treatment can be accurately predicted. Chen and colleagues were the first to identify an 18-gene response signature that predicts treatment response with a positive prediction value (PPV) of 96% [20]. Most of these 18 genes are ISGs and all of them are more highly expressed in the pretreatment NR liver tissues, prompting the hypothesis that pre-activation of type I interferon signaling leading to the increased expression of a subset of ISGs contributes to interferon resistance and treatment failure [20]. ISGs are the down-stream effective molecules following type I interferon signaling activation and they have long been recognized as anti-viral molecules to restrict viral replication. We and others have found that not all ISGs have anti-viral function. At least for ISG15 and USP18, activation of them not only stimulates HCV production but also blunts interferon anti-HCV activity [36]. The pro-viral activity of some ISGs can explain why increased expression of a subset of ISGs contributes to treatment non-response and treatment failure.

Single Nucleotide Polymorphism (SNP) of IL28B

Most recently, four groups using genome-wide association study (GWAS), reported that a single-nucleotide change at roughly 3kb upstream of the promoter region of IL28B gene (encodes a type III IFNλ3) is associated with both treatment-induced and spontaneous Hepatitis C Virus clearance [46-49]. Ge et al., [49] scanned the genomic DNA sequences from more than 1,600 treatment-naïve HCV genotype 1 (G1)-infected patients and found that SNP rs12979860 was strongly associated with treatment-induced SVR. This SNP was also found to be associated with virus spontaneous clearance [46]. Although SNP rs12979860 is indeed the strongest hit that was found to be associated with treatment-induced or spontaneous HCV clearance from these GWAS, the authors did find that other SNPs were also associated, albeit the associations were statistically weaker. However, from the top 100 SNPs that were associated with SVR, none of them replicated previous reported SNPs (MxA, OAS-1, PKR, IFNγ, and IL-10) that were associated with treatment response [49]. Failure to replicate these previous findings suggests that either those associations were too weak to be picked up by GWAS, or those previous SVR-associated SNPs were not real. To support this notion, from 637 non-Hispanic Caucasian patients infected with genotype 1, a study by Morgan et al., [50] indicated that none of the 8 previously reported treatment response-associated SNPs was associated with SVR. Taking all these together, further cross-validation using even larger patient populations for GWAS is needed to confirm these associations.

Conclusion

HCV infection is a major health burden for many countries, with world-wide prevalence of more than 170 million. Current treatment regime with pegylated interferon alpha and ribavirin achieves SVR in 50% of patients. Although adding DAA to the existing regimen as "triple therapy" dramatically increases SVR up to 90%, there are still a considerable portion of patients (and a large number world-wide) that do not respond. Therefore, further understanding of the molecular mechanisms of treatment resistance is needed. Using cDNA microarray gene expression profiling, we identified an 18-gene signature that differentiates treatment responders and non-responders. A majority of these 18 genes are ISGs that are more highly expressed in the treatment non-responder pretreatment liver tissue; therefore, we propose that pre-activation of the endogenous interferon signaling leading to increased expression of a subset of ISGs contributes to treatment non-response and treatment failure. Based on the expression levels of these 18 genes, we can accurately predict whether a given patient will respond to treatment or not with a positive prediction value of 96%, providing a foundation for HCV personalized medicine. Three of these 18 genes are involved in the same ubiquitin-like signaling pathway, so activation of this pathway is related to treatment non-response. Two of the ISGs, ISG15 and USP18 have pro-HCV activity and they offer potential drug targets for the better treatment of HCV.

Acknowledgment

Financial support from grants by Canadian Institutes of Health Research (CIHR) and by Sichuan Provincial Science and Technology department (2011SZ001 to Dr. L Chen, 2013JY0048 to Dr. S Li through the Institute of Blood Transfusion, Chinese Aademy of Medical Sciences/Peking Union Medical College and 2013SZ0066 to Dr. J Yuan throug Sichuan Mingri Pharmaceutical Research Institute of New Technology) is gratefully acknowledged.

References

[1] Alter, MJ. Epidemiology of hepatitis C virus infection. *World J Gastroenterol*, 2007, 13, 2436–2441.

[2] Choo, QL; Kuo, G; Weiner, AJ; Overby, LR; Bradley, DW; Houghton, M. Isolation of a cDNA clone derived from a blood-borne non-A, non-B viral hepatitis genome. *Science*, 1989, 244, 359–362.

[3] Gerlach, JT; Diepolder, HM; Zachoval, R; Gruener, NH; Jung, MC; Ulsenheimer, A; et al. Acute hepatitis C: high rate of both spontaneous and treatment-induced viral clearance. *Gastroenterology*, 2003, 125, 80-88.

[4] Shepard, CW; Finelli, L; Alter, MJ. Global epidemiology of hepatitis C virus infection. *Lancet Infect Dis*, 2005, 5, 558-567.

[5] Thimme, R; Lohmann, V; Weber, F. A target on the move: innate and adaptive immune escape strategies of hepatitis C virus. *Antiviral Res*, 2006, 69, 129-141.

[6] Pileri, P; Uematsu, Y; Campagnoli, S; Galli, G; Falugi, F; Petracca, R; et al. Binding of hepatitis C virus to CD81. *Science*, 1998, 282, 938-941.

[7] Scarselli, E; Ansuini, H; Cerino, R; Roccasecca, RM; Acali, S; Filocamo, G; et al. The human scavenger receptor class B type I is a novel candidate receptor for the hepatitis C virus. *EMBO J*, 2002, 21, 5017-5025.

[8] Evans, MJ; von Hahn, T; Tscherne, DM; Syder, AJ; Panis, M; Wölk, B; et al. Claudin-1 is a hepatitis C virus co-receptor required or a late step in entry. *Nature*, 2007, 446, 801-805.

[9] Ploss, A; Evans, MJ; Gaysinskaya, VA; Panis, M; You, H; de Jong, YP; et al. Human occludin is a hepatitis C virus entry factor required for infection of mouse cells. *Nature*, 2009, 457, 882-886.

[10] Pöhlmann, S; Zhang, J; Baribaud, F; Chen, Z; Leslie, GJ; Lin, G; et al. Hepatitis C Virus Glycoproteins Interact with DC-SIGN and DC-SIGNR. *J Virol*, 2003, 77, 4070-4080.

[11] Fried, MW; Shiffman, ML; Reddy, KR; Smith, C; Marinos, G; Gonçales, FL Jr; et al. Peginterferon alfa-2a plus ribavirin for chronic hepatitis C virus infection. *N Engl J Med*, 2002, 347, 975-982.

[12] Feld, JJ; Hoofnagle, JH. Mechanism of action of interferon and ribavirin in treatment of hepatitis C. *Nature*, 2005, 436, 967-972.

[13] Randall, G; Panis, M; Cooper, JD; Tellinghuisen, TL; Sukhodolets, KE; Pfeffer, S; et al. Cellular cofactors affecting hepatitis C virus infection and replication. *Proc Natl Acad Sci USA*, 2007, 104, 12884-12889.

[14] Poordad, F; McCone, J Jr; Bacon, BR; Bruno, S; Manns, MP; Sulkowski, MS; et al. Boceprevir for untreated chronic HCV genotype 1 infection. *N Engl J Med*, 2011, 364, 1195–1206.

[15] Zeuzem, S; Andreone, P; Pol, S; lawitz, E; Diago, M; Roberts, S; et al. Telaprevir for retreatment of HCV infection. *N Engl J Med*, 2011, 364, 2417–2428.

[16] Margeridon, S; Le Pogam, S; Liu, TF; Hanczaruk, B; Simen, BB; Shulman, N; et al. No detection of variants bearing NS5B S282T mericitabine (MCB) resistance mutation in DAA treatment-naive HCV genotype 1-infected patients using ultra-deep pyrosequencing (*UDPS*). Presented at 62nd Annual Meeting of the American Association for the Study of Liver Diseases, San Francisco, CA, Nov. 6–9 (2011).

[17] Asselah, T; Marcellin, P. Direct acting antivirals for the treatment of chronic hepatitis C: one pill a day for tomorrow. *Liver Int*, 2012, 32(Suppl 1), 88–102.

[18] Kiser, JJ; Flexner, C. Direct-acting antiviral agents for hepatitis C virus infection. *Annu Rev Pharmacol Toxicol*, 2013, 53, 427-449 .

[19] Hopkins, S; Gallay, P. Cyclophilin inhibitors: an emerging class of therapeutics for the treatment of chronic hepatitis C infection. *Viruses*, 2012, 4(11), 2558-2577.

[20] Chen, L; Borozan, I; Feld, J; Sun, J; Tannis, LL; Coltescu, C; et al. Hepatic gene expression discriminates responders and nonresponders in treatment of chronic hepatitis C viral infection. *Gastroenterology*, 2005, 128, 1437–1444.

[21] Feld, JJ; Nanda, S; Huang, Y; Chen, W; Cam, M; Pusek, SN; et al. Hepatic gene expression during treatment with peginterferon and ribavirin: identifying molecular pathways for treatment response. *Hepatology*, 2007, 46, 1548–1563.

[22] Asselah, T; Bieche, I; Narguet, S; Sabbagh, A; Laurendeau, I; Ripault, MP; et al. Liver gene expression signature to predict response to pegylated interferon plus ribavirin combination therapy in patients with chronic hepatitis C. *Gut*, 2008, 57, 516-524.

[23] Ritchie, KJ; Hahn, CS; Kim, KI; Yan, M; Rosario, D; Li, L; et al. Role of ISG15 protease UBP43 (USP18) in innate immunity to viral infection. *Nat Immunol*, 2004, 10, 1374-1378.

[24] Yuan, W; Krug, RM. Influenza B virus NS1 protein inhibits conjugation of the interferon (IFN)-induced ubiquitin-like ISG15 protein. *EMBO J*, 2001, 20, 362–371.

[25] Furman, MH; Ploegh, HL. Lessons from viral manipulation of protein disposal pathways. *J Clin Invest*, 2002, 110, 875-879.

[26] Lenschow, DJ; Giannakopoulos, NV; Gunn, LJ; Johnston, C; O'Guin, AK; Schmidt, RE; et al. Identification of interferon-stimulated gene 15 as an antiviral molecule during Sindbis virus infection in vivo. *J Virol*, 2005, 79, 13974–13983.

[27] Kumura, G; Lu, I; Pitha-Rowe, Pitha, PM. Innate antiviral response targets HIV-1 release by the inducion of ubiquitin-like protein ISG15. *Proc Natl Acad Sci USA*, 2006, 103, 1440–1445.

[28] Lu, G; Reinert, JT; Pitha-Rowe, I; Okumura, A; Kellum, M; Knobeloch, KP; et al. ISG15 enhances the innate antiviral response by inhibition of IRF-3 degradation. *Cell Mol Biol*, 2006, 52, 29-41.

[29] Durfee, LA; Lyon, N; Seo, K; Huibregtse, JM. The ISG15 conjugation system broadly targets newly synthesized proteins: implications for the antiviral function of ISG15. *Mol Cell*, 2010, 38, 722-732.

[30] Recht, M; Borden, EC; Knight, E Jr. A human 15-kDa IFN-induced protein induces the secretion of IFN-gamma. *J Immunol*, 1991, 147, 2617-2623.

[31] Ritchie, KJ; Malakhov, MP; Hetherington, CJ; Zhou, L; Little, MT; Malakhova, OA; et al. Dysregulation of protein modification by ISG15 results in brain cell injury. *Genes Dev*, 2002, 16, 2207–2212.

[32] Osiak, AO; Utermöhlen, S; Niendorf, I; Horak, I; Knobeloch, KP. ISG15, an interferon-stimulated ubiquitin-like protein, is not essential for STAT1 signaling and responses against vesicular stomatitis and lymphocytic choriomeningits virus. *Mol Cell Biol*, 2005, 25, 6338-6345.

[33] Lenschow, DJ; Lai, C; Frias-Staheli, N; Giannakopoulos, NV; Lutz, A; Wolff, T; et al. IFN-stimulated gene 15 functions as a critical antiviral molecule against influenza, herpes, and Sindbis viruses. *Proc Natl Acad Sci USA*, 2007, 104, 1371–1376.

[34] Knobeloch, KP; Utermohlen, O; Kisser, A; Prinz, M; Horak, I. Re-examination of the role of ubiquitin-like modifier ISG15 in the phenotype of UBP43-deficient mice. *Mol Cell Biol*, 2005, 25, 11030–11034.

[35] Kim, MJ; Hwang, SY; Imaizumi, T; Yoo, JY. Negative feedback regulation of RIG-I-mediated antiviral signaling by interferon-induced ISG15 conjugation. *J Virol*, 2008, 82, 1474-1483.

[36] Chen, L; Sun, J; Meng, L; Heathcote, J; Edwards, AM; McGilvray, ID. ISG15, a ubiquitin-like interferon stimulated gene, promotes hepatitis C production in vitro: Implications for chronic infection and response to treatment. *J Gen Virol*, 2010, 91, 382-388.

[37] Broering, R; Zhang, X; Kottilil, S; Trippler, M; Jiang, M; Lu, M; et al. The interferon stimulated gene 15 functions as a proviral factor for the hepatitis C virus and as a regulator of the IFN response. *Gut*, 2010, 59, 1111-1119.

[38] Malakhov, MP; Malakhova, OA; Kim, KI; Ritchie, KJ; Zhang, DE. UBP43 (USP18) specifically removes ISG15 from conjugated proteins. *J Biol Chem*, 2002, 277, 9976-9981.

[39] Malakhova, OA; Yan, M; Malakhov, MP; Yuan, Y; Ritchie, KJ; Kim, KI; et al. Protein ISGylation modulates the JAK–STAT signaling pathway. *Genes Dev*, 2003, 17, 455–460.

[40] Ritchie, KJ; Hahn, CS; Kim, KI; Yan, M; Rosario, D; Li, L; et al. Role of ISG15 protease UBP43 (USP18) in innate immunity to viral infection. *Nat Immunol*, 2004, 10, 1374-1378.

[41] Ritchie, KJ; Malakhov, MP; Hetherington, CJ; Zhou, L; Little, MT; Malakhova, OA; et al. Dysregulation of protein modification by ISG15 results in brain cell injury. *Genes Dev*, 2002, 16, 2207–2212.

[42] Osiak, AO; Utermöhlen, S; Niendorf , I; Horak, I; Knobeloch, KP. ISG15, an interferon-stimulated ubiquitin-like protein, is not essential for STAT1 signaling and responses against vesicular stomatitis and lymphocytic choriomeningits virus. *Mol Cell Biol*, 2005, 25, 6338-6345.

[43] Kim, KI; Yan, M; Malakhova, O; Luo, JK; Shen, MF; Zou, W. Ube1L and protein ISGylation are not essential for alpha/beta interferon signaling. *Mol Cell Biol*, 2006, 26, 472-479.

[44] Malakhova, OA; Kim, KI; Luo, JK; Zou, W; Kumar, KG; Fuchs, SY; et al. UBP43 is a novel regulator of interferon signaling independent of its ISG15 isopeptidase activity. *EMBO J*, 2006, 25, 2358-2367.

[45] Duex, JE; Sorkin, A. RNA interference screen identifies Usp18 as a regulator of epidermal growth factor receptor synthesis. *Mol Biol Cell*, 2009, 20, 1833-1844.

[46] Thomas, DL; Thio, CL; Martin, MP; Qi, Y; Ge, D; O'Huigin, C; et al. Genetic variation in IL28B and spontaneous clearance of hepatitis C virus. *Nature*, 2009, 461, 798-801.

[47] Suppiah, V; Moldovan, M; Ahlenstiel, G; Berg, T; Weltman, M; Abate, ML; et al. IL28B is associated with response to chronic hepatitis C interferon-alpha and ribavirin therapy. *Nat Genet*, 2009, 41, 100-1104.

[48] Tanaka, Y; Nishida, N; Sugiyama, M; Kurosaki, M; Matsuura, K; Sakamoto, N; et al. Genome-wide association of IL28B with response to pegylated interferon-alpha and ribavirin therapy for chronic hepatitis C. *Nat Genet*, 2009, 41, 1105-1109.

[49] Ge, D; Fellay, J; Thompson, AJ; Simon, JS; Shianna, KV; Urban, TJ; et al. Genetic variation in IL28B predicts hepatitis C treatment-induced viral clearance. *Nature*, 2009, 461, 399-401.

[50] Morgan, TR; Lambrecht, RW; Bonkovsky, HL; Chung, RT; Naishadham, D; Sterling, RK; et al; HALT-C Trial Group. DNA polymorphisms and response to treatment in patients with chronic hepatitis C: results from the HALT-C trial. *J Hepatol*, 2008, 49, 548-556.

In: Microarrays: Principles, Applications and Technologies
Editor: James V. Rogers
ISBN: 978-1-62948-669-7

Chapter 9

Microarray Diagnostics for Orthopoxviruses, Filoviruses, Arenaviruses and Type A Influenza Virus

Alexander N. Sinyakov* and Vladimir A. Ryabinin
Institute of Chemical Biology and Fundamental Medicine,
Siberian Branch, Russian Academy of Sciences, Novosibirsk, Russia

Abstract

This chapter describes the methods for microarray-based analysis of orthopoxviruses using hybridization and sequencing microchips. These microchips are able to detect the orthopoxviruses pathogenic for humans and distinguish between the latter and the herpesviruses VZV, HSV-1, and HSV-2, which cause human infections with similar clinical manifestations. The advantages and shortcomings of hybridization and sequencing microarrays for typing influenza A virus are considered. The microarrays able to detect almost all existing hemagglutinin and neuraminidase subtypes of this virus are described. A new method for selecting hybridization probes for the microchip based on analysis of the peptide fragments of influenza A virus hemagglutinin and neuraminidase is briefed. Examples of successful application of this method to select hybridization probes when designing microarrays for typing influenza A virus, filoviruses, and arenaviruses are discussed.

Introduction

Microarray DNA technology has been designed for the parallel assay of complex nucleic acid mixtures. First and foremost, this technology is intended for analyzing the changes in

* Corresponding author: asinyakov1951@gmail.com.

gene expression in different organisms. However, a diagnostic perspective of microarrays is acquiring an ever increasing importance. Elaboration of diagnostic microarray chips is a promising and rapidly developing field in the virus diagnostics. There are two types of microarrays, namely, sequencing and hybridization microchips. Hybridization microchips are more widespread. This type of microarray contains a set of oligonucleotide probes to form duplexes with the target DNA carrying a label, for example, a fluorophore [1–8]. The label present in the formed duplex is recorded by special devices, thus making the specified DNA fragments detectable within an analyzed specimen. This chapter describes the microarray-based analysis of viral genomes as a case study of orthopoxviruses, filoviruses, and arenaviruses as well as typing of influenza A virus.

Microarray Analysis of Orthopoxviruses

Orthopoxviruses are among the most abundant virus families and affect numerous representatives of various taxa from insects to humans [9]. Of the orthopoxviruses, four species are pathogenic for humans, causing diseases of different severities, such as, variola virus (VARV), monkeypox virus (MPXV), cowpox virus (CPXV), and vaccinia virus (VACV) [9–11]. Currently, humans are regarded as the only VARV carrier, since no other wildlife reservoir of this virus has been found. VARV is the agent causing smallpox, one of the most harmful infectious diseases. The mortality rate for smallpox reached 30–40%, indicating that smallpox alone may have killed more people than all the remaining infectious diseases in the history of humankind.

The vaccine against smallpox has been known for over 200 years [9]. Owing to the mass vaccination under the aegis of the World Health Organization (WHO), no human smallpox cases have been recorded since 1977, which allowed the WHO to officially declare worldwide smallpox eradication in 1980 [12, 13]. Since then, the antismallpox vaccination was ceased in many countries, and currently the major part of human population has no immunity against poxvirus infections. This sensitivity of this growing non-immune human population to orthopoxviruses implicates the use of VARV as a bioterrorism agent [11].

Currently, orthopoxviruses are detected by morphological (electron microscopy), biological (virus isolation using chick embryos and cell cultures), and serological methods. All these assays are sufficiently sensitive and adequate, but laborious and time-consuming [14–16]. The advent of polymerase chain reaction (PCR) considerably simplified and accelerated VARV detection in a much safer manner [17, 18]. Restriction fragment length polymorphism (RFLP) can be also used to detect VARV; however, mutations in the orthopoxvirus genome [19] considerably decrease the diagnostic value of this assay. The diagnostic kits based on PCR are highly sensitive and specific for typing orthopoxviruses [20, 21]. However, in the case of a genetically-modified or spontaneously-mutated virus, a single species-specific probe may appear insufficient. In these cases, a microarray-based assay utilizing a set of typing probes is likely to be much more reliable due to several probes being used for assaying one genetic marker. Such redundancy guarantees a significant increase in reliability of virus identification despite possible mutations. This particular specific feature of microarray-based assays has stimulated elaboration of several microchip-based detection methods and typing of orthopoxviruses allowing these pathogens to be distinguished from other infections with similar clinical symptoms, such as chickenpox.

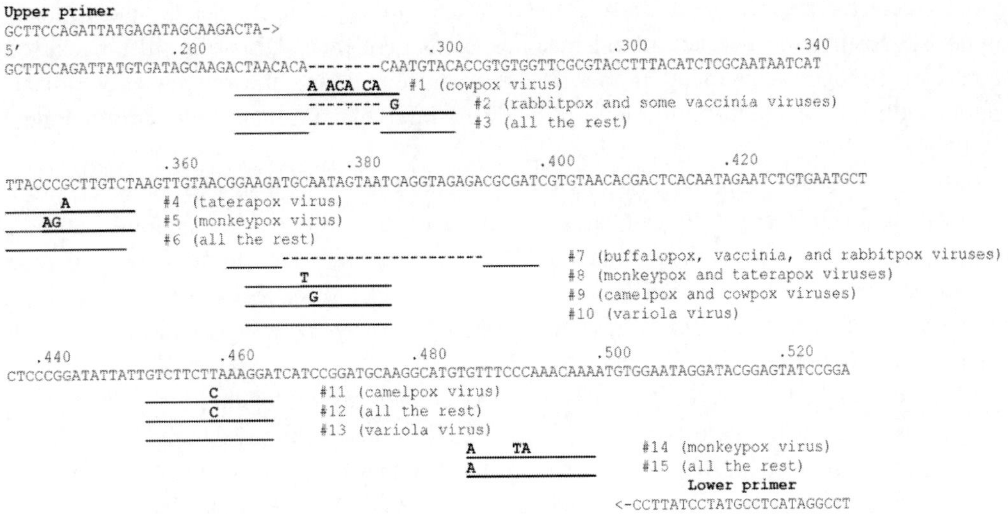

Figure 1. Structure of the amplicon of vTNFR gene used for microarray-based assay of orthopoxviruses. Hybridization probes are shown as solid lines; the mutations in typing probes relative to the variola virus genome (capital letters); deletions are represented by with dashed lines.

Hybridization Microarrays for Assaying Orthopoxviruses

The *CrmB* [22], *C23L/B29R* [23], and *B19R* [24] genes are used as genetic markers in microarray diagnostics of orthopoxviruses. The very first method used for detection and amplification of orthopoxviruses was a technique utilizing an oligonucleotide MAGIChip (for microarray of gel-immobilized compounds on a chip) [25]. The method is based on hybridization of a fluorescently-labeled amplified DNA specimen with the oligonucleotide DNA probes immobilized on three-dimensional polyacrylamide gel microchip. The *crmB* gene was chosen as a target for species-specific typing of orthopoxviruses [22]. This gene is also referred to as the gene of viral tumor necrosis factor receptor (vTNFR) and is a homolog of the cellular TNFR. The vTNFR is rarely subject to mutations [26], thereby allowing this gene to be used for species-specific typing of orthopoxviruses [27]. This microarray utilizes 15 oligonucleotide probes homologous to five species-specific regions in the vTNFR gene for identifying orthopoxviruses. The coordinates for the probes in vTNFR gene sequence of VARV (EMBL accession no. X69198) are shown in Figure 1.

In this microarray chip, the vertical columns contain probes to five species-specific regions in the vTNFR gene (Figure 2). Note that only one probe in each column is able to form a perfect duplex with the studied amplicon; therefore, the hybridization patterns for different orthopoxvirus species should theoretically differ. Figure 2 shows the expected and experimental patterns for different orthopoxvirus species. The spots containing perfect DNA duplexes display a stronger fluorescence signal as compared with the spots containing unpaired bases. By assaying the relative intensities of fluorescence signals within one column and selecting the most intense, it is possible to identify the species of an analyzed orthopoxvirus. For example, when analyzing the DNA of camelpox virus (CMLV) strain CP-1 (Figure 2f), all spots in the column containing probes 7, 8, 9, and 10 yield a fluorescence signal except for the last one. However, the spot containing probe 9 gives the most intensive signal, suggesting the absence of mismatches in its sequence. Discarding the background signals of the spots containing probes 7 and 8 and comparing the obtained hybridization pattern with theoretical patterns, it is possible to infer that the analyzed virus belongs to CMLV.

Estimation of the selectivity of this typing method has shown that the selected primers (TNFR1f and TNFR3r) amplify the target fragment of the orthopoxvirus genome and do not amplify human genomic DNA; cellular DNA of uninfected Vero culture, DNA of related ectromelia virus (ECTV), which lacks the target vTNFR gene, and DNAs of the viruses belonging to other genera, such as myxoma virus (genus *Leporipoxvirus*) and fowlpox virus (genus *Avipoxvirus*).

The authors have analyzed 59 different specimens of six orthopoxvirus species [25]. In general, the specimens were correctly identified. The only virus presenting a problem for identification was VACV. It has been found that the analyzed locus in VACV genome is variable (Figure 1). The vTNFR amplicons of various VACV strains give two types of hybridization patterns (Figure 2b, d). Nine of the 13 hybridization patterns matched VACV, while the remaining patterns were similar to the rabbitpox virus. Sensitivity of this method was assessed using dilutions of VACV strain LIVP and counting of virus particles using electron microscopy, which yielded 10^2–10^3 particles per sample. The time required for assay is 6 h [25]. The main disadvantages of this method are a long hybridization process as determined by a slow diffusion of the labeled amplicons, inward the gel units of microchip, and insufficient selectivity of hybridization probes as mainly determined by single nucleotide substitutions (Figure 1).

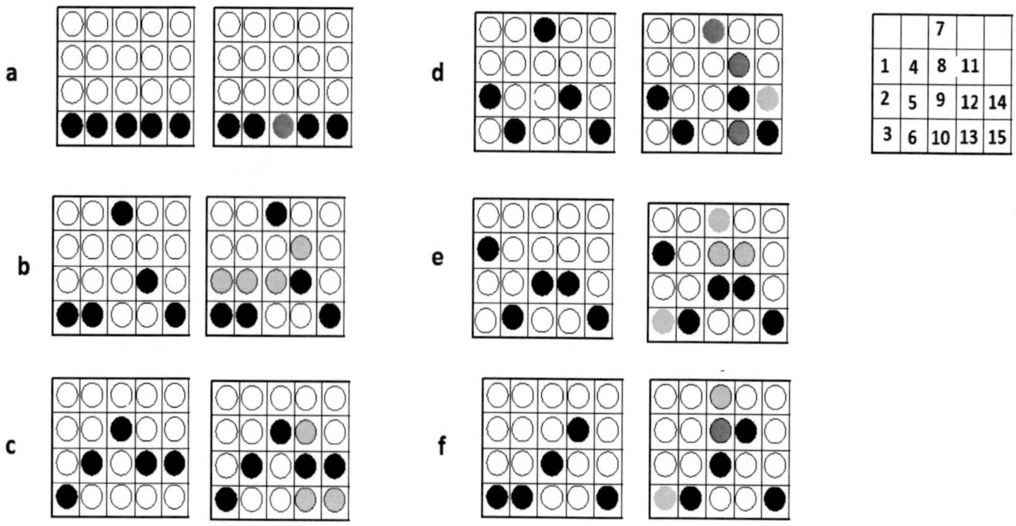

Figure 2. Hybridization patterns for the amplicons of the following orthopoxviruses: (a) variola minor virus strain Garcia-1966, (b) vaccinia virus strain Wyeth, (c) monkeypox virus strain Zaire-96, (d) vaccinia virus strain rabbitpox Utrecht, (e) cowpox virus strain GRI-90, (f) and camelpox virus strain CP-1. The left panels show the theoretical hybridization patterns and the right panels show experimental data. The rightmost panel shows the location of typing probes detailed in Figure 1.

The microarray chip designed by Laassri et al., [28] is a glass slide with covalently attached typing probes. The viral gene *C23L/B29R,* encoding the CC chemokine binding protein [23] is used as a target.

Along with the probes for detecting orthopoxviruses, this microchip also contains the probes for herpesvirus HHV-3, since this infection is frequently confused with smallpox. The probes immobilized to the slide surface are arranged as a single layer, where the targets readily and rapidly hybridize with probes. The time required for assay [28] is reduced to 3 h versus 6 h in the above variant [25].

Use of the orthopoxvirus *C23L/B29R* gene for typing these viruses makes it possible to considerably increase the number of discriminating probes for identification of each individual species of this genus. Thus, the on-slide microarray involving the *C23L/B29R* gene, unlike the gel variant [25], may be used for concurrent typing of several orthopoxvirus species. Supplementing of the microchip with the specific probes for chickenpox virus allows for a rapid diagnosing of each virus even when they are simultaneously present in the same analyzed specimen. A rapid and simultaneous detection of the herpesvirus and orthopoxviruses is especially important for the mixed outbreaks of monkeypox and chickenpox in the same population [29].

Despite significant improvements in diagnosing orthopoxviruses, the microarray by Laassri et al., [28] also has certain shortcomings. This microarray can type only the four orthopoxvirus species, VARV, MPXV, CPXV, and VACV. In addition, the probes for VACV may cross-hybridize to the amplicon synthesized from ECTV and CMLV DNAs. This is associated with the fact that it was impossible to find a sufficient number of selective probes in the sequence of only one gene (*C23L/B29R*) [23]. It is typical for this microarray that the perfect and imperfect duplexes formed by probe and target differ only by one mismatch.

To overcome the above mentioned problems, Ryabinin et al., [30] have designed a microarray involving the two genetic markers, *C23L/B29R* and *B19R* [24, 31]. A concurrent typing using several genes instead of only one gene increases the accuracy and reliability of assay. This microchip contains not only the probes specific for different orthopoxvirus species, but also the probes detecting herpesviruses 1 (HSV-1), 2 (HSV-2), and 3 (VZV), since these viruses cause the diseases with clinical manifestations resembling smallpox. The probes and PCR primers for the identification of herpesviruses were selected using the conserved regions in ORF31, *US4*, and *US5* genes of VZV, HSV-1, and HSV-2, respectively. The primers and typing probes for species identification of orthopoxviruses were based on analysis of 86 *C23L/B29R* gene variants and 72 *B19R* gene variants [9] (Table 1) using the OligoScan program.

The orthopoxvirus strains used for designing probes were isolated at different times and in different geographic areas of the world, which is likely to elevate the practical value of the resulting probes. The designed microarray chip contains the probes to six orthopoxvirus species, namely, VARV, MPXV, CPXV, VACV, CMLV, and ECTV. The sequences of all typing probes are listed in Table 2. Involvement of almost all data on the target gene structures available at the moment this microarray was deigned allowed the authors to construct highly selective probes for detection of six orthopoxvirus species [30]. A high selectivity of the calculated probes is determined by the fact that most of them form perfect duplexes with their targets, which differ from the imperfect duplexes by at least two mismatches. This was almost impossible when using only one gene for designing the typing probes.

Table 1. Orthopoxvirus and human herpesvirus strains used for designing oligonucleotide probes and testing microarray for orthopoxviruses

Orthopoxvirus	Strain	Strain abbreviation[a]	Isolated in	Source of strain[b]	GenBank accession number	
					Gene B29R	Gene B19R
Monkeypox virus (MPXV)	CDC# v79-I-005	**M/Z79**	Zaire	1	AF531450	AY315880
	CDC# v97-I-004	**M/Z97**	Zaire	1	AF531451	AY315881
	CDC# v70-I-187	**M/LIB**	Liberia	2	AF531452	AY315882
	CDC# v78-I-3945	**M/BEN**	Benin	2	AF531453	AY315883
	Congo 8	**M/CNG8**	Congo	2	AY102985	AY315879
	Zaire 96-I-16	**M/Z96**	Zaire	2	–	AF380138
	CDC#77-666	**M/Z77**	Zaire	2	AY102986	–
Vaccinia virus (VACV)	WR	**V/WR**	United States	4	Y17730	Y17730
	Elstree 3399	**V/EL33**	Netherlands	1	AY102939	–
	Elstree 6664 (Vac Utrecht)	**V/EL66**	Netherlands	1	AY102940	–
	CVI-78	**V/CVI**	Netherlands	1	AY102947	AY315833
	Patwadanger (1273)	**V/PAT**	India	1	AY102945	AY315831
	Chambon St-Yves Menard (11-7-73)	**V/CHA**	France	1	–	AY315830
	LIVP	**V/LIVP**	Russia	1	AY102942	AY315829
	Wyeth	V/WYE	Netherlands	2	AY102943	–
	Buffalopox virus BP-1	V/BP1	India	1	AY102948	AY315835
	Copenhagen, VC-2	V/COP-2	Denmark	4	M35027	M35027
	Copenhagen	V/COP	Denmark	2	AY102938	AY315828
	Tian Tan	V/TIA	China	4	AF095689	AF095689
	EM-63	V/EM63	Russia	1	AY102941	AY315832
	MVA	V/MVA	Germany	4	U94848	U94848
	Rabbitpox virus strain Utrecht	V/RPU	Netherlands	4	U64724	U64724
Variola virus (VARV)	M-Abr-60	S/ABR	Russia	1	AY102975	AY315869
	M-Sok-60	**S/SOK**	Russia	1	AY102976	AY315870
	M-Sur-60	S/SUR	Russia	1	AY102977	AY315871
	M-N-60	S/MN	Russia	1	AF531446	–
	Ind-3a	**S/IND3**	India	1	AY102972	AY315866
	India 164*	**S/I164**	India	1	AF531444	AY315859
	India 378*	**S/I378**	India	1	AY102973	AY315867
	Kuw-5	**S/KUW5**	Kuwait	1	AY102974	AY315868
	Aziz	S/AZI	Pakistan	1	AY102968	AY315862
	Aslam*	**S/ASL**	Pakistan	1	AY102967	AY315861
	Khateen*	**S/KHA**	Pakistan	1	AF531445	–
	Wsim Ahmed	**S/WSI**	Pakistan	1	AF531449	–
	Butler	**S/BUT**	United Kingdom	1	AY102969	AY315863
	Brazil 128*	**S/B128**	Brazil	1	AF531442	AY315857
	Brazil 131*	**S/B131**	Brazil	1	AF531443	AY315858
	Congo-2	**S/CNG2**	Congo	1	AY102970	AY315864

Orthopoxvirus	Strain	Strain abbreviation[a]	Isolated in	Source of strain[b]	GenBank accession number	
					Gene B29R	Gene B19R
	Congo-9	**S/CNG9**	Congo	1	AY102971	AY315865
	Semat	**S/SEM**	Tanzania	1	AY102979	AY315873
	Ngami	**S/NGA**	Tanzania	1	AY102978	AY315872
	12/62	**S/12**	Tanzania	1	AY102966	AY315860
	22/62	**S/22**	Tanzania	1	AF531441	AY315855
	Helder	**S/HEL**	Tanzania	1	AF531440	–
	Rw18	**S/RW18**	Ruanda	1	AF531447	–
	Taj Barin	**S/TAJ**	Pakistan	1	AF531448	–
	Somalia-1977	S/SOM	Somalia	4	U18341	U18341
	Bangladesh-1975	S/BSH	Bangladesh		L22579	L22579
	Garcia-1966	S/GAR	Brazil		Y16780	Y16780
	India-1967	S/IND67	India	4	–	X67117
	6/58	S/6	Tanzania	1	–	AY315856
	India	S/IND	India		X69198	X69198
Ectromelia virus (mousepox virus; ECTV)	MP-1	**E/MP1**	Germany	3	AY102987	–
	MP-2	**E/MP2**	Germany	3	AY102988	–
	#4908	**E/4908**	Netherlands	3	AY102989	–
	#33221	**E/33221**	Netherlands	3	AY102991	–
	K1	**E/K1**	Russia	1	AY102990	AY315884
	K1-2	E/K1-2	Russia	1	–	AY315885
	Moscow	E/MOS	Russia	1	AF012825	–
	Naval	E/NAV	United States	5	–	–
	Mill Hill	E/MIL	United Kingdom	4	AJ277112	AJ277112
Camelpox virus (CMLV)	CP-1	**CP/1**	Iran	3	AY102980	AY315874
	CP-5	CP/5	IAE	3	AY102983	AY315877
	CP-1260	**CP/1260**	IAE	3	AY102982	AY315876
	CP-1231	CP/1231	IAE	3	AY102981	AY315875
	CP Saudi	CP/SAU	IAE	3	AY102984	AY315878
	CMS	CP/CMS	Iran	4	AY009089	–
	M-96	CP/M96	Kazakhstan	4	AF438165	–
Cowpox virus (CPXV)	Rev-91	**C/REV**	Russia	1	AY102952	AY315840
	Ham-85	**C/HAM**	Germany	1	AY102950	AY315838
	Turk-74	**C/TUR**	Turkmenia	1	AY102953	AY315841
	EP-1	**C/EP1**	Germany	1	AY102954	AY315843
	EP-2	**C/EP2**	Germany	1	AY102955	AY315844
	EP-3	**C/EP3**	Germany	1	AY102957	AY315846
	EP-4	**C/EP4**	Germany	1	AY102958	AY315847
	EP-5	**C/EP5**	Austria	1	AY102959	AY315847
	EP-6	C/EP6	Netherlands	1	AY102960	AY315849
	EP-7	C/EP7	Netherlands	1	AY102961	AY315850
	EP-8	C/EP8	Germany	1	AY102962	AY315851
	EP-267	C/EP267	Germany	1	AY102956	AY315845
	RP	C/RP	Germany	1	AY102964	–
	RP-1	**C/RP1**	Germany	1	AY102965	AY315853
	RP-9	C/RP9	Germany	1	AF531439	AY315854
	EXP-GR	**C/EXP**	United Kingdom	1	AY102963	AY315852
	Puma M-73	**C/PUM**	Russia	1	AY102951	AY315839
	Gri90	**C/GRI**	Russia	1	Y15035	Y15035

Table 1. (Continued)

Orthopoxvirus	Strain	Strain abbreviation[a]	Isolated in	Source of strain[b]	GenBank accession number	
					Gene B29R	Gene B19R
	Brighton	C/BRI	United Kingdom	3	AY315836	–
	OPV-88	C/OPV88	Germany	3	AY315842	–
	Gruziya	C/GRU	Georgia	1	AY315837	–
Human herpesvirus 1 (HSV-1)[c]	192 strains and isolates	**HSV-1**	Kenya, Sweden, Japan, China, South Korea, United Kingdom			
	4	AY240646–AY240838				
Human herpesvirus 2 (HSV-2)	17 strains and isolates	**HSV-2**	Sweden	4	AJ252251–AJ252267	
Human herpesvirus 3 (VZV)	20 strains and isolates	**HSV-3**	–	4	AF314216; AH010213; AH010254; AH010255; AH010733; AY016445; AY016451; AY016457; AY016462; AY253673; AY253679; AY253685; AY253691; AY253697; AY253703; AY253709; AY253720; AY548170; AY548171; NC_001348	

[a] The orthopoxvirus strains tested using this microarray are boldfaced and underlined.

[b] Sources of strains: (1) collection of the State Research Center of Virology and Biotechnology Vector (Russia); (2) DNA specimens were kindly provided by Dr. J. Esposito (Atlanta, United States); (3) DNA specimens were kindly provided by Dr. H. Meyer (Munich, Germany); (4) DNA sequences were extracted from GenBank; and (5) DNA sequences are available at http://www.sanger.ac.uk/Projects/Ectromelia_virus/.

[c] The *US4* and *US5* genes of HSV-1 encode glycoproteins G and O, respectively; *US4* gene of HSV-2 encodes glycoprotein G-2; and ORF31 encodes glycoprotein B. Reproduced with permission from the [30]; copyright 2006 John Willey and Sons.

The first group of 14 oligonucleotides comprises universal probes for detecting the *C23L/B29R* and *B19* genes. The last three universal probes for detecting the *B29R* gene can distinguish between a G–T mismatch in the duplexes when hybridizing with the target. To level the effect of such mismatch on the duplex melting temperature, the G base in these probes was replaced with 5-nitroindole (Table 2).

Table 2. Oligonucleotide probes used for detection of and discrimination between the orthopoxvirus (OPV) species and human herpesviruses of 1, 2, and 3

No.	OPV and herpesvirus species	Oligoprobe	Sequence (5' → 3')	Position of oligoprobe on microarray[b]
1	Common to all OPV species (B29R gene)	Pox_B29_1	ATATGGGAATCGATGTTATT	A1
2		Pox_B29_2	ACTTTGGATTCACCAAAT	B1
3		713POX_B29[a]	ATGTAGCGAAGAAGAGAAAGACAGC	C1
4		900POX2_B29[a]	CTTGAGGTCAAGGATGGATT	D1
5		B29_11pox_Ind[c]	CGTAT**X**GGAGACATGTG	E1
6		B29_12pox_Ind	TGTAC**X**GAAGAAGAAAACA	F1
7		B29_13pox_Ind	ATGTAG**X**GAAGAAGAGAAA	G1
8	Common to all OPV species (B19R gene)	Pox_B19_3	ACTCTAAATGGTTGAATCC	H1
9		Pox_B19_4	GAACCTGCCAATATAACAT	I1
10		Pox_B19_5	TGGATAGCCAACTATACA	J1
11		Pox_B19_6	CTGCCAATATAACATGCA	A2
12		Pox_B19_7	GGAGAACCTGCCAATA	B2
13		B19_25pox	CGACGTTAGAATCAAGA	C2
14		B19_26pox	ACAACTACAGTAGTATTGG	D2
15	Ectromelia virus (B29R gene)	Ect_B29_4	TAGCGTCTCTCCAATG	E2
16		Ect_B29_5	TGTACGGCATTGAATC	F2
17		Ect_B29_6	TGCGGTGTCTATCAAT	G2
18		Ect_B29_11_1mut	TTCGCATCCTCGT	H2
19		Ect_B29_12_1mut	ATGCTCGGTTGTC	I2
20		Ect_B29_13_1mut	TGCTCGGTTGTCT	J2
21		Ect_B29_14_1mut	CAACGACTGTGCG	A3
22		Ect_B29_15_1mut	AACGACTGTGCGG	B3
23		Ect_B29_16_1mut	TGAATCTCCAGCCA	C3
24		Ect_B29_17_1mut	AGTGAGGTACGAAGAT	D3
25	Ectromelia virus (B19R gene)	Ect_B19_1	TCACAATCTGGTCAGA	E3
26		Ect_B19_2	AGTATTCACAATCTGGTC	F3
27		Ect_B19_3	AAGAACTCATTATCGATGG	G3
28		Ect_B19_11_1mut	TCCATAGCTACGCCA	H3
29		Ect_B19_12_1mut	AGTGTCCGCCTATTG	I3
30		Ect_B19_13_1mut	AGTAGTGGCGGTATC	J3
31	Variola virus (B29R gene)	Var_B29_5	ATCGTGTGTCTGAATAAAT	A4
32		Var_B29_13_1mut	AATCCGCTAATGGAAA	B4
33		Var_B29_14_1mut	ACTGTGAGCTGTCTA	C4
34		Var_B29_15_1mut	TGAAGGTGTAACCGAT	D4
35		Var_B29_17_1mut	AGTGTACCTGACACTA	E4
36	Variola virus (B19R gene)	Var_B19_1	AACGTAGTTAACTGGGA	F4
37		Var_B19_2	AGTGTCGTATGATGAAGA	G4
38		Var_B19_3	AGACAACGTAGTTAACTG	H4
39		Var_B19_4	AACATGTATATTTGGAGGC	I4
40		Var_B19_11_1mut	AGGAAAAACTAGACGAC	J4
41		Var_B19_12_1mut	AATGGCGACTGTATTCA	A5
42		Var_B19_13_1mut	CGGAAGATATGACTGTT	B5
43		Var_B19_14_1mut	CACTGCTGTATCAACG	C5
44	Monkeypox virus (B29R gene)	Mpx_B29_3	TGCCTACTAGTCTTCAA	D5
45		Mpx_B29_4	CGGGTCTTAACATGAAC	E5

Table 2. (Continued)

No.	OPV and herpesvirus species	Oligoprobe	Sequence (5' → 3')	Position of oligoprobe on microarray[b]
46		Mpx_B29_5	TAACATGAACTTTGGATTCA	F5
47		Mpx_B29_11_1mut	TGTGCCTAGTGGC	G5
48		Mpx_B29_14_1mu	AATGGATGAGTTGGG	H5
49		Mpx_B29_16_1mut	CCAGTACTTGGGTCTA	I5
50		Mpx_B29_17_1mut	TTAGCGTACGTATCGG	J5
51		Mpx_B29_18_1mut	ACATGTGTGAGGAATC	A6
52		Mpx_B29_19_1mut	TCAAAGCATGTGTCTG	B6
53	Monkeypox virus (B19R gene)	Mpx_B19_1	AAGATGAAGATGATGGTC	C6
54		Mpx_B19_2	ACCTATGATGAAGATGAAG	D6
55		Mpx_B19_11_1mut	TCACAAGCGGGAAA	E6
56		Mpx_B19_12_1mut	ACATGCAGTGCTGT	F6
57		Mpx_B19_13_1mut	TCTCATGTGATGTGTG	G6
58		Mpx_B19_14_1mut	AATCACCGAATTCTTCA	H6
59		Mpx_B19_15_1mut	TTCAGGGTGTAGTTAGA	I6
60	Vaccinia virus (B19R gene)	Vac_B19_1	GTTGTATTTTTCTCATGCG	J6
61		Vac_B19_2	GTATTTTTCTCATGCGATG	A7
62		Vac_B19_11_1mut	ATCGCACAGATATAAAGAC	B7
63		Vac_B19_12_1mut	ACCGTAACTACAAAGAAT	C7
64	Vaccinia virus strain CVI-78	V/CVI_19_1	GAATGAATGAGGATACGC	D7
65	Camelpox virus (B29R gene)	Camel_B29_2	ATCAAGACTCACCCAG	E7
66		Camel_B29_3	GACTCACCCAGTACTC	F7
67		Camel_B29_11_1mut	AGCAAGACAAAACACC	G7
68		Camel_B29_12_1mut	AACGGAAAATACAGAGTC	H7
69		Camel_B29_13_1mut	AGATGATTTCACATCAGT	I7
70	Camelpox virus (B29R gene)	Camel_B19_1	ATTCTACCAACTAAAGACTC	J7
71		CM3_B19a	GTGTACGGCGATTACAA	A8
72		Camel_B19_11_1mut	GGCACAACGAATGATA	B8
73		Camel_B19_12_1mut	AAAGACTAAACGACCTC	C8
74		Camel_B19_13_1mut	GTTGAATCCTGCATGT	D8
75		Camel_B19_14_1mut	TGTGTACGGCGATTAC	E8
76		Camel_B19_16_1mut	GGTTACACCGTCACA	F8
77	Cowpox virus (B29R gene)	Cpv_B29_1	GCACACCACAAATTAAA	G8
78		Cpv_B29_2	GCGATCACTAGTGAAGA	H8
79	Cowpox virus subgroup A1 (B19R gene)	Cpv_B19_1	CCGGTATGTCGAAGAT	I8
80		Cpv_B19_2	TGGTCAGCGATGTAAT	J8
81		Cpv_B19_3	AGCGATGTAATGGTTG	A9
82		Cpv_B19_4	AACTTACGCATCCATC	B9
83	Cowpox virus subgroup A2 (B19R gene)	Cpv_B19_5	TTTTTATCACGCGATGT	C9
84		Cpv_B19_6	AGACGACAACAGATTTC	D9
85		Cpv_B19_7	AAGAATAGACGACAACAGA	E9
86	Cowpox virus group B (B19R gene)	Cpv_B19_8	ATTTTTTTCTCATGCGATG	F9
87		Cpv_B19_9	TTTGTATTTTTTTCTCATGCG	G9
88	Cowpox virus, strains BRI, Turk-74, Gruzia (B29R gene)	C/BRI_TUR_B29_1	CTGAAGACGAGTCAGAA	H9
89		C/BRI_TUR_B29_2	AAGACGAGTCAGAAGAA	I9
90		C/BRI_TUR_B29_3	TCACTCATCAATTCAGC	J9

No.	OPV and herpesvirus species	Oligoprobe	Sequence (5' → 3')	Position of oligoprobe on microarray[b]
91	Cowpox virus, strain EP-2	C_EP2_RP9_B29_2	AGAAAGTGAACTACGAAG	A10
92	Herpesvirus 1	HSV1_1	GACTTGGCGTTCTG	B10
93		HSV1_2	GTGTCGCGATGTCT	C10
94		HSV1_3	AGTCTGGCATCTGG	D10
95		HSV1_4	TTTGGGAAGCCTCG	E10
96		HSV1_5	CAACGGACCCCTTA	F10
97		HSV1_6	CTGCTTGAGCTCCT	G10
98	Herpesvirus 2	HSV2_1	CGCCAGGAATTCTTG	H10
99		HSV2_2	TAAGGACATGCCCTC	I10
100		HSV2_3	CCTGTTCTGGTTCCT	J10
101		HSV2_4	TCCCCTGCTCTAGAT	A11
102		HSV2_5	CGTTCGTTTGTCTGG	B11
103		HSV2_6	AGCACAACTTTGGC	C11
104	Herpesvirus 3 (VZV)	HHV3_1	GCGCGGAACTTTTT	D11
105		HHV3_2	AACGGTTGGTTGGA	E11
106		HHV3_3	GCCAAAACGAACGG	F11
107		HHV3_4	TTGTCAAGTGGCGT	G11
108		HHV3_5	AAAGGAGGAAGCCC	H11
109		HHV3_6	GCACACAATTTTCGC	I11
110		HHV3_7	ACACTTTCTACCACGT	J11

[a] Oligonucleotide probe sequences were taken from [28].

[b] Positions of oligonucleotide probes are given according to spot numbers in microarray (Figure 3).

[c] Oligonucleotide probes contained 5-nitroindole. Reproduced with permission from the [30]; copyright 2006 John Willey and Sons.

As mentioned above, the *C23L/B29R* gene sequence contains the typing probes allowing for identification of four orthopoxvirus species—VARV, MPXV, ECTV, and CMLV (Table 2). In the case of CPXV and VACV, the intraspecific divergence considerably hinders finding probes for identification of these orthopoxvirus species. In particular, only two probes specific for most but not all CPXV strains have been found. One of these probes, Cpv_29_1 identified all CPXV strains except for C/BRI, C/TUR, C/OPV-88, C/RP-1, and C/GRI, while Cpv_29_2 is unable to detect C/EP-2, C/EP-5, and C/RP-9. Therefore, the microarray was supplemented with the set of probes (C/BRI_T-UR_B29_1, C/BRI_TUR_B29_2, C/BRI_TUR_B29_3, and CEP2_RP9_B29_2) to identify the CPXV strains undetectable by Cpv_29_1 and Cpv_29_2.

The *B19R* gene sequence contains the probes for typing the six OPV species, VARV, MPXV, ECTV, CMLV, VACV, and CPXV. The main problems associated with the selection of probes from the *B19R* gene sequence results from heterogeneity of CPXV. To overcome this, all CPXV strains were divided into three subgroups—A, B, and C (Table 2)—and the separate typing probes were constructed for each group. The designed microarray chip was tested using 52 OPV strains, two HSV-1 strains, three HSV-2 strains, and two HSV-3 strains (Table 1). All the strains used in testing were unambiguously identified (Figure 3). Testing of this microarray has demonstrated that the probes forming the duplexes with targets containing two or more mismatches have a weak effect on hybridization patterns unlike the probes forming duplexes with one mismatch only. For example, the hybridization pattern in the case of ECTV amplicons displays several additional signals resulting from cross-hybridization with the probes for other orthopoxvirus species (Figure 3a, *B19R* gene). All these subsidiary hybridizations are determined by the probes with single nucleotide substitutions. Nonetheless, such a cross-hybridization does not influence the correctness in interpreting hybridization pattern (Figure 3a). The probes with two and more mismatches display a high specificity and almost never cross-hybridize with heterologous specimens. All the probes for typing herpesviruses displayed a high specificity and did not hybridize with orthopoxvirus amplicons (Figure 3b).

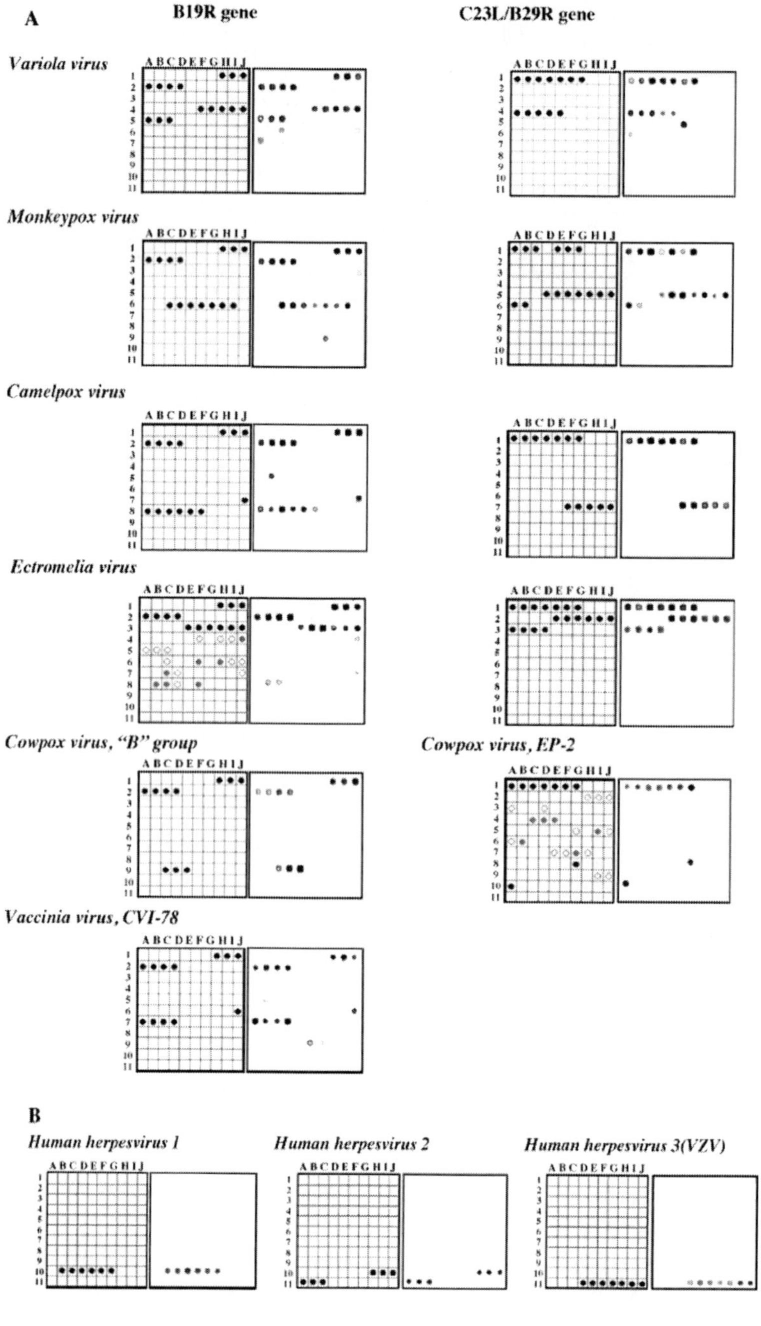

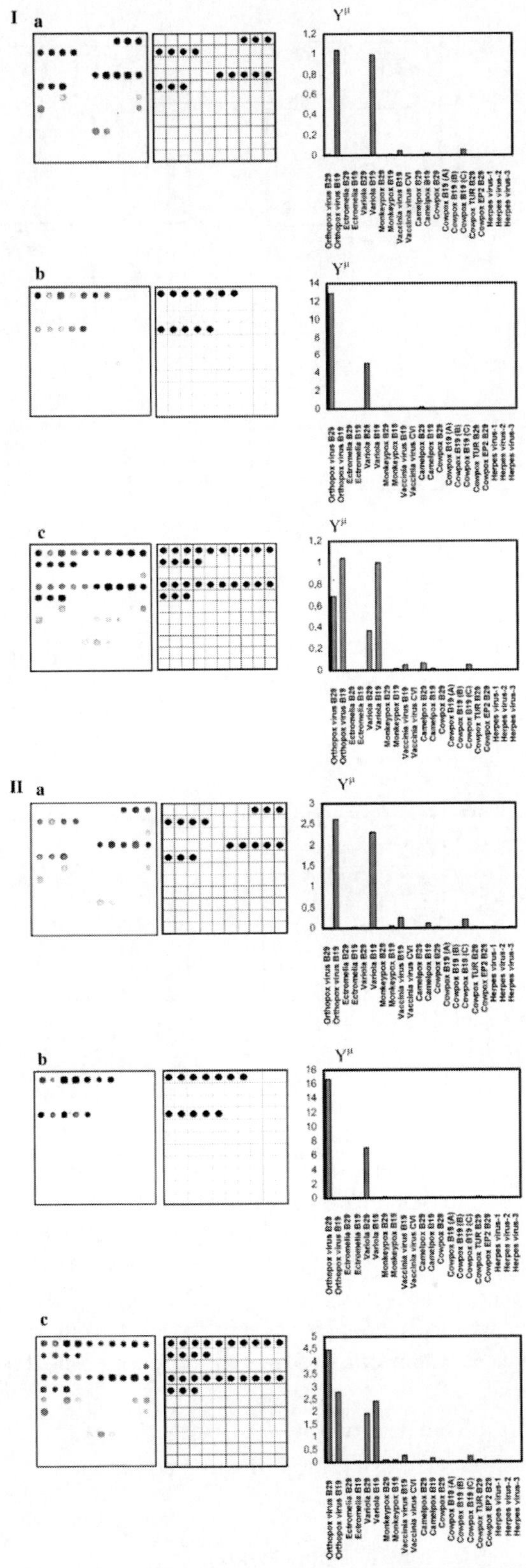

Figure 3. (Continued)

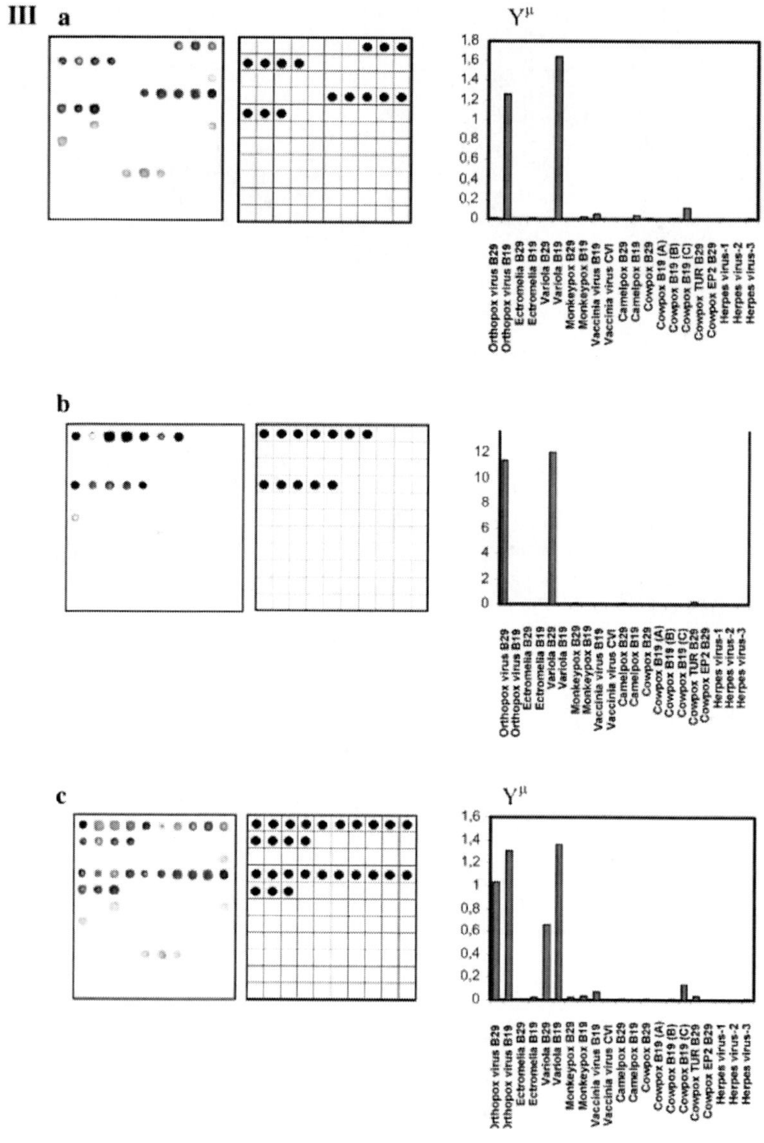

Figure 3. Theoretical (left) and experimental (right) patterns of microarray hybridization: column letters and row numbers show the positions of oligoprobes according to Table 2. (a) Hybridization of amplified *B19R* and *C23/B29R* gene fragments of orthopoxviruses. (b) Hybridization of amplified *US4–US5* gene fragments for HSV-1 and HSV-2 and ORF31 gene for HSV-3 (VZV). For theoretical patterns of ectromelia virus (*B19R* gene) and cowpox virus EP-2 (*C23L/B29R* gene), the spots containing the oligonucleotide probes with one mismatch are denoted by (●) and the oligoprobes with two mismatches, by (○). Reproduced with permission from the [30]; copyright 2006 John Willey and Sons.

Discrimination ability of the microarray was estimated as a mean square of normalized fluorescence signal intensities (Y^{μ}) using the following equation:

$$Y^{\mu}=(\sum_{i}^{N}(I_{i}^{\mu})^{2}/N,$$

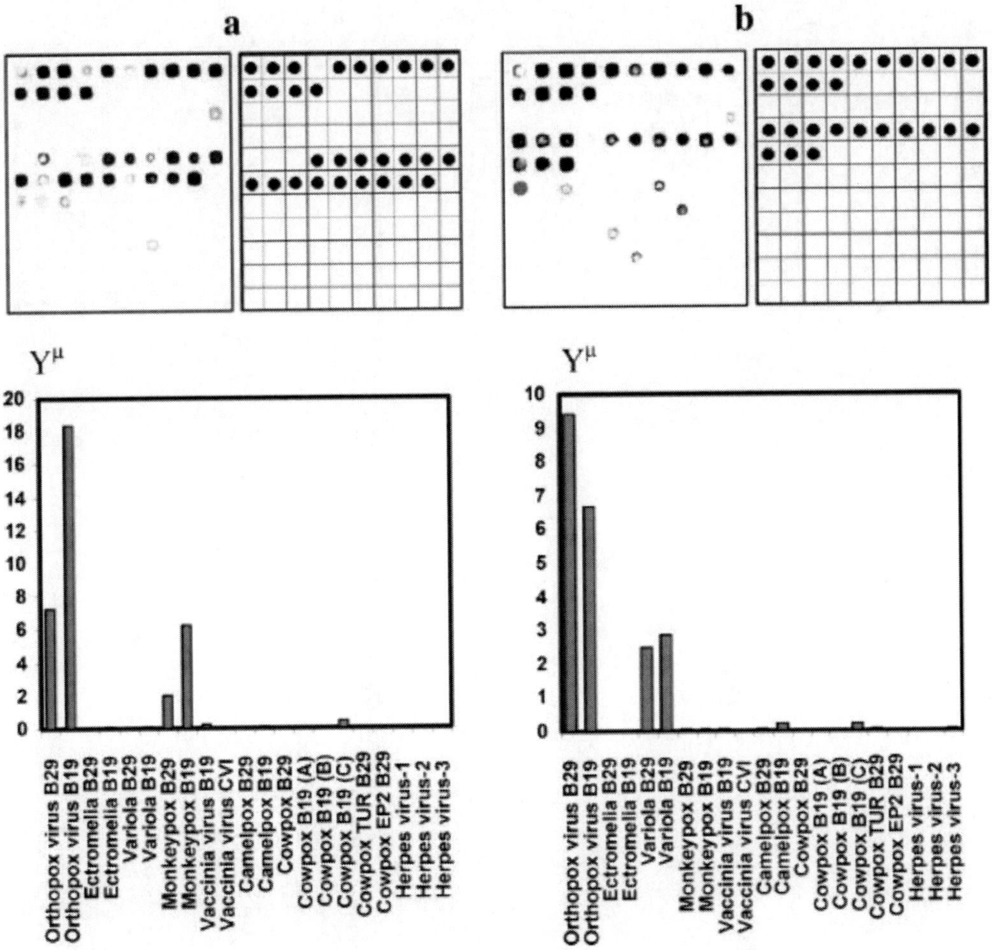

Figure 4. Detection of and discrimination between OPV strains using the mixture of *C23L/B29R* and *B19R* amplicons produced by multiplex PCR: experimental (left) and theoretical (right) patterns for (a) monkeypox virus M/LIB and (b) variola virus S/SOK. Reproduced with permission from [30]; copyright 2006 John Willey and Sons.

where μ is a group of spots specifically discriminating orthopoxvirus species (or herpesvirus species); I_i^{μ} is the normalized signal from each spot belonging to μ; and N, the number of specific oligoprobes for detection the of orthopoxvirus species (Table 2). The signal quantification using the mean square of normalized fluorescence signals (Y_{μ}) appeared useful for accurate and reliable identification of orthopoxvirus species (Figure 4).

The designed microarray has been tested using the VARV strains Ind-4A (isolated in India in 1967), M-Gavr-60 (isolated in Russia in 1960), and 13/62 (isolated in Tanzania in 1962). These strains were not sequenced; thus, their *B19R* and *C23/B29R* gene sequences were not used when computing the probe structures for the microarray. Nonetheless, these VARV strains were correctly identified by typing involving only one of the genes (*B19R* or *C23/B29R*) or both genes at once (Figure 5).

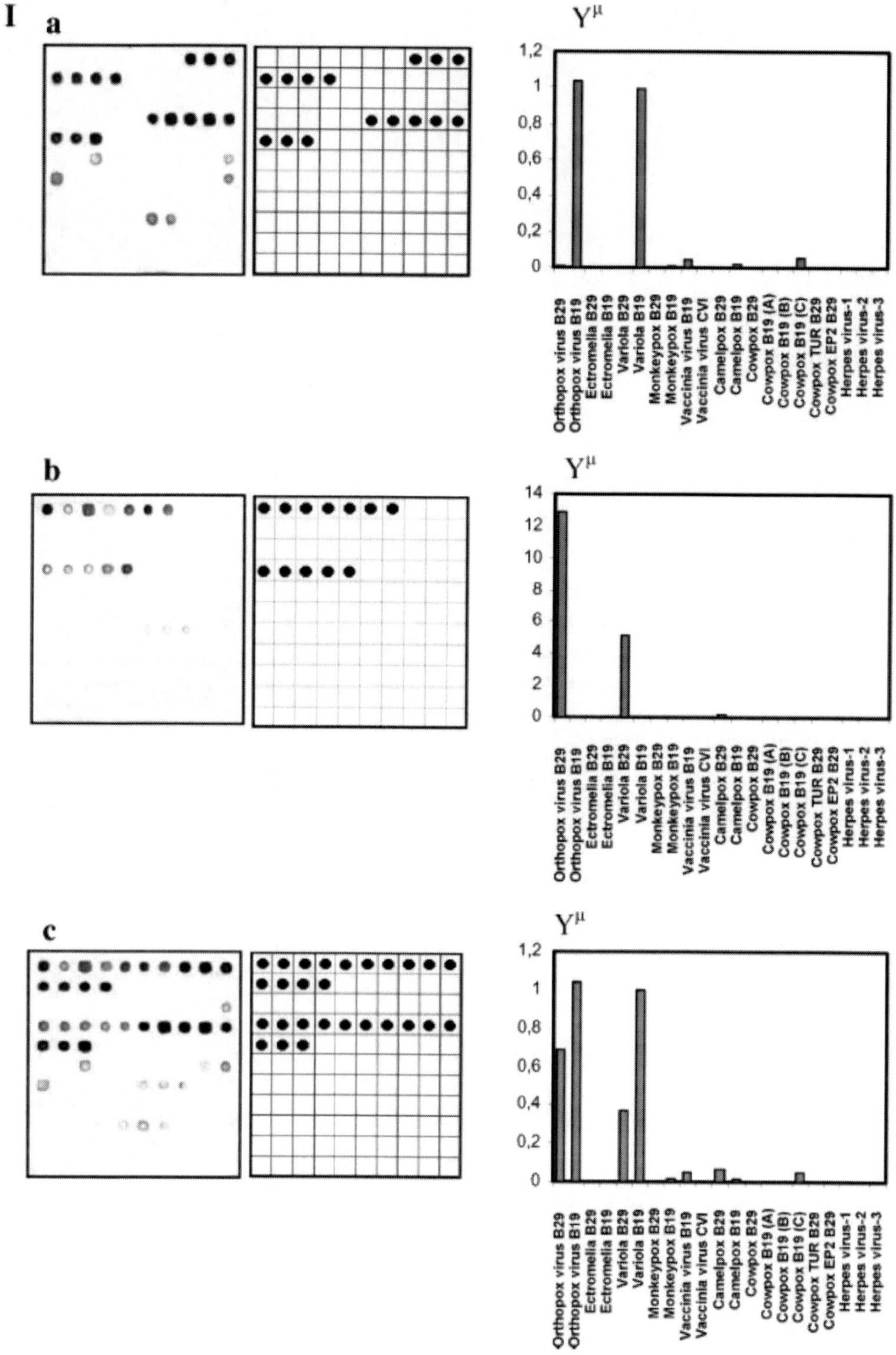
I
a
Yμ
1,2
1
0,8
0,6
0,4
0,2
0
Orthopox virus B29
Orthopox virus B19
Ectromelia B29
Ectromelia B19
Variola B29
Variola B19
Monkeypox B29
Monkeypox B19
Vaccinia virus B19
Vaccinia virus CVI
Camelpox B29
Camelpox B19
Cowpox B29
Cowpox B19 (A)
Cowpox B19 (B)
Cowpox B19 (C)
Cowpox TUR B29
Cowpox EP2 B29
Herpes virus-1
Herpes virus-2
Herpes virus-3
b
Yμ
14
12
10
8
6
4
2
0
Orthopox virus B29
Orthopox virus B19
Ectromelia B29
Ectromelia B19
Variola B29
Variola B19
Monkeypox B29
Monkeypox B19
Vaccinia virus B19
Vaccinia virus CVI
Camelpox B29
Camelpox B19
Cowpox B29
Cowpox B19 (A)
Cowpox B19 (B)
Cowpox B19 (C)
Cowpox TUR B29
Cowpox EP2 B29
Herpes virus-1
Herpes virus-2
Herpes virus-3
c
Yμ
1,2
1
0,8
0,6
0,4
0,2
0
Orthopox virus B29
Orthopox virus B19
Ectromelia B29
Ectromelia B19
Variola B29
Variola B19
Monkeypox B29
Monkeypox B19
Vaccinia virus B19
Vaccinia virus CVI
Camelpox B29
Camelpox B19
Cowpox B29
Cowpox B19 (A)
Cowpox B19 (B)
Cowpox B19 (C)
Cowpox TUR B29
Cowpox EP2 B29
Herpes virus-1
Herpes virus-2
Herpes virus-3

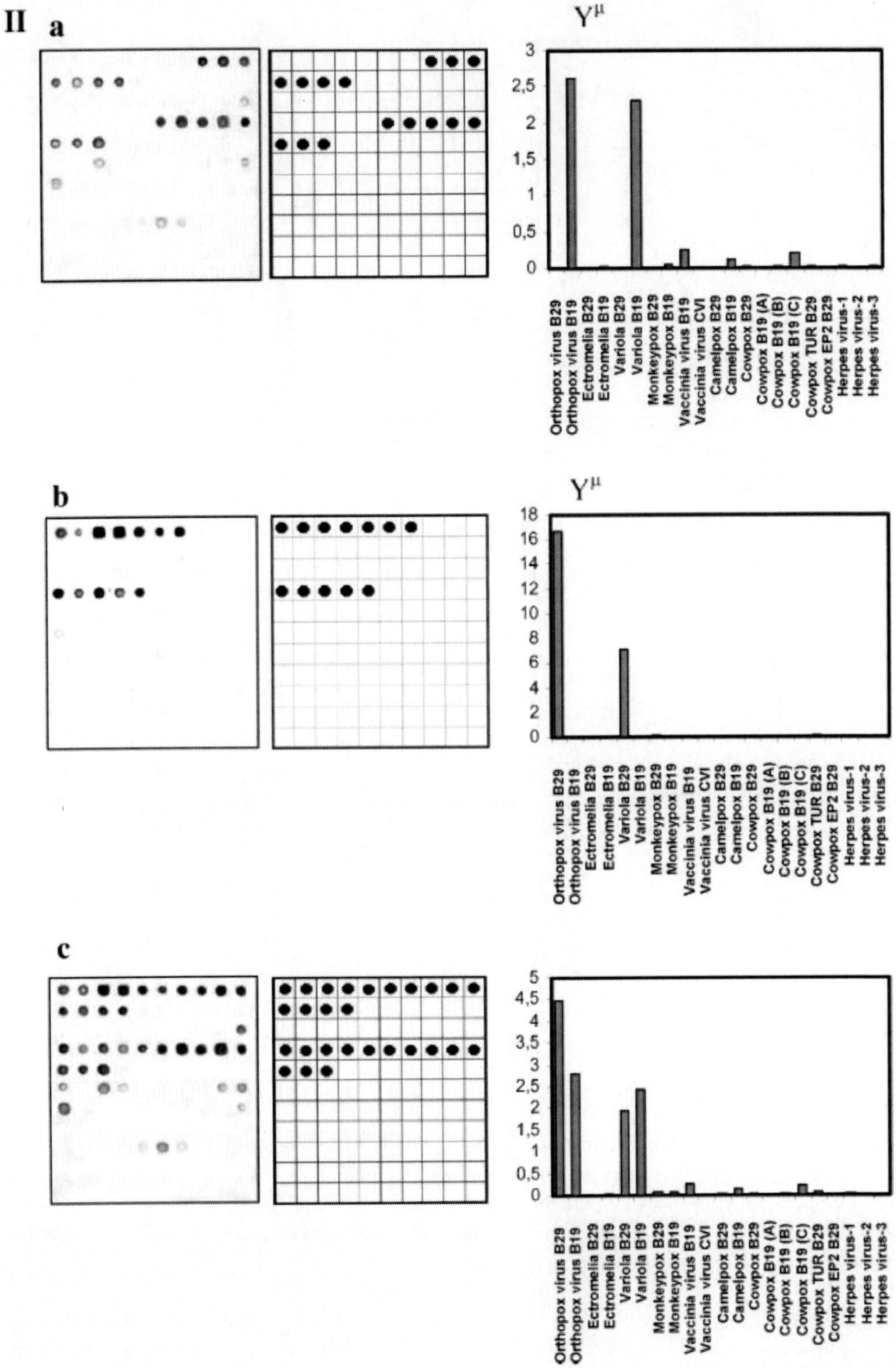

Figure 5. (Continued)

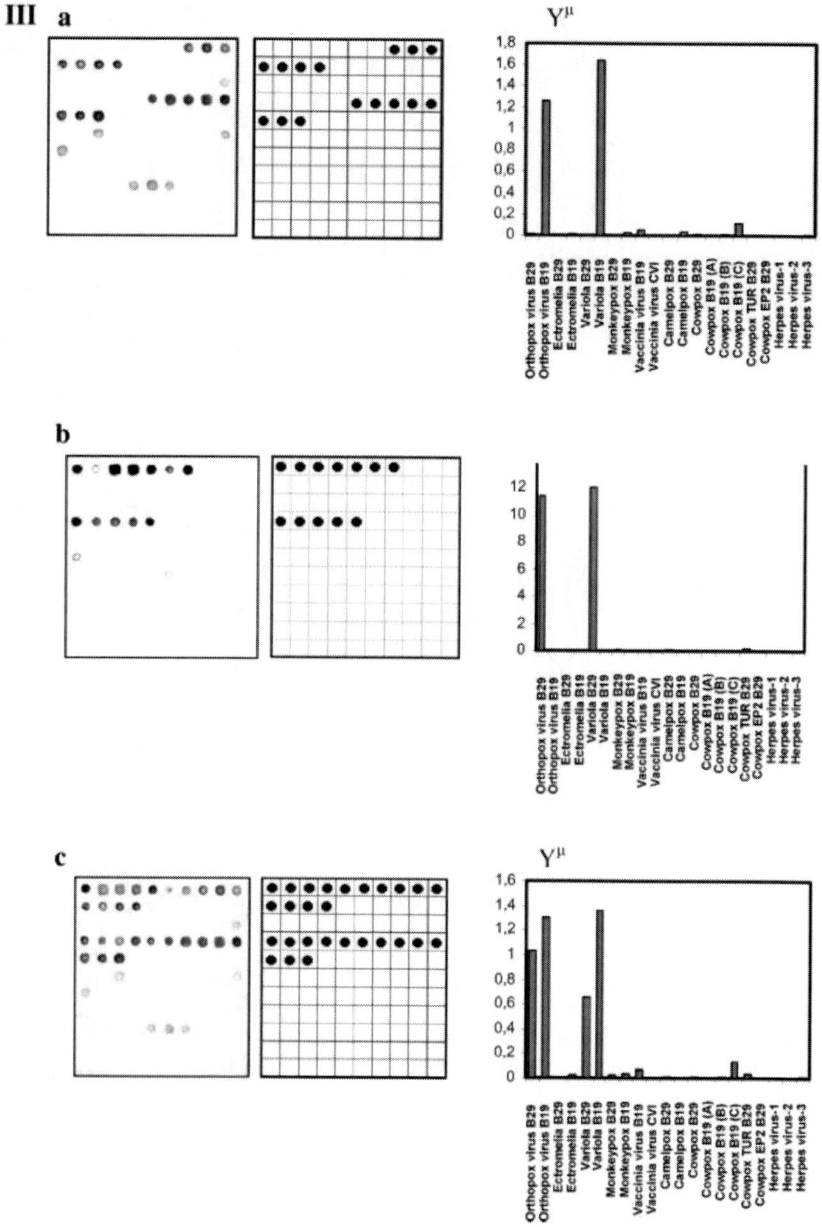

Figure 5. Detection of and discrimination between unknown variola strains using (a and b) *B19R* and *C23L/B29R* amplicons separately and (c) together with the help of 19-U1/19-L1 and B29-U1var/B29-L1var primers: experimental (left) and theoretical (right) patterns for (I) VARV Ind-4A, (II) VARV M-Gavr-60, and (III) VARV 13/62. Reproduced with permission from [30]; copyright 2006 John Willey and Sons.

Microarray-based typing of orthopoxviruses demonstrates that this diagnostic assay is a promising tool. Use of the probes for several genes within the same microarray considerably increases the probability for detecting recombinant structures carrying genes originating from different viruses.

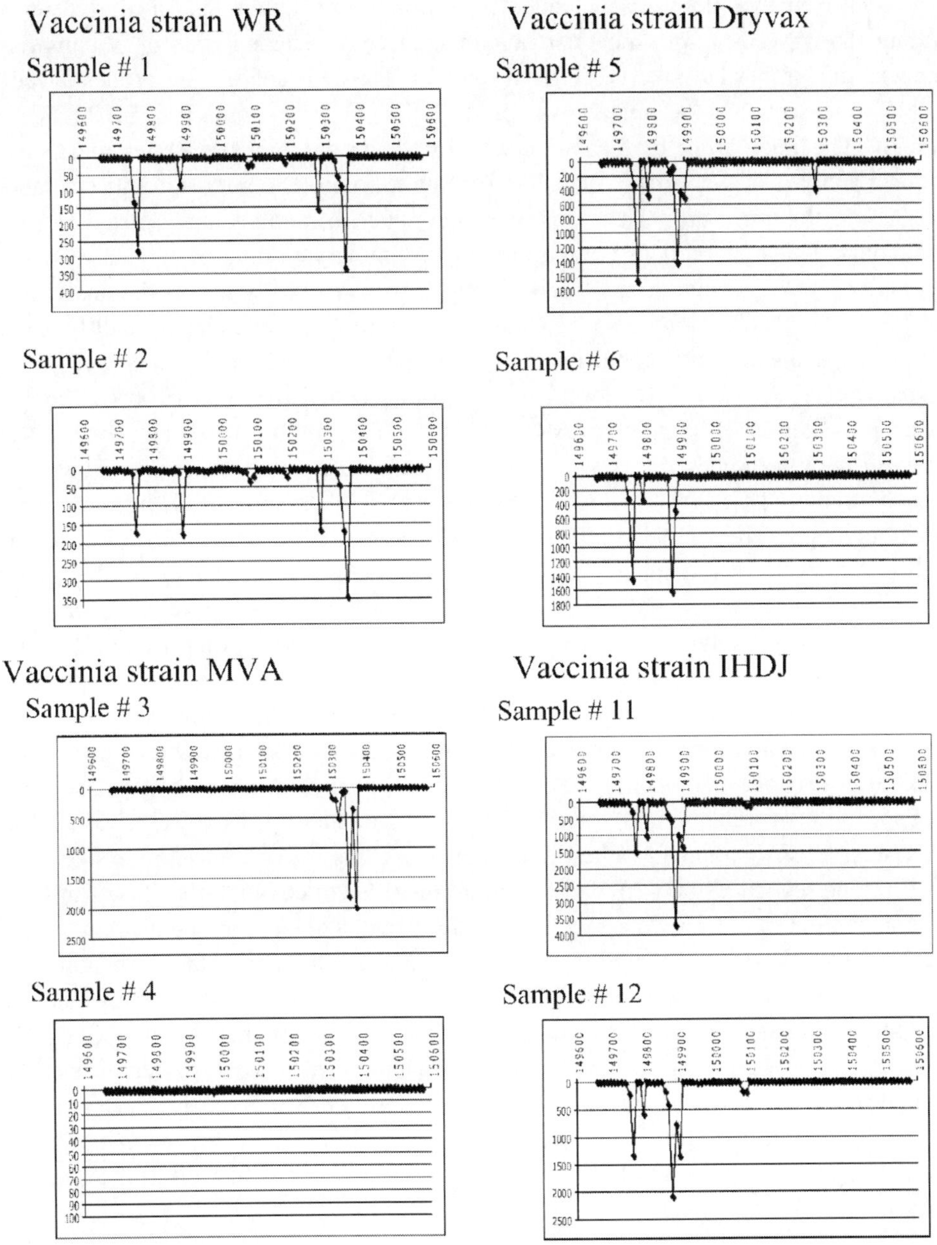

Figure 6. Mutation screening of the B5R gene from different vaccinia virus strains. The B5R clones from vaccinia virus strains (1 and 2) WR, (3 and 4) MVA, (5 and 6) NYCBH/Dryvax, and (11 and 12) IHD-J were assayed for the presence of mutations by microarray. Reproduced with permission from [32]; copyright 2006 John Willey and Sons.

Sequencing Microarrays for Analysis of Orthopoxvirus Genome

Laassri et al., [32] have designed a microarray chip for analysis of the VACV B5R gene. This microchip, which utilizes MARSH (microarrays for resequencing and sequence heterogeneity) methodology, is intended for monitoring mutations in this gene when passaging the VACV strain Ankara in mammalian cells. The microarray is comprised of oligonucleotide probes half overlapping one another. Thus, the set of these probes cover twice the B5R gene sequence, which allows for detection of point mutations in this sequence (Figure 6). The microarray by Laassri et al., [32] considerably accelerates and simplifies mutational analysis of the virus in question, making it possible to put off traditional laborious assays. Using this microarray, it is possible to conduct in parallel many assays during just several hours. This method does not require any preliminary cloning of the analyzed sample; correspondingly, the results of such analysis much more precisely reflect the analyzed virus population. This method may also be useful for controlling the vaccine quality, screening virus isolates, and analyzing genetic changes in this pathogen during chemotherapy.

Sulaiman et al., [33] have designed a set of resequencing microarrays (GeneChips), intended for analysis of complete VARV genomes. The authors used the sequences of 24 complete VARV genomes for constructing this microarray. The genomic sequences were aligned with the CLUSTAL_X and DnaSP programs in order to find the conserved and variable segments. The CHN48_horn strain (GenBank acc. no. DQ437582) was used as a reference sequence. All deletions and inserts were computed relative to this reference genome. Each microarray chip manufactured according to Affymetrix technology by photolithography and solid-phase oligonucleotide synthesis [34] is comprised of 240,000 25-mer oligonucleotides. Each position in the reference sequence is represented by four identical 25-mer oligonucleotides differing at the 13th position. Each of the seven sequencing microchips is intended for assaying a viral genome segment with a length of approximately 30,000 bp. In general, the coverage of the reference sequence provided by this microarray is about 99% of the overall genome, since 1,779 bases belong to repeats and escaped analysis.

Fourteen VARV strains have been analyzed using this set of GeneChips. Note that only seven genomic sequences of VARV strains were involved to compute the GeneChips probes. All genomes used in testing were sequenced beforehand. The data on the genomic structures of analyzed strains were compared to the conventional sequencing data. Six genomes of the analyzed VARV strains appeared identical according to the microarray assay and conventional sequencing. An analogous result was obtained for the reference CHN48_horn strain, sequenced twice using two different sets of GeneChips. Five virus genomes contained a single error as compared to conventional sequencing; two genomes, three errors each; and the SOM77_ali genome differed by 12 bases. Undoubtedly, the constructed GeneChips is an efficient tool for analyzing possible VARV genetic modifications as well as the evolution of this virus.

A Microarray Assay for the Influenza Virus

Type A influenza virus, circulating in the human population as well as among feral and domestic animals, presents a significant medical and economic problem due to annual epidemics of seasonal influenza. Since the influenza virus rapidly changes its antigenic properties, it is impossible to design any single anti-influenza vaccine. Once in several decades, influenza virus strains emerge that display unique antigenic and biological properties, allowing them to spread worldwide, leading to tens of millions of deaths. There has yet been a successful attempt to forecast such a pandemic. Prevention of influenza epidemics and pandemics is based on monitoring of the influenza viruses circulating in the human population as well as among domestic and feral animals and assessment of their pandemic potential. Recently, the candidates for pandemic strains have been mutants of avian H5N1 and swine H1N1 influenza viruses.

Since the neutralizing antibodies to the viral proteins hemagglutinin and neuraminidase are the particular tool providing protection against influenza A virus, it is of paramount importance to determine the subtypes of these proteins in new influenza virus isolates. To date, 16 hemagglutinin (HA) subtypes and 9 neuraminidase (NA) subtypes are known [2, 3]. As a rule, the protocol for influenza A virus typing includes virus isolation and its immunochemical or real-time PCR identification [4] with subsequent partial or complete sequencing of the hemagglutinin and neuraminidase genes [35–38]. Currently, microarrays are rather widely used as a diagnostic tool for this purpose [2, 4, 5, 39, 40–50]. Both types of microarrays—sequencing and hybridization—are appropriate for solving this problem. In the former case, microarray or microreactor devices allow for determination of hemagglutinin and neuraminidase gene sequences to further attribute the isolates to certain serotypes using phylogenetic analysis [44, 48, 49]. The latter variant of microarrays, which utilizes the oligonucleotides specific for subtypes of the analyzed genes, is more widespread [2, 4, 5, 43, 50].

Hybridization Microchips for Typing Influenza A Virus

Some specialized microchips have been also used to identify a limited number of influenza virus types. For example, MChip [5, 45] consists of 15 distinct oligonucleotides designed to target only a segment of the influenza A virus matrix gene. For automated analysis of hybridization pattern, the authors developed a specialized computer program. The MChip allows for detection of only H3N2, H1N1, and H5N1 subtypes of influenza A virus.

Evans at al. [43] used 24 cDNA sequences with a length of 500 nt encoding fragments of HA, NA, and MP genes to design another microarray capable of detecting H1–H3, N1, and N2 influenza A virus subtypes. The assay protocol involves obtaining of fluorescently-labeled cDNA and its hybridization with typing probes on microarrays. Using this microarray, the authors succeeded in detecting three previously unknown influenza virus strains despite the fact that the rate of mismatches in hybridization duplexes of these strains amounted to approximately 14%.

When constructing a three-dimensional microarray (Flow-Thru Chip, MetriGenix), Kessler et al., [41] applied the probes with a length of 45 to 65 bp, attempting to increase assay specificity. These probes were immobilized into microchannels of a silicon wafer. The double-stranded cDNA produced using biotin-labeled random primers was used as a target. The double-stranded cDNA was denatured by heating and hybridized to the microarray with subsequent development of the hybridization pattern by chemiluminescence method in which nine virus isolates were analyzed. This microarray was used to type H1N1, H3N2, H1N2, and H5N1 influenza A virus strains as well as to detect influenza B virus. Unfortunately, the authors do not discuss the rate of correct and incorrect detections when assaying the tested virus samples; thus, the practical value of this microarray is unclear. Another low-density microarray, FluChip-55, designed to detect H1N1, H3N2 and H5N1 influenza A virus subtypes. [40].

Sengupta et al., [2] attempted to construct a microarray capable of typing all influenza A virus subtypes. For this purpose, they searched the VirOligo, a database of virus-specific oligonucleotides, and selected a set of 476 specific probes with an average length of 21 nt. The authors assayed fluorescently-labeled cDNA after reverse transcription and subsequent amplification. Unfortunately, this study lacks any data on the correct identification of virus subtypes by the designed microarray. It is likely that this microarray appeared unable to correctly detect many virus subtypes; nonetheless, the idea of using a set of relatively short oligonucleotides for typing influenza virus undoubtedly deserves attention.

The efficiency of any hybridization microarray is determined by the quality of the typing probes used for its construction. It is easy enough to find specific probes for genomic sequences considerably differing in their structures, such as with genus-specific probes. In particular, it is no trouble to distinguish between VARV and a herpesvirus [30]. However, the problem is much more complex when analyzing genomic sequences close in their structures. In this case, selection of probes specific for a certain microbial subtype is rather a hard problem. Often the selection of probes is difficult when the number of assayed sequences is several hundreds or thousands; therefore, the strategy for probe selection and data analysis is of paramount importance.

Several programs have been elaborated for analysis of genomic sequences and computation of microarray probes [51–58]. Choice of selective probes using traditional approaches for nucleotide sequence analysis frequently fails to give satisfactory results. It is possible to enhance the search for selective probes by reducing the search area to the regions encoding conserved peptides. Such an approach has been proposed [53, 58] but not fully implemented. Presumably, this method would be appropriate when searching for the microarray probes in a set of DNA sequences rather close in their structures, such as cDNA of influenza A and B viruses.

The search for the probes that would allow for typing of all influenza A virus subtypes is a nontrivial task. First, the number of subtypes to be identified is rather big—16 hemagglutinin subtypes and 9 neuraminidase subtypes. Second, the amino acid and nucleotide sequences of hemagglutinin subtypes depend on the neuraminidase subtypes of a virion and vice versa. For example, a virus particle with H1 subtype may contain N1–N9 neuraminidase subtypes (H1N1, H1N2, H1N3, H1N4, H1N5, H1N6, H1N7, H1N8, and H1N9). Note that the nucleotide probes detecting the H1 subtype in H1N1 may differ in their structure from the probes detecting H1 subtype in H1N7. This specific feature of influenza virus increases the number of typing probes necessary to identify each of the hemagglutinin and neuraminidase subtypes. The need to find all the necessary typing probes within relatively short sequences of the HA and NA genes (1700 and 1400 bp, respectively) further complicates the problem.

Researchers [59–63] have developed a method for selecting the typing oligonucleotide probes suitable for microarray assay of influenza A virus based on analysis of hemagglutinin and neuraminidase amino acid sequences. Specialized software searching for such probes was also created. This software detects the peptides characteristic of only a certain subtype in the hemagglutinin and neuraminidase amino acid sequences. Then, based on the data for the structure of such typing peptides, the oligonucleotide probes for microarray assay are computed and selected. This approach simulated the search for antigenic determinants in immunological typing of influenza virus. The proposed scheme for selection of the probes for microarray typing is illustrated by the example of the influenza A virus neuraminidase gene and comprises the following stages (Figure 7):

- Compilation of a database of protein sequences for the influenza virus neuraminidase;
- Detection of the peptides common for the neuraminidase subtype to be identified;
- Elimination of the peptides that are present in protein sequences of other neuraminidase subtypes from the obtained peptide set;
- Reverse transcription of the unique peptides characteristic of the neuraminidase subtype in question to get the oligonucleotide sequences encoding these peptides;
- Selection of the most representative probes from the obtained set of oligonucleotide sequences;
- Elimination of the oligonucleotide probes capable of forming perfect duplexes with the DNAs of other neuraminidase subtypes after introducing one or two mutations into their sequences; and
- Optimization of the probe structures in order to unify melting temperatures of the duplexes formed with the target.

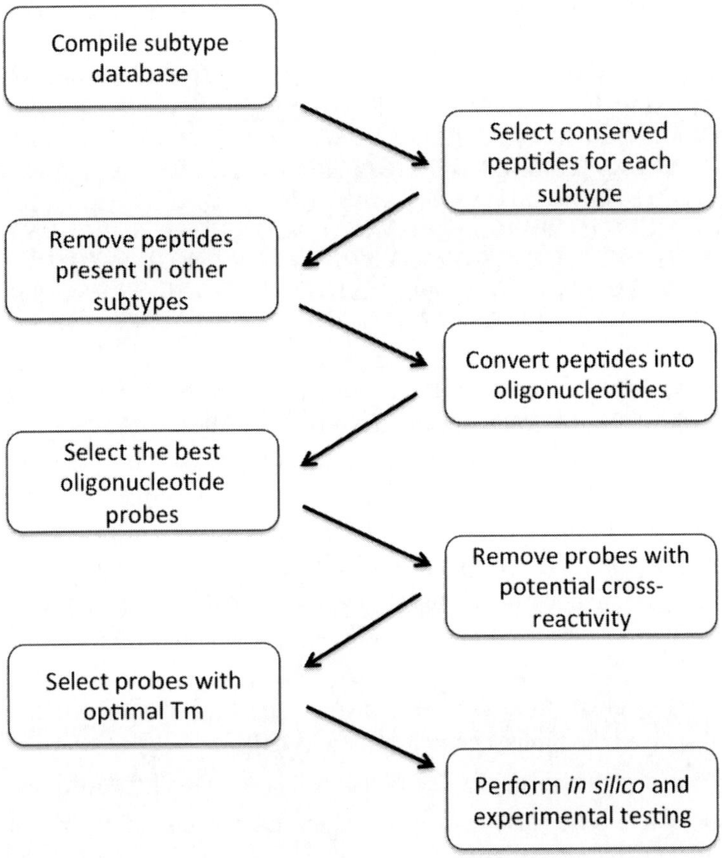

Figure 7. Scheme for the selection of oligonucleotide probes.

In the databases on amino acid sequences of influenza virus hemagglutinin and neuraminidase, the structures of these sequences are incomplete. The protein fragments with a length over 200 amino acid residues were used for computing the probes. In the compiled database, identical sequence segments with a length of 7 residues were found for each neuraminidase subtype (the set for all hemagglutinin subtypes). A preliminary analysis has shown that use of longer peptides decreases the number of oligonucleotide probes for the subtype to be identified, while shorter peptides are encoded by insufficiently specific oligonucleotides. Since it is impossible to find common peptides for absolutely all proteins of a particular influenza virus subtype, the selected peptides were contained in no less than 90–96% of all proteins. This is associated with both variability in the target proteins and the fact that GenBank contains many incomplete sequences for these proteins.

At the next stage, the peptide sets that characterized the neuraminidase serotypes were selected. For this purpose, the peptides with the same structures belonging to sets for different serotypes were discarded. This stage resulted in obtaining of nine sets of 7-mer peptides characterizing all nine subtype of influenza virus neuraminidase. As an example, see below the amino acid sequence of neuraminidase of the isolate A/California/UR0660552/2007(H1N1). The boldfaced peptide sequences are characteristic of 90% of the analyzed N1 sequences but are absent in all the remaining neuraminidase subtypes (N2–N9); 89 7-mer peptides have been found.

MNPNQKIITIGSISIAIGIISLMLQIGNIISIWASHSIQTGSQNHTGICNQRIITYEN
STWVNHTYVSINNTNVVAGKDKTSVT<u>***LAGNSSLC***</u>SISGWAIYTKDNSIRIGSKGD
VFVI**RE*****PFISCSHLECRTFFLTQGALLNDKHSNGTVKDRSP***YRALMSCPL***GEAPS***
PYNSKFESVAWSASACHDGMG***WLTIGISGPD***N***GAVAVLKYNGIIT***ETIKSWKKR***I***
LRTQESECVCVNGSCFTIMTDGPSNGAASYKIFKIEKGKVTKSIELNAPNF***HYEEC***
SCYPDTGTVM***CVCRDNWHGSNRPWVSFNQNL***D***YQIGYICSGVFGDNPR***PKDGK
GSCNPVTVDGADGVKGFSYKY***GNGVWIGRTKS***NRLRK***GFEMIWDPNGWT***DTD
SDFSVKQDVVAI***TDWSGYSGSFVQHPELTG***LDCI***RPCFWVEL***VRGLPRENA***TIW***
TSGSSISFCGVDSDTAN***WSWPDGA***ELPFTIDK

Each unique set of 7-mer peptides were then converted by reverse transcription into the corresponding sets of 21-mer oligonucleotides encoding these peptides. Note that if a set contains peptides that differ by a shift in one to two amino acid residues, the corresponding sets of oligonucleotide probes will give almost the same information about the analyzed neuraminidase subtype. In such cases, one of the peptides was discarded.

Consider the peptide LAGNSSLC, containing two 7-mer peptides—LAGNSSL and AGNSSLC, which are encoded by 51 and 43 oligonucleotide probes, respectively (Table 3). Only one of these two oligonucleotide sets was chosen based on the following criteria:

(1) For several sets of oligonucleotide probes, the set where the first oligonucleotide probe displayed the highest content was selected;
(2) If the contents of oligonucleotide probes (1) were equal, then the set where another oligonucleotide probe also displayed content higher than all the remaining probes in addition to (1) was selected;

(3) If the contents (1) and contents (2) were equal, than the set containing smaller probes was selected; and
(4) If the contents (1), contents (2), and number of oligonucleotides of sets were equal, than a random set of oligonucleotides probes was used.

In Table 3, the peptides LAGNSSL and AGNSSLC are encoded by 51 and 43 oligonucleotides, respectively. Note that probe #1 (TTAGCGGGCAATTCATCTCTT) for the peptide LAGNSSL detects 1,043 influenza virus strains, while probe #1 (GCGGGCAATTCATCTCTTTGC) for the peptide AGNSSLC detects 1,022 strains. Therefore, the set of oligonucleotides for the peptide LAGNSSL was chosen for microarray diagnostics, whereas the oligonucleotides for the other peptide were discarded. In cases when the length of a conserved peptide exceeded 7 amino acid residues, such as REPFISCSHLEECRTFFLTQGALLNDKHSNGTVKDRSP from A/California/UR0660552/2007(H1N1), then such a sequence was cut into several peptides so that no more than four amino acid sequences overlapped. The basic principles for selecting the sets of conserved peptides for the other neuraminidase subtypes (N2–N9) were the same.

Table 3. Oligonucleotide probes for N1 neuraminidase generated for LAGNSSL and AGNSSLC peptides. The number of virus cDNA sequences containing each probe is in parentheses

#	LAGNSSL	#	AGNSSLC
1	TTAGCGGGCAATTCATCTCTT (1043)	1	GCGGGCAATTCATCTCTTTGC (1022)
2	TTGGCCGGCAATTCATCTCTT (708)	2	GCCGGCAATTCATCTCTTTGT (769)
3	TTAGCCGGCAATTCATCTCTT (121)	3	GCAGGCAATTCCTCTCTCTGT (94)
4	CTAGCAGGCAATTCCTCTCTC (98)	4	GCCGGCAATTCATCTCTTTGC (76)
5	CTAGCCGGCAATTCCTCTCTT (62)	5	GCCGGCAATTCCTCTCTTTGC (69)
6	TTAGCGGGCAATTCCTCTCTC (61)	6	GCGGGCAATTCCTCTCTCTGC (67)
7	TTGGCAGGCAATTCGTCTCTT (37)	7	GCGGGCAATTCATCTCTTTGT (41)
	…………………………		………………………
42	CTAGCAGGTAATTCCTCTCTT (1)	42	GCGGGTAATTCCTCTCTCTGC (1)
43	CTAGCAGGCAATTCTTCTCTC (1)	43	GCGGGAAATTCATCTCTTTGC (1)
44	CTAGCAGGCAACTCCTCTCTC (1)		
45	TTGGCCGGTAATTCATCTCTT (1)		
46	TTAGCCGGCAATTCATCTCTC- (1)		
47	CTAGCGGGCAACTCCTCTCTC (1)		
48	TTGGCCGGCAATTCATCTCTC (1)		
49	CTAGCGGGCAATTCATCTCTC (1)		
50	CTAGCAGGCAATTCCTCTCTA (1)		
51	TTAGCGGGAAATTCATCTCTT (1)		

Reverse translation of the peptides unique for each subtype gives a large number of oligonucleotides encoding them. Many of these oligonucleotides are able to type the target protein in only one virus strain or isolate (Table 3). Correspondingly, the next stage in computing probes was the search of initial complete set of probes and selection of the oligonucleotide set most optimal in its size (24–30 probes) and able to identify the neuraminidase subtype with a highest probability. At the first glance, it seems sufficient to choose the probes that identify the largest number of virus strains; however, this would lead to selection of the probes for neuraminidase with the most abundant hemagglutinin subtypes. This is a result of coevolution of the influenza A virus proteins, in which changes in one gene are accompanied by changes in other genes [64, 65]. This brings about the dependence of neuraminidase gene structure on the hemagglutinin subtype. Thus, in the case of N1 neuraminidase subtype (Table 4), mainly represented by subtype H1N1 (42% of all virus strains) and H5N1 (48% of all virus strains), this approach will give the probes detecting the neuraminidase DNA of influenza virus subtypes H1N1 and H5N1. Analogously, the most abundant hemagglutinin subtypes are H1, H3, and H5, and the combinations with different neuraminidase subtypes also display different frequencies (Tables 4 and 5).

The following scheme was used when searching for the probes able to identify rare subtypes on the background of abundant subtypes (Figure 8). From the complete set of probes (Probe_Total), a probe that detects the maximal number of strains of a target neuraminidase subtype (Probe_1-1, database on nucleotide sequences of all available influenza virus neuraminidase genes Seq_0) was selected. At the next stage, the sequences of the genes detected by Probe_1-1 were eliminated from Seq_0 to get the set Seq_1. In the set Seq_1, the probe detecting the maximal number of neuraminidase gene sequences (Probe_1-2) was selected. In an analogous manner, the sequences containing Probe_1-2 were discarded from the set Seq_1 to obtain the set Seq_2. Subsequently, the described algorithm was used to select Probe_1-3, Probe_1-4, Probe_1-5, and Probe_1-6. At the next stage, the first six probes (Probe_1-1 to Probe_1-6) were discarded from the Probe_Total set and an additional six probes—Probe_2-1 to Probe_2-6—were selected from the remaining oligonucleotide probes using the complete DNA set (Seq_0) as the initial set. This selection was repeated three additional times to obtain the final set of 30 probes for the microarray. If the number of analyzed sequences is small, the procedure should be repeated more than five times to obtain the necessary number of probes. The selected oligonucleotide probes intended for identification of a certain subtype of the target gene were supplemented with one or two point substitutions to estimate whether they would react with any other subtypes of this gene. If such mutations changed an oligonucleotide into a perfect probe for identification of another subtype, such probe was discarded from the set for microarray. The melting temperatures for duplexes formed by the probes and assayed DNAs were unified by varying the lengths of typing probes. The performed computations allowed for selection of 191 probes able to type all nine neuraminidase subtypes. Each subtype was identified by a set of 19–24 probes.

Table 4. The number of influenza A virus isolates with different subtypes of neuraminidase analyzed. [59–63]

Sub-type of Neuraminidase	Sub-type of Hemagglutinin																
	H1	H2	H3	H4	H5	H6	H7	H8	H9	H10	H11	H12	H13	H14	H15	H16	Total
N1	1,14	12	14	4	1,295	150	61	0	9	2	8	1	0	0	0	0	2,696
N2	140	96	2,394	14	108	86	89	0	661	0	8	0	3	0	0	0	3,599
N3	2	23	10	4	43	5	80	0	3	2	7	0	1	0	0	6	186
N4	0	1	0	4	0	1	3	12	0	2	7	0	1	0	0	0	31
N5	2	9	3	4	0	14	1	1	3	4	0	18	0	1	0	0	60
N6	4	0	50	72	1	10	0	0	3	3	7	0	12	0	0	0	162
N7	1	3	0	1	3	0	32	0	0	21	0	0	0	0	0	0	61
N8	1	3	105	15	2	0	0	3	1	6	2	0	1	0	2	0	141
N9	2	16	2	2	1	3	2	0	0	2	32	1	2	0	3	0	68
																Total	7,004

Table 5. The number of influenza A virus isolates with different subtypes of hemagglutinin analyzed. [59–63]

Subtype of Hemagglutinin	Subtype of Neuraminidase									
	N1	N2	N3	N4	N5	N6	N7	N8	N9	Total
H1	1,303	133	1	0	1	3	1	0	4	1,446
H2	13	115	24	1	9	0	2	5	18	187
H3	13	4,226	11	1	2	37	0	150	1	4,441
H4	4	15	3	3	2	73	1	18	1	120
H5	1,874	234	52	0	1	4	7	5	14	2,191
H6	69	100	6	5	21	6	1	46	9	263
H7	121	109	93	5	1	0	69	1	2	401
H8	0	1	0	12	0	0	0	0	0	13

Table 5. (Continued)

Subtype of Hemagglutinin	Subtype of Neuraminidase									
H9	9	831	2	0	5	4	1	1	0	853
H10	2	0	1	2	0	3	18	3	2	31
H11	3	9	6	1	0	5	0	0	31	55
H12	1	1	0	1	15	0	0	0	1	19
H13	0	4	0	0	0	11	0	1	1	17
H14	0	0	0	0	2	0	0	0	0	2
H15	0	2	0	0	0	0	0	3	4	9
H16	0	0	8	0	0	0	0	0	0	8
									Total	10,056

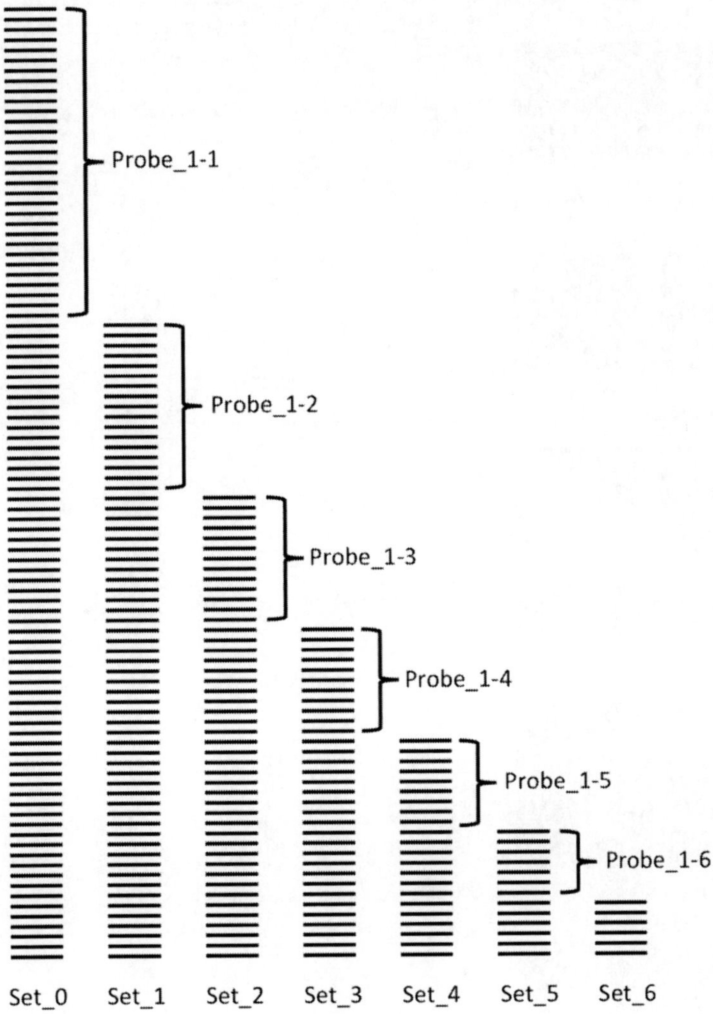

Figure 8. Selection of the most representative oligonucleotide.

The probes for typing influenza virus hemagglutinin gene were selected using an analogous scheme. However, the probes for H14 subtype were not determined at that time, since only two sequences of this gene were available in GenBank. Special probes were selected for the H1 subtype of H1N1 swine influenza virus. This is associated with distinction of the swine H1 subtype in both amino acid and nucleotide sequences from the other H1 subtypes compiled in GenBank. The N1 subtype H1N1 swine influenza virus was rather similar to the earlier known N1 neuraminidase subtypes.

The microchip designed by Ryabinin et al., [61] was tested using the specimens representing different HA and NA subtypes. Totally, 40 samples of HA genes (H1–H7, H9–H11, and H13) and 40 samples of NA genes (N1–N3 and N6–N8) were tested, and 36 of them unambiguously identified HA and NA subtypes of influenza A virus (Figure 9). Note that several incorrectly typed influenza A virus strains were detected during construction of this microarray, in particular, the strain A/mallard duck/Minnesota/1979 (H3N2) appeared to belong to H3N1 subtype and A/equine/New Market/79 (H3N3) was identified as H3N8.

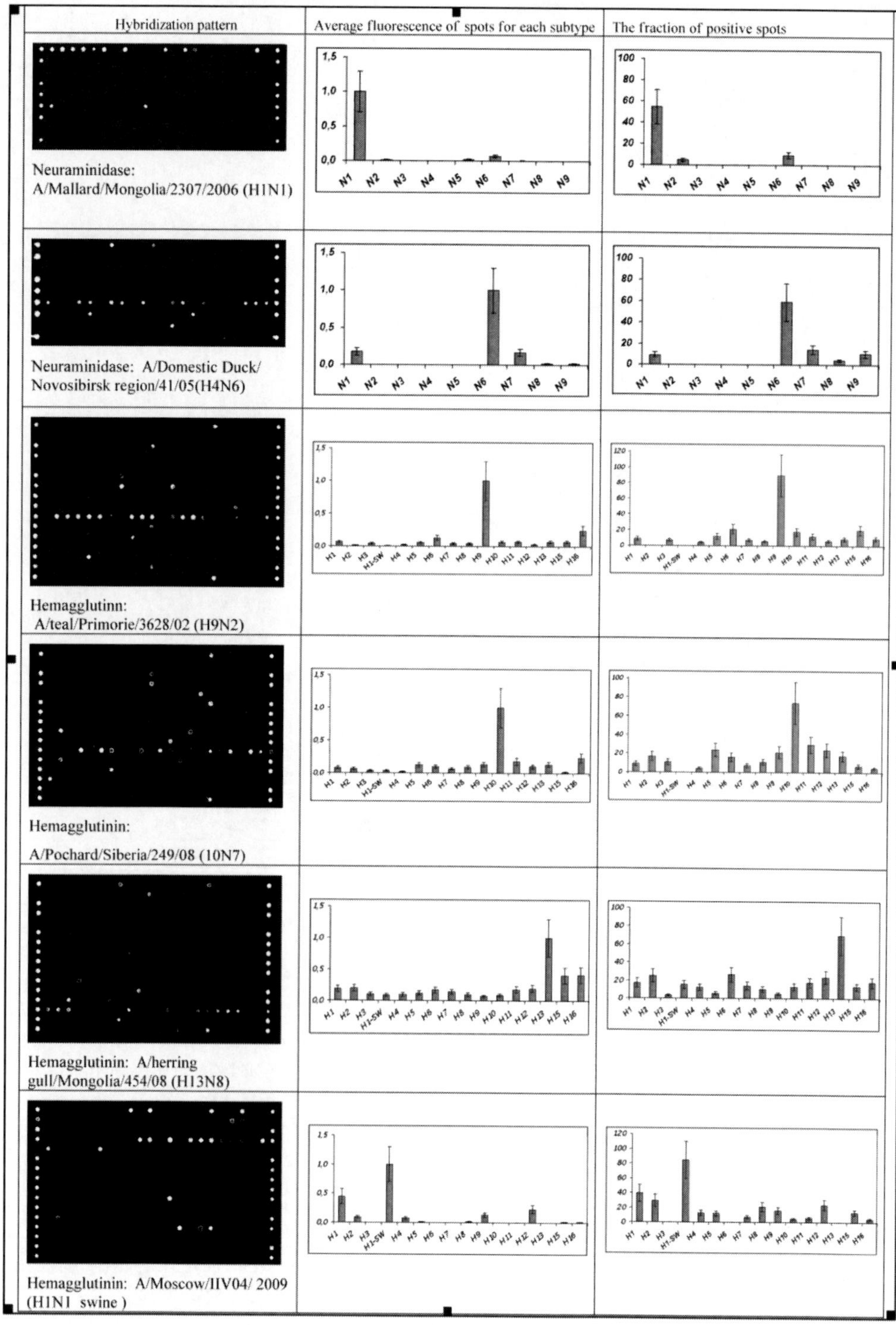

Figure 9. Microarray hybridization patterns for amplicons of different influenza A virus neuraminidase and hemagglutinin subtypes and the corresponding histograms showing the average fluorescence of individual microchip elements.

Fesenko et al. [66] designed a microarray able to identify 15 hemagglutinin subtypes and 2 neuraminidase subtypes (N1 and N2) of influenza A virus. This microarray contained 3D gel-based elements with 41 immobilized typing probes. These probes are specific; however, it remains unclear whether all influenza A virus strains are detectable by such a small set of probes.

Sequencing Microarray for Typing Influenza A Virus

Lodes et al. [67] designed a microarray able to identify 15 hemagglutinin subtypes and 9 neuraminidase subtypes of influenza A virus. This sequencing microarray utilized 7,354 probes for typing hemagglutinin and 4,646, for neuraminidase. The typing procedure itself was rather complex and was comprised of cDNA amplification of an analyzed sample and ligation of microarray probes with one of the strands of the labeled amplicon. In the process of typing, it was necessary to obtain and analyze 177,264 observations recorded after performing hybridization, enzymatic reactions, and washing of the slide. The assay reliability was high and the assay cost using this sequencing microarray was 40–50-fold higher as compared with hybridization microarrays.

Microarray Typing of Arenaviruses and Filoviruses

The computation of hybridization oligonucleotide probes based on sequence analysis of viral proteins has appeared efficient for highly variable viral genomes. This method was successfully applied for the first time for typing influenza A virus [59–63]. The same approach was used for calculation of hybridization probes when designing an oligonucleotide microarray for detection of arena- and filoviruses (unpublished data). When developing this microarray, the data for the nucleocapsid protein (NP) and RNA-dependent RNA polymerase for Junin (JUNV), Machupo (MACV), Guanarito (GTOV), and Lassa (LASV) viruses were extracted from GenBank (http://www.ncbi.nlm.nih.gov) and Protein Sequence Database for Pathogenic Arenaviruses (http://epitope.liai.org:8080/projects/arena) [68]. The data for RNA-dependent RNA polymerase of Ebola (EBOV) and Marburg (MARV) viruses were obtained from GenBank. Selection of the protein to be analyzed in all the cases was determined according to the best representation of the corresponding amino acid sequences in the databases.

The six sets of 7-mer peptides obtained by analyzing the nucleocapsid protein and RNA-dependent RNA polymerase specific for JUNV, GTOV, MACV, LASV, EBOV, and MARV were converted into sets of 21-mer oligonucleotides encoding these peptides. The resulting sets of oligonucleotide sequences were used for the selection of the probes following the described scheme [61] to obtain the sets of oligonucleotides capable of discriminating between the target viruses. Thus, according to the above described principles, six sets of oligonucleotide probes were selected for identification of JUNV, GTOV, MACV, LASV, EBOV, and MARV. The number of probes for individual target viruses varied from 7 to 25 (94 probes total).

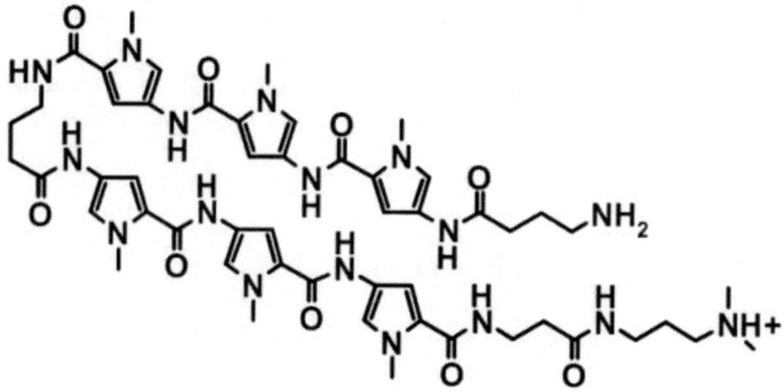

Figure 10. Structure of the carboxamidopyrrole ligand Y-γ-Py-Py-Py-γ-Py-Py-Py-β-Dp.

Because of a high heterogeneity of the genomic sequences of target viruses, the primers were selected in several stages similar to the scheme used in probe selection. First, a conserved 7-mer peptide was found based on the analysis of the corresponding amino acid sequences for all strains and isolates of each target virus species. Then, each peptide was converted into the degenerate nucleotide sequence corresponding to a set of 21-mer oligonucleotides encoding this peptide. As in the case of probe selection, the total number of oligonucleotide structures in each of the sets was very large. As an example, consider the conserved 7-mer peptide QLSGSCL from the amino acids of LASV RNA-dependent RNA polymerase. This heptapeptide may be encoded by the degenerate nucleotide sequence 5'-CA**RY**T**R**TC**W**GG**N**AG**Y**TG**YY**T**V**-3' (**R** = G/A, **Y** = T/C, **W** = A/T, **N** = A/T/G/C, and **V** = A/G/C). It is evident that a set of $2 \times 2 \times 2 \times 2 \times 4 \times 2 \times 2 \times 2 \times 3 = 1536$ oligonucleotides corresponds to this structure; however, only eight of them are actually primers.

Construction of a microarray-based diagnostic kit that would allow for discrimination between the target viruses is purposeful only if multiplex PCR can be performed at a certain stage. The critical point in this situation is the number of primers used in the assay. In the case of a large number of primers, the probability of their undesirable interactions drastically increases, which significantly decreases the efficiency of multiplex PCR. The total number of primers can be reduced by decreasing their length and the number of degenerate positions. However, use of this approach decreases the melting temperature for heteroduplexes between the primers and the template, which also has a negative effect on PCR efficiency.

To resolve this problem, the conjugates of oligodeoxyribonucleotides with a minor groove carboxamidopyrrole ligand, Y-γ-Py-Py-Py-γ-Py-Py-Py-β-Dp (Figure 10), which stabilizes an AT-tract [69], were used as the PCR primers. This approach allowed the length and total number of the oligonucleotides used as primers to be considerably decreased (owing to a significant increase in the thermal stability of hybrid DNA duplexes). Thus, shortened (15 to 22 nucleotides) oligodeoxyribonucleotides containing a quadruplet of three A or T bases and one G or C base at their 5' ends were used for synthesizing conjugates. The produced conjugates were divided into two groups corresponding to forward (33 oligonucleotides) and reverse (32 oligonucleotides) primers. The derivatives of oligodeoxyribonucleotides belonging to the same group were pooled in equimolar amounts and used in PCR. Amplification (and concurrent fluorescent labeling with Cy5) was conducted in an asymmetric mode to produce predominantly single-stranded products for further hybridization on microarrays. Configuration of the microarray used in this work and containing the probes for JUNV, GTOV, MACV, LASV, EBOV, and MARV is shown in Figure 11. At the 3' end, the oligonucleotide probes contained an aminohexyl linker, used to immobilize these probes on the surface of a glass slide modified with phenyldiisothiocyanate. The marker oligodeoxyribonucleotide $(dT)_8$ with a fluorescent label at its 5' end and an aminohexyl group as a linker at its 3' end was used as c control when printing the slides. This microarray was tested using amplified fragments of the nucleocapsid protein genes of JUNV (strain XJ#44), GTOV (strain VHF-1750), and MACV (strain Carvallo) and amplified fragments of the RNA-dependent RNA polymerase genes of LASV (strain CSF), MARV (strain Popp), and Zaire EBOV (strain Mayinga). Mean fluorescence value was used to estimate the results of microarray-based identification. Figure 12 shows examples of hybridization patterns obtained using this microarray and the corresponding histograms of mean fluorescence distribution for the spots.

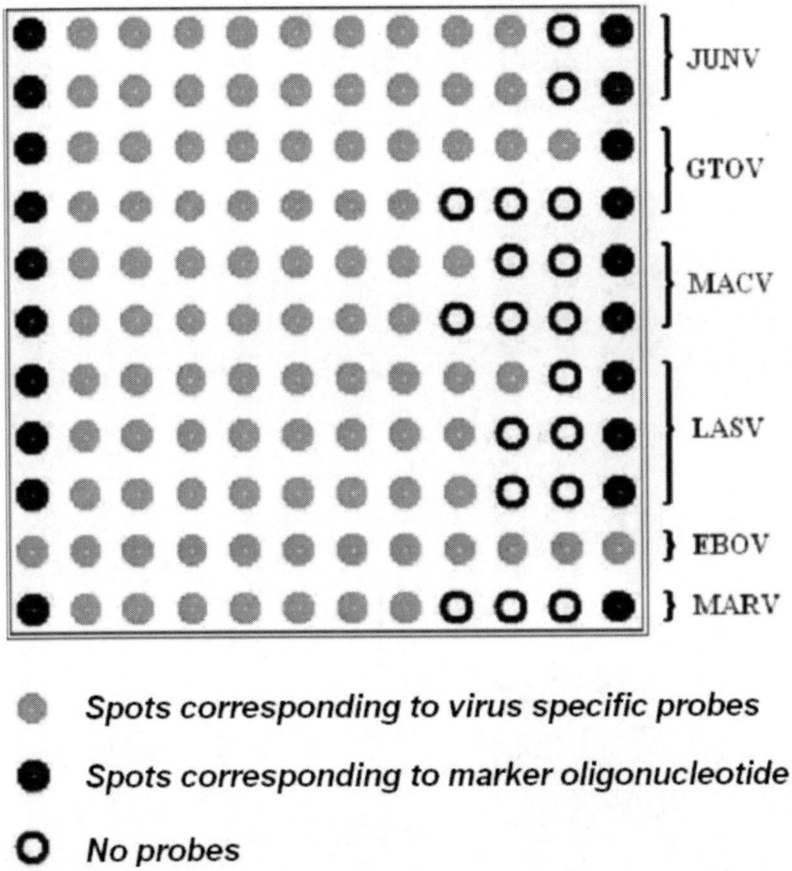

Figure 11. A microarray with 12 × 11 spots for identification of the BSL-4 pathogens belonging to the families Arenaviridae and Filoviridae. Each row contains the probes for detection of the corresponding virus (JUNV, GTOV, MACV, LASV, EBOV, or MARV).

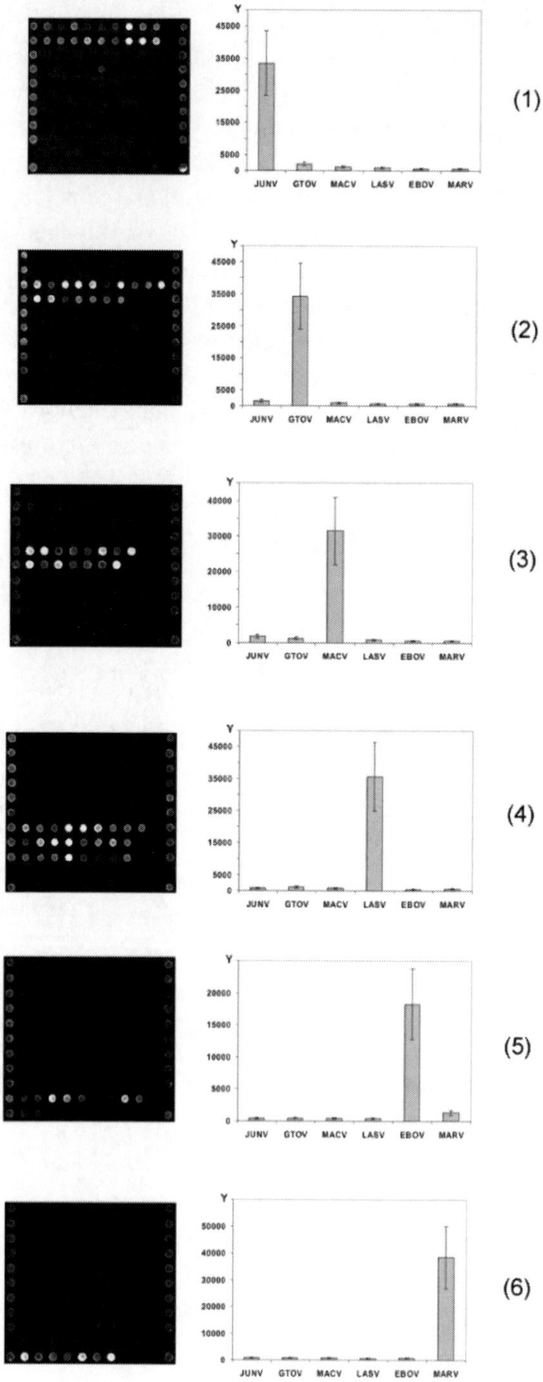

Figure 12. Hybridization patterns for the amplified fragments of nucleocapsid protein genes of the (*1*) Junin (JUNV, strain XJ#44), (*2*) Guanarito (GTOV, strain VHF-1750), and (*3*) Machupo (MACV, strain Carvallo) viruses and the amplified fragments of RNA-dependent RNA polymerase genes of the (*4*) Lassa (LASV, strain CSF), (*5*) Zaire Ebola (ZEBOV, strain Mayinga, and (*6*) Marburg (MARV, strain Popp) viruses on the microarray (67°C, 1.5 h) and the corresponding histograms of mean fluorescence distribution for the spots (Y, relative units).

It is evident from these examples that the designed assay allows for an unambiguous discrimination between the target viruses. Cross-hybridization with nonspecific probes, observed in some cases, does not interfere with data interpretation owing to a considerable (by an order of magnitude) difference between the mean fluorescence values of spots.

Conclusion

In general, the microarray diagnostics for orthopoxviruses is sufficiently developed for a reliable identification of the viruses that are pathogenic in humans. The microarrays allowing single nucleotide substitutions to be detected in both individual viral genes and the overall genome have been designed. This is a convenient tool for monitoring the evolution of virus strains and detecting genetically-modified viruses. Microarray diagnostics provides for a reliable, rapid identification of orthopoxviruses and their discrimination from the viruses causing diseases with similar clinical manifestations.

As for the microarray diagnostics of influenza virus, microarrays have been constructed that allow for identification of influenza virus subtypes potentially pathogenic in humans. Such microarrays are already used for constant monitoring of virus subtypes aiming to determine the vaccine strains for influenza prevention. Universal microarrays making it possible to identify any influenza virus subtypes affecting animals, birds, and humans have been designed. These microarrays are extremely useful for analyzing molecular evolution of this virus.

References

[1] Li, J; Chen, S; Evans, DH. Typing and subtyping influenza virus using DNA microarrays and multiplex reverse transcriptase PCR. *J Clin Microbiol*, 2001, 39, 696-704.

[2] Sengupta, S; Onodera, K; Lai, A; Melcher, UJ. Molecular detection and identification of influenza viruses by oligonucleotide microarray hybridization. *Clin Microbiol*, 2003, 41, 4542-4550.

[3] Han, X; Lin, X; Liu, B; Hou, Y; Huang, J; Wu, S; Liu, J; Mei, L; Jia, G; Zhu, QJ. Simultaneously subtyping of all influenza A viruses using DNA microarrays. *Virol Methods*, 2008, 152, 117-121.

[4] Gall, A; Hoffmann, IB; Harder, T; Grund, C; Hoper, D; Beer, MJ. Design and validation of a microarray for detection, hemagglutinin subtyping, and pathotyping of avian influenza viruses. *Clin Microbiol*, 2009, 47, 327-334.

[5] Dawson, ED; Moore, CL; Smagala, JA; Dankbar, DM; Mehlmann, M; Townsend, MB; Smith, CB; Cox, NJ; Kuchta, RD; Rowlen, KL. MChip: a tool for influenza surveillance *Anal Chem*, 2006, 78, 7610-7615.

[6] Laassri, M; Chizhikov, V; Mikheev, M; Shchelkunov, S; Chumakov, K. Detection and discrimination of orthopoxviruses using microarrays of immobilized oligonucleotides. *J Virol Methods*, 2003, 112, 67-78.

[7] Laassri, M; Dragunsky, E; Enterline, J; Eremeeva, T; Ivanova, O; Lottenbach, K; Belshe, R; Chumakov, K. Genomic analysis of vaccine-derived poliovirus strains in stool specimens by combination of full-length PCR and oligonucleotide microarray hybridization. *J Clin Microbiol*, 2005, 43, 2886-2894.

[8] Wilson, WJ; Strout, CL; DeSantis, TZ; Stilwell, JL; Carrano, AV; Andersen, GL; Sequence-specific identification of 18 pathogenic microorganisms using microarray technology. *Mol Cell Probes*, 2002, 16, 119-127.

[9] Shchelkunov, SN ; Marennikova, SS ; Moyer, RW. Orthopoxviruses pathogenic for humans. Berlin, Heidelberg, New York: Springer, 2005.

[10] Kesson, AM; Ferguson, JK; Rawlinson, WD; Cunningham, AL. Progressive vaccinia treated with ribavirin and vaccinia immune globulin. *Clin Infect Dis*, 1997, 25, 911-914.

[11] Henderson, DA. Countering the posteradication threat of smallpox and polio. *Clin Infect Dis*, 2002, 34, 79-83.

[12] Fenner, F. The eradication of smallpox. *Prog Med Virol*, 1977, 23, 1-21.

[13] Henderson, DA. The history of smallpox eradication. *Henry E Sigerist Suppl Bull Hist Med*, 1980, 4, 99-114.

[14] Obijeski, JF; Palmer, EL; Gafford, LG; Randall, CC. Polyacrylamide gel electrophoresis of fowlpox and vaccinia virus proteins. *Virology*, 1973, 51, 512-516.

[15] Arita, M; Tagaya, I. Structural polypeptides of several strains of orthopoxvirus. *Microbiol Immunol*, 1977, 21, 343-346.

[16] Mukinda, VB; Mwema, G; Kilundu, M; Heymann, DL; Khan, AS; Esposito, JJ. Re-emergence of human monkeypox in Zaire in 1996. *Monkeypox Epidemiologic Working Group. Lancet*, 1997, 349, 1449-1450.

[17] Ibrahim, MS; Esposito, JJ; Jahrling, PB; Lofts, RS. The potential of 50 nuclease PCR for detecting a single-base polymorphism in Orthopoxvirus. *Mol Cell Probes*, 1997, 11, 143-147.

[18] Shchelkunov, SN; Gavrilova, EV; Babkin, IV. Multiplex PCR detection and species differentiation of orthopoxviruses pathogenic to humans. *Mol Cell Probes*, 2005, 19, 1-8.

[19] Loparev, VN; Massung, RF; Esposito, JJ; Meyer, H. Detection and differentiation of old world orthopoxviruses: Restriction fragment length polymorphism of the crmB gene region. *J Clin Microbiol*, 2001, 39, 94-100.

[20] Sofi Ibrahim, M; Kulesh, DA; Saleh, SS; Damon, IK; Esposito, JJ; Schmaljohn, AL; Jahrling, PB. Real-time PCR assay to detect smallpox virus. *J Clin Microbiol*, 2003, 41, 3835-3839.

[21] Olson, VA; Laue, T; Laker, MT; Babkin, IV; Drosten, C; Shchelkunov, SN; Niedrig, M; Damon, IK; Meyer, H. Real-time PCR system for detection of orthopoxviruses and simultaneous identification of smallpox virus. *J Clin Microbiol*, 2004, 42, 1940-1946.

[22] Cunnion, KM. Tumor necrosis factor receptors encoded by poxviruses. *Mol Genet Metab*, 1999, 67, 278-282.

[23] Graham, KA; Lalani, AS; Macen, JL; Ness, TL; Barry, M; Liu, LY; Lucas, A; Clark-Lewis, I; Moyer, RW; McFadden, G. The T1/35kDa family of poxvirus-secreted proteins binds chemokines and modulates leukocyte influx into virus-infected tissues. *Virology*, 1997, 229, 12-24.

[24] Symons, JA; Alcami, A; Smith, GL. Vaccinia virus encodes a soluble type I interferon receptor of novel structure and broad species specificity. *Cell*, 1995, 81, 551-560.

[25] Lapa, S; Mikheev, M; Shchelkunov, S; Mikhailovich, V; Sobolev, A; Blinov, V; Babkin, I; Guskov, A; Sokunova, E; Zasedatelev, A; Sandakhchiev, L; Mirzabekov, A. Species-level identification of Orthopoxviruses with an oligonucleotide microchip. *J Clin Microbiol*, 2002, 40, 753-757.

[26] Smith, TD; Smolak, PJ; Friend, D; Hagen, H; Gerhart, M; Park, L; Pickup, DJ; Torrance, D; Mohler, K; Schooley, K; Goodwin, RG. Poxvirus genomes encode a secreted, soluble protein that preferentially inhibits b-chemokine activity yet lacks sequence homology to known chemokine receptors. *Virology*, 1997, 236, 316-327.

[27] Loparev, VN; Massung, RF; Esposito, JJ; Meyer, H. Detection and differentiation of Old World orthopoxviruses: restriction fragment length polymorphism of the crmB gene region. *J Clin Microbiol*, 2001, 39, 94-100.

[28] Laassri, M; Chizhikov, V; Mikheev, M; Shchelkunov, S; Chumakov, K. Detection and discrimination of orthopoxviruses using microarrays of immobilized oligonucleotides. *J Virol Methods*, 2003, 112, 67-78.

[29] Hutin, YJ; Williams, RJ; Malfait, P; Pebody, R; Loparev, VN; Ropp, SL; Rodriguez, M; Knight, JC; Tshioko, FK; Khan, AS; Szczeniowski, MV; Esposito, JJ. Outbreak of human monkeypox, Democratic Republic of Congo, 1996 to 1997. *Emerg Infect Dis*, 2001, 7, 434-438.

[30] Ryabinin, VA; Shundrin, LA; Kostina, EB; Laassri, M; Chizhikov, V; Shchelkunov, SN; Chumakov, KM; Sinyakov, AN. Microarray assay for detection and discrimination of orthopoxvirus species. *J Med Virol*, 2006, 78, 1325-1340.

[31] Mikheev, MV; Feshchenko, MV; Shchelkunov, SN. Phylogenetic analysis of chemokine-binding protein gene from orthopoxviruses. *Mol Gen Mikrobiol Virusol*, 2004, 1, 29-36.

[32] Laassri, M; Meseda, CA; Williams, O; Merchlinsky, M; Weir, JP; Chumakov, K. Microarray assay for evaluation of the genetic stability of modified vaccinia virus ankara B5R gene. *J Med Virol*, 2007, 79, 791-802.

[33] Sulaiman, IM; Tang, K; Osborne, J; Sammons, S; Wohlhueter, RM. GeneChip resequencing of the smallpox virus genome can identify novel strains: a biodefense application. *J Clin Microbiol*, 2007, 45, 358-363.

[34] McGall, G; Labadie, J; Brock, P; Wallraff, G; Nguyen, T; Hinsberg, W. Light-directed synthesis of high-density oligonucleotide arrays using semiconductor photoresists. *Proc Natl Acad Sci USA*, 1996, 93, 13555-13560.

[35] Hoffmann, B; Depner, K; Schirrmeier, H; Beer, M. A universal heterologous internal control system for duplex real-time RT-PCR assays used in a detection system for pestiviruses. *J Virol Methods*, 2006, 136, 200-209.

[36] Hoffmann, E; Stech, J; Guan, Y; Webster, RG; Perez, DR. Universal primer set for the full-length amplification of all influenza A viruses. *Arch Virol*, 2001, 146, 2275-2289.

[37] Wright, KE; Wilson, GA; Novosad, D; Dimock, C; Tan, D; Weber, JM. Typing and subtyping of influenza viruses in clinical samples by PCR. *J Clin Microbiol*, 1995, 33, 1180-1184.

[38] Phipps, LP; Essen, SC; Brown, IH. Genetic subtyping of influenza A viruses using RT-PCR with a single set of primers based on conserved sequences within the HA2 coding region. *J Virol Methods*, 2004, 122, 119-122.

[39] Fesenko, EE; Kireyev, DE; Gryadunov, DA; Mikhailovich, VM; Grebennikova, TV; L'vov, DK; Zasedatelev, AS. Oligonucleotide microchip for subtyping of influenza A virus. *Influenza Other Respi Viruses*, 2007, 1, 121-129.

[40] Townsend, MB; Dawson, ED; Mehlmann, M; Smagala, JA; Dankbar, DM; Moore, CL; Smith, CB; Cox, NJ; Kuchta, RD; Rowlen, KL. Experimental evaluation of the FluChip diagnostic microarray for influenza virus surveillance. *J Clin Microbiol*, 2006, 44, 2863-2871.

[41] Kessler, N; Ferraris, O; Palmer, K; Marsh, W; Steel, A. Use of the DNA flow-thru chip, a three-dimensional biochip, for typing and subtyping of influenza viruses. *J Clin Microbiol*, 2004, 42, 2173-2185.

[42] Li, H; McCormac, MA; Estes, RW; Sefers, SE; Dare, RK; Chappell, JD; Erdman, DD; Wright, PF; Tang, YW. Simultaneous detection and high-throughput identification of a panel of RNA viruses causing respiratory tract infections. *J Clin Microbiol*, 2007, 45, 2105-2109.

[43] Li, J; Chen, S; Evans, DH. Typing and subtyping influenza virus using DNA microarrays and multiplex reverse transcriptase PCR. *J Clin Microbiol*, 2001, 39, 696-704.

[44] Lin, B; Blaney, KM; Malanoski, AP; Ligler, AG; Schnur, JM; Metzgar, D; Russell, KL; Stenger, DA. Using a resequencing microarray as a multiple respiratory pathogen detection assay. *J Clin Microbiol*, 2007, 45, 443-452.

[45] Moore, CL; Smagala, JA; Smith, CB; Dawson, ED; Cox, NJ; Kuchta, RD; Rowlen, KL. Evaluation of MChip with historic subtype H1N1 influenza A viruses, including the 1918 ''Spanish Flu'' strain. *J Clin Microbiol*, 2007, 45, 3807-3810.

[46] Wang, LC; Pan, CH; Severinghaus, LL; Liu, LY; Chen, CT; Pu, CE; Huang, D; Lir, JT; Chin, SC; Cheng, MC; Lee, SH; Wang, CH. Simultaneous detection and differentiation of Newcastle disease and avian influenza viruses using oligonucleotide microarrays. *Vet Microbiol*, 2008, 127, 217-226.

[47] Gall, A; Hoffmann, B; Harder, T; Grund, C; Ehricht, R; Beer, V. Rapid haemagglutinin subtyping and pathotyping of avian influenza viruses by a DNA microarray. *J Virol Methods*, 2009, 160, 200-205.

[48] Liu, RH; Lodes, MJ; Nguyen, T; Siuda, T; Slota, M; Fuji, HS; McShea, A. Validation of a fully integrated microfluidic array device for influenza A subtype identification and sequencing. *Anal Chem*, 2006, 78, 4184-4193.

[49] Lodes, MJ; Suciu, D; Elliott, M; Stover, AG; Ross, M; Caraballo, M; Dix, K; Crye, J; Webby, RJ; Lyon, WJ; Danley, DL; McShea, A. Use of semiconductor-based oligonucleotide microarrays for influenza a virus subtype identification and sequencing. *J Clin Microbiol*, 2006, 44, 1209-1218.

[50] Han, X; Lin, X; Liu, B; Hou, Y; Huang, J; Wu, S; Liu, J; Mei, L; Jia, G; Zhu, Q. Simultaneously subtyping of all influenza A viruses using DNA microarrays. *J Virol Methods*, 2008, 152, 117-121.

[51] Mehlmann, M; Dawson, ED; Townsend, MB; Smagala, JA; Moore, CL; Smith ,CB; Cox, NJ; Kuchta, RD; Rowlen, KL. Robust sequence selection method used to develop the FluChip diagnostic microarray for influenza virus. *J Clin Microbiol*, 2006, 44, 2857-2862.

[52] Wernersson, R; Juncker, AS; Nielsen, HB. Probe selection for DNA microarrays using *OligoWiz. Nat Protoc*, 2007, 2, 2677-2691.

[53] Jabado, OJ; Liu, Y; Conlan, S; Quan, PL; Hegyi, H; Lussier, Y; Briese, T; Palacios, G; Lipkin, WI. Comprehensive viral oligonucleotide probe design using conserved protein regions. *Nucleic Acids Res*, 2008, 36, e3.
[54] Hall, TA. BioEdit: a user – friendly biological sequence alignment editor and analysis program for Windows 95/98/NT. *Nucleic Acids Symp Ser*, 1999, 41, 95-98.
[55] Neverov, AA; Riddell, MA; Moss, WJ; Volokhov, DV; Rota, PA; Lowe, LE; Chibo, D; Smit, SB; Griffin, DE; Chumakov, KM; Chizhikov, VE. Genotyping of measles virus in clinical specimens on the basis of oligonucleotide microarray hybridization patterns. *J Clin Microbiol*, 2006, 44, 3752-3759.
[56] Array Designer. Premier Biosoft International. Available: http://www.premierbiosoft.com/dnamicroarray.
[57] Oligos4Array. Ocimum Biosolutions. Available: http://www.ocimumbio.com/web/arrays/array_design.asp.
[58] SF. Amino acid substitution matrices from an information theoretic perspective. *J Mol Biol*, 1991, 219, 555-565.
[59] Ryabinin, VA; Kostina, EV; Neverov, AA; Maksakova, GA; Sinyakov, AN. *J Bioorg Chem*, 2010, 36, 634-644.
[60] Ryabinin, VA; Kostina, EV; Maksakova, GA; Sinyakov, AN. Subtyping of influenza virus A hemagglutinin with a hybridization microarray. *J Bioorg Chem*, 2010, 36, 779-782.
[61] Ryabinin, VA; Kostina, EV; Maksakova, GA; Neverov, AA; Chumakov, KM; Sinyakov, AN. Universal oligonucleotide microarray for sub-typing of influenza A virus. *PLoS ONE*, 2011, 6(4), e17529.
[62] Kostina, EV; Ryabinin, VA ; Agafonov, AP ; Ternovoi, VA ; Sinyakov, AN. Microarray for diagnostics of human pathogenic influenza A virus subtypes. *J Bioorg Chem*, 2011, 37, 644-646.
[63] Kostina, EV; Ryabinin, VA; Maksakova, GA; Sinyakov, AN. A Second Generation Universal Microarray for Subtyping Influenza A Virus. *Russian J Bioorg Chem*, 2012, 38, 599-604.
[64] Ghedin, E; Sengamalay, NA; Shumway, M; Zaborsky, J; Feldblyum, T; Subbu, V; Spiro, DJ; Sitz, J; Koo, H; Bolotov, P; Dernovoy, D; Tatusova, T; Bao, Y; St George, K; Taylor, J; Lipman, DJ; Fraser, CM; Jeffery, K; Taubenberger, JK; Salzberg, SL. Large-scale sequencing of human influenza reveals the dynamic nature of viral genome evolution. *Nature*, 2005, 437, 1162-1166.
[65] Obenauer, JC; Denson, J; Mehta, PK; Su, X; Mukatira, S; Finkelstein, DB; Xu, X; Wang, J; Ma, J; Fan, Y; Rakestraw, KM; Webster, RG; Hoffmann, E; Krauss, S; Zheng, J; Zhang, Z; Naeve, CW. Large-scale sequence analysis of avian influenza isolates. *Science*, 2006, 311, 1576-1580.
[66] Fesenko, EE; Kireyev, DE; Gryadunov, DA; Mikhailovich, VM; Grebennikova, TV; L'vov, DK; Zasedatelev, AS. Oligonucleotide microchip for subtyping of influenza A virus. *Influenza Other Respi Viruses*, 2007, 1, 121-129.
[67] Lodes, MJ; Suciu, D; Elliott, M; Stover, AG; Ross, M; Caraballo Dix, K; Crye, J; Webby, RJ; Lyon, WJ; Danley, DL; McShea, A. Use of semiconductor-based oligonucleotide microarrays for influenza A virus subtype identification and sequencing. *J Clin Microbiol*, 2006, 44, 1209-1218.

[68] Bui, HH; Botten, J; Fusseder, N; Pasquetto, V; Mothe, B; Buchmeier, MJ; Sette, A. Protein sequence database for pathogenic arenaviruses. *Immunome Research*, 2007, 3, 1-8.

[69] Sinyakov, AN; Lokhov, SG; Kutyavin, IV; Gamper, HB; Meyer, RB. Exceptional and selective stabilization of A-T rich DNA-DNA duplexes by N-methylpyrrole carboxamide peptides conjugated to oligodeoxynucleotides. *JACS*, 1995, 117, 4995-4996.

In: Microarrays: Principles, Applications and Technologies
Editor: James V. Rogers
ISBN: 978-1-62948-669-7

Chapter 10

Using DNA Microarray Technology to Monitor the Distribution of Antimicrobial Resistance Genes in *Escherichia Coli* Isolated from Animal Colibacillosis Outbreaks

Pietro Badagliacca[1*], *Luke Masson*[2], *Alfreda Tonelli*[1], *Giacomo Migliorati*[1] *and Josée Harel*[3]
[1]Istituto Zooprofilattico Sperimentale dell'Abruzzo e del Molise "G. Caporale", Campo Boario, Teramo, Italy
[2]National Research Council of Canada, Montreal, QC, Canada
[3]Centre de Recherche en Infectiologie Porcine, Université de Montréal, QC, Canada

Abstract

A customized DNA microarray was designed to detect *Escherichia coli* virulence genes in addition to Gram-negative genes conferring resistance to antimicrobials belonging to the aminoglycoside, β-lactam, tetracycline, sulfonamide, phenicol, and quinolone families as well as to rifampin, erythromycin, trimethoprim, olaquindox, quaternary ammonium compound resistant determinant genes, and mobile genetic elements such as class 1, 2, 3 integrons and transposon Tn21. This microarray was applied in the genotyping of two collection strains coming from colibacillosis outbreaks in rabbit and cattle facilities located in Italy and Iran, respectively.

For each collection, the positive gene distribution and frequency per encoded enzyme or protein involved in antimicrobial resistance was first determined. Thereafter, strains positive for integrons were matched with the presence of transposon Tn21 (*tnpM*

* Corresponding author. Mailing address: Campo Boario, 64100 Teramo, Italy. Phone 0039 0861332407; Fax 0039 0861332251; e-mail p.badagliacca@izs.it.

gene) in order to highlight the genetic potential of resistance (multi-resistance) transferability in the examined strains.

Fourteen of 26 strains in our rabbit *E. coli* collection were positive for a class 1 integron, six of which were associated with the *tnpM* gene and multidrug resistance-related genes to the aminoglycoside, tetracycline, and sulfonamide-trimethoprim antimicrobial families, prevalently.

Class 1 and/or class 2 and 3 integrons were also detected in 23 out of 51 strains in our cattle *E. coli* collection, 13 of which were also positive for *tnpM* gene. The presence of multiple drug resistance involving the aminoglycoside, tetracycline, and sulfonamide antimicrobial families are positively associated with this *tnpM* gene group. Moreover, strains having positive genes for integrons and transposons were also carriers, to a lesser extent, of class A and/or class C β-lactamase (11 strains) and quinolone (9 strains) resistant genes. The DNA microarray used in this chapter has proven to be a powerful tool in determining the genetic profile of antimicrobial resistance in *E. coli* field strains. The DNA microarray technology may be an appropriate technology for inclusion in antimicrobial resistance monitoring programs because of its adaptability to and ease of data collection for any particular monitoring program. Furthermore, this technology could be used to assess the effectiveness of antimicrobial use-reduction plans or in comparative studies between traditional and antibiotic-free animal production systems.

Introduction

There is a growing consensus in the scientific community that the steady increase in antimicrobial resistance (AMR) is rapidly becoming a global public health emergency involving both medical and veterinary sectors, to the extent that it is now a primary focus of the 'One World, One Health' approach to counteract health risks [1, 2]. The phenomenon, if we exclude natural resistance to specific antimicrobials, finds its biological basis in the genetic selective pressure exerted by the (ab)use of antibiotics (acquired resistance, AR). As antibiotic resistant bacteria have been observed in natural environments, in humans, numerous animals and food, it cannot be excluded that animal-human-environment interactions lead to the circulation of these microorganisms according to the known models of the infectious diseases [3-6].

AMR is recognized as an underlying cause of failure in rehabilitation therapy and as a factor in nosocomial mortality rates due to resistant organisms, especially in the treatment of post-operative infections in intensive care units and in the treatment of immune-compromised individuals. In the veterinary field, the use of antimicrobials has been consolidated over the years, both as a therapeutic tool and growth promoter in the control of pathogens and undesirable microorganisms in the food chain. The appearance of antibiotic-resistant strains in food-producing animals induces a self-sustaining mechanism of progressive increases in the use of other antimicrobials or combination of antimicrobials inevitably resulting in the selection of multi-resistant strains. Moreover, the pharmaceutical industry has reached a stage where the development of new or modifications of existing antibiotics to replace ineffective compounds is decreasing, and it is uncertain that sufficient time and resource investments needed for new drug discoveries will continue in the future as in past decades.

Acquired multi-drug resistance (MDR) has been found in pathogenic and non-pathogenic *E. coli* of human, animal, food and environmental origin [7-9]. The major encountered resistance to aminoglycosides in *E. coli* is through bacterial enzymatic modification of aminoglycoside transferases whose plasmid genes are grouped into three classes: *aac* (aminoglycoside-acetyltransferases), *aad* (-adeniltransferases) and *aph* (-phosphotransferases) family genes. Acquired resistance to tetracycline in *E. coli* results from the acquisition of *tet* genes functionally distributed into energy-dependent efflux pumps (*tetA, tetB, tetC, tetD, tetE, tetG, tetJ, tetL, tetY*) or ribosomal protection (*tetM*) categories. Although few macrolide-licosamide-streptogramin B antibiotics are targeted at *E. coli* infections, an acquired resistance to erythromycin by inactivating estererases (*ereA, ereB* genes) or phosphorylases (*mph* family genes) was found. Resistance to chloramphenicol is due to inactivating chloramphenicol acetyltransferases (CAT) encoded by *cat* genes and an efflux system controlled by the *cmlA* gene. However, chloramphenicol resistant strains whose mechanism is based on the activity of CAT are susceptible to florfenicol. Resistance to both phenicols was found on another efflux system based on the *floR* gene. Specific plasmid-mediated AR to sulfonamides was found in the acquisition of *sul* family genes encoding resistant variants of bacterial dihydropteroate syntase whose natural homolog is competitively inhibited by the drug. Similarly, specific plasmid-mediated *dhfr* family genes encoding resistant variants of dihydrofolate reductase were found in AR to trimethoprim. Resistance to rifampin is expressed by inactivating ADP ribosylase, encoded by *arr-3* gene. Although the clinical resistance to quinolones in *E. coli* is achieved by a combination of chromosomal mutations and plasmid-mediated pathways, a specific plasmid AR was identified in the *qnr* determinant which encodes for a protein protecting the DNA gyrase and topoisomerase IV from quinolone inhibition [10-13]. The β-lactam antibiotics have been effectively used for decades against Gram-positive bacteria and more recently against Gram-negatives (mostly second, third, fourth generation of cephalosporins, cephamycins, and carbapenems). While in Gram-positives the resistance to β-lactams is primarily through modification of penicillin-binding-proteins (PBPs), in Gram-negative bacteria the major AR mechanism is the production of β-lactamases, encoded by genes that are widely distributed by mobile genetic elements. This distribution has recently risen to alarming levels with *E. coli* strains producing extended-spectrum β-lactamases (ESBL) conferring resistance to penicillins, cephems and monobactams [14-16]. At the molecular level, β-lactamases are grouped into four so called Ambler classes A to D. Class A, C and D β-lactamases utilize serine at the active site while class B requires zinc ions to facilitate the lactam hydrolysis [17]. Based on targeted substrates, class A serine β-lactamases range from penicillinases, cephalosporinases to ESBL and carbapenemases. A large number of derivatives by amino acid substitution within $bla_{\text{TEM-1}}$, $bla_{\text{SHV-1}}$ and $bla_{\text{CTX-M}}$ genes were found in strains resistant to different spectra of β-lactams comprising bla_{PER} ESBL-producing Gram-negative bacteria as well as the bla_{KPC} family genes which encode class A carbapenemases. Class B metallo β-lactamase (MBL) includes carbapenemases whose main genotypes are bla_{VIM} and bla_{IMP} which spread via mobile elements in *Klebsiella pneumoniae* and *Pseudomonas aeruginosa*. Class C β-lactamase includes *ampC*-type extended-spectrum cephalosporinases not inhibited by clavulanic acid, commonly detected in the *Enterobacteriaceae* family. A class C cephamycinases encoded by bla_{CMY} and $bla_{\text{FOX-2}}$ were also described in *Enterobacteriaceae*. Type bla_{OXA} conferring resistance to cloxa- and oxacillin is the most representative class D β-lactamase, whose ESBL and carbapenemase phenotypes have been described [6, 18-22].

The spread of antibiotic-resistant organisms is predicated on the horizontal gene transfer (HGT) of HGT-encoding proteins (enzymes) through a variety of mechanisms such as plasmid gene transmission, bacteriophage-transduced DNA, and/or mobile genetic elements as integron-inserted gene cassettes and transposons mostly in Gram-negative bacteria. Based on the primary sequence of integrases, integrons were classified into three classes. Class 1, 2 and 3 integrons exist primarily in Gram-negative bacteria. In particular, class 1 integrons are widely distributed among *Enterobacteriaceae* and are associated with a variety of gene cassettes including aminoglycoside, trimethoprim, sulfonamide, quinolone, chloramphenicol, rifampicin and β-lactam determinants of genetic resistance as well as the *qac* gene encoding for a specific protein efflux pump involved in the quaternary ammonium compound resistance [23-27]. HGT of class 1 integron-inserted cassettes mediated by transposon Tn21, whose genetic determinant is *tnpM*, is well reported [28-30].

The monitoring and reporting of AMR are routinely carried out within national and transnational surveillance programs by government networks in human and animal health predominantly in North American and European countries [31-33]. There is a very large body of harmonized data available on drug consumption (as mg active ingredient/animal biomass unit) per food producing animal/country/antimicrobial class [34] and on the occurrence of pathogenic resistant strains and indicator bacteria such as *E. coli* per origin (animal/food), animal species and category [35].

Although rabbit chain production isn't usually included in these programs, data on antimicrobial consumption and awareness on the spreading AMR phenomenon in rabbits has been well documented [36]. Similarly, in Iran the antibiotic resistance related to human and animal health is poorly monitored by institutional programs. However, emerging data on AMR in the human health area has rapidly raised the visibility of this issue in this country to that found in European and American countries [37, 38]. Genotyping of antimicrobial resistant strains is occasionally carried out for epidemiological investigations or special studies related to single antimicrobial categories or a panel of AMR genes, but the integration of genotyping tools in the AMR surveillance programs remains poor [39].

Among the various molecular techniques currently utilized to assess the molecular composition of AMR resistance in bacteria, DNA microarray with its strong parallel processing capabilities is a highly discriminatory, robust candidate to be considered in an integrated system within larger AMR surveillance programs [40]. An oligonucleotide microarray initially developed by Bruant et al. [41] to assess the virulence potential of *E. coli* isolates, also possesses an exhaustive list of antimicrobial resistance genes found among Gram-negative bacteria. This array, or its upgraded versions, have been useful in assessing the virulence and AMR gene composition of *E. coli* isolated from water and wastewater, [9, 42-43], animals [44-47] and humans [47]. The aim of this chapter was to develop a methodology for defining the AMR gene profiles of *E. coli* strains and their usefulness in integrated monitoring programs, using the above mentioned *E. coli* microarray. For this purpose, we examined two collections of strains from outbreaks of colibacillosis in rabbit and cattle facilities located in two different production regions, namely a European region (northern Italy) with high intensive rabbit production, and a non-intensive cattle production region in a densely populated area of a developing country (Tehran province, Iran).

Materials and Methods

Strain Collections

Two collections of *E. coli* consisting of 26 rabbit enteropathogenic strains (hereinafter termed the rabbit collection or RC) coming from 16 rabbitries located in Veneto and Piemonte Regions (Northern Italy) and 51 calf diarrheagenic strains (hereinafter termed the cattle collection or CC) from 14 farms located in the Province of Tehran (Iran) were selected on the basis of their virulence gene profiles. Microbiological detection, biochemical identification, pathotyping and DNA extraction were described in Badagliacca et al. [48] and Staji et al. [49], respectively.

Microarray Design and Antimicrobial Resistance Genes

The microarray version used in this chapter was validated as previously described [43,47] by performing hybridizations with DNA from *E. coli* reference strains representing all known pathotypes and AMR genes by classical PCR. Only data related to AMR gene hybridizations were considered. The microarray was composed of 70-mer oligonucleotide probes (Table 1) targeting AMR family genes to aminoglycosides, β-lactams, tetracyclines, sulfonamides, phenicols, quinolones, rifampin, as well as trimethoprim, erythromycin, olaquindox, quaternary ammonium compound resistant determinants and mobile genetic elements such as class 1, 2, 3 integrons and transposon Tn21. Oligonucleotides were printed in duplicate on Corning Ultra GAPS slides (Corning Canada, Whitby, Ontario) at the National Research Council (Montreal, QC, Canada).

Table 1. List of the antimicrobial resistance-related genes on the microarray used in this chapter

Oligo	Gene[a]	Encoded protein/enzyme	AM preferred substrate[b, c, d]
70-aacC(2)200	*aac(3)-IIa (aacC2)*	Aminoglycoside acetyltransferase	Dib, Gent, Net, Sis, Tob
70-aac3(IV)380	*aac(3)-IV (aacC4)*	Aminoglycoside acetyltransferase	Gent, Dib, Net, Sis, Tob, Apr
RB_aac3VI	*aac(3)-VI (aacC6)*	Aminoglycoside acetyltransferase	Gent, Sis
RB_aac6Ib	*aac(6')-Ib (aacA4)*	Aminoglycoside acetyltransferase	Dib, Sis, Net, Tob, Isep, Amk
70-aadA(1)290	*aadA1 (ant(3")-Ia)*	Aminoglycoside adenyltransferase	Streptomycin, spectinomycin
F_aadA7	*aadA7*	Aminoglycoside adenyltransferase	Streptomycin, spectinomycin
CM_aadA14	*aadA14*	Aminoglycoside adenyltransferase	Streptomycin, spectinomycin
70-aadB1778	*ant(2")-Ia (aadB)*	Aminoglycoside adenyltransferase	Dib, Gent, Kan, Sis, Tob
70-aac(6)91	*ant(6)-Ia (aadE)*	Aminoglycoside adenyltransferase	Streptomicin
70-aphA(1)1310	*aph(3')-Ia (aphA1)*	Aminoglycoside phosphotransferase	Kan, Neo, Paro, Rib, Liv, Gent

Table 1. (Continued)

Oligo	Gene[a]	Encoded protein/enzyme	AM preferred substrate[b, c, d]
70-aphA(2)220	*aph(3')-IIa (aphA2)*	Aminoglycoside phosphotransferase	Kan, Neo, Paro, Rib, But, Gent
F_aphA3	*aph(3')-IIIa (aphA3)*	Aminoglycoside phosphotransferase	Kan, Amk, Isep, Neo, Paro, Rib, Liv, But, Gent
F_aph N/A	*aph(3')-IV*	Aminoglycoside phosphotransferase	Kan, Neo, Paro, Rib, But
F_aph3strA	*aph(3'')-Ib (strA)*	Streptomycin phosphotransferase	Streptomicin
F_aph6 N/A	*aph(6)-Id (strB)*	Streptomycin phosphotransferase	Streptomicin
CM_tet30	*tet(30)*	Tetracycline pump efflux	Tetracycline
70-tetA1390	*tet(A)*	Tetracycline pump efflux	Tetracycline
70-tetB190	*tet(B)*	Tetracycline pump efflux	Tetracycline
70-tetC130	*tet(C)*	Tetracycline pump efflux	Tetracycline
70-tetD1770	*tet(D)*	Tetracycline pump efflux	Tetracycline
70-tetE370	*tet(E)*	Tetracycline pump efflux	Tetracycline
RB_tetG	*tet(G)*	Tetracycline pump efflux	Tetracycline
CM_tetH	*tet(H)*	Tetracycline pump efflux	Tetracycline
70-tetY1770	*tet(Y)*	Tetracycline pump efflux	Tetracycline
70-tetM575	*tet(M)*	Ribosomal tetracyline-protecting protein	Tetracycline
70-tetQ694	*tet(Q)*	Ribosomal tetracyline-protecting protein	Tetracycline
F_tetR	*tet(R)*	Tetracycline repressor protein	Tetracycline
CM_tetX	*tet(X)*	Tetracycline-inactivating enzyme	Tetracycline
F_ampC	*ampC* (bla_{DHA-1})	Class C β-lactamase	BJM Group 1
F_ampR	*ampR* ($bla_{DHA-1\ repressor}$)	Transcriptional regulator	BJM Group 1
F_b-a_cmy2	bla_{CMY-2}	Class C β-lactamase	BJM Group 1
F_b-a_FOX2	bla_{FOX-2}	Class C β-lactamase	BJM Group 1
F_b-a_PSE1	bla_{PSE-1}	Class A β-lactamase	Pen, Carbenecillin
F_b-a_OXY_K1	bla_{OXY-K1}	Class A β-lactamase	Pen, monobactams
70-tem8674	bla_{TEM}	Class A β-lactamase	BJM group 2b
F_b-a_ROB1	bla_{ROB-1}	Class A β-lactamase	BJM group 2b
F_b-a_SHV37	bla_{SHV-37}	Class A β-lactamase	BJM group 2be
70-shv86	bla_{SHV-86}	Class A β-lactamase	BJM group 2be
F_b-a_CTX_M12	$bla_{CTX-M-12}$	Class A β-lactamase	BJM group 2be
CM_ctxM14	$bla_{CTX-M-14}$	Class A β-lactamase	BJM group 2be
70-ctx143	$bla_{CTX-M-1}$	Class A β-lactamase	BJM group 2be
CM_ctxM2	$bla_{CTX-M-2}$	Class A β-lactamase	BJM group 2be
F_b-a_PER2	bla_{PER-2}	Class A β-lactamase	BJM group 2be

Oligo	Gene[a]	Encoded protein/enzyme	AM preferred substrate[b, c, d]
70-oxa(1)256	*bla*$_{OXA-1}$	Class D β-lactamase	BJM group 2d
F_b-a_OXA2	*bla*$_{OXA-2}$	Class D β-lactamase	BJM group 2d
70-oxa(7)295	*bla*$_{OXA-7}$	Class D β-lactamase	BJM group 2d
F_b-a_OXA9	*bla*$_{OXA-9}$	Class D β-lactamase	BJM group 2d
F_b-a_OXA2b	*bla*$_{OXA-2b}$	Class D β-lactamase	BJM group 2d
b-a_OXA26	*bla*$_{OXA-26}$	Class D β-lactamase	BJM group 2df
F_b-a_OXA27	*bla*$_{OXA-27}$	Class D β-lactamase	BJM group 2df
F_b-a_OXA61	*bla*$_{OXA-61}$	Class D β-lactamase	BJM group 2df
F_bla-KPC3	*bla*$_{KPC-3}$	Class A β-lactamase	BJM group 2f
F_bla_SME1	*bla*$_{SME-1}$	Class A β-lactamase	BJM group 2f
F_b-a_IMP2	*bla*$_{IMP-2}$	Class B β-lactamase	BJM group 3a
F_bla_VIM2	*bla*$_{VIM-2}$	Class B β-lactamase	BJM group 3a
F_cat	*cat*	Chloramphenicol acetyltransferase	Chloramphenicol
CM_catB2	*catB2*	Chloramphenicol acetyltransferase	Chloramphenicol
70-cat550	*catI*	Chloramphenicol acetyltransferase	Chloramphenicol
70-cat(2)300	*catII*	Chloramphenicol acetyltransferase	Chloramphenicol
70-cat(3)370	*catIII*	Chloramphenicol acetyltransferase	Chloramphenicol
RB_cmlA1	*cmlA1*	Chloramphenicol pump efflux	Chloramphenicol
70-floR384	*floR*	Florfenicol/chloramphenicol pump efflux	Phenicols
70-sul(1)960	*sul1 (sulI)*	Dihydropteroate synthase type 1	Sulfonamides
70-sul(2)420	*sul2 (sulII)*	Dihydropteroate synthase type 2	Sulfonamides
GM_sulIII	*sul3 (sulIII)*	Dihydropteroate synthase type 3	Sulfonamides
70-dhfr(1)490	*dfrA1 (dhfrI)*	Dihydrofolate reductase	Trimethoprim
70-dhfr(9)830	*dfrA9 (dhfrIX)*	Dihydrofolate reductase	Trimethoprim
70-dhfr(5)1560	*dfrA5 (dhfrV)*	Dihydrofolate reductase	Trimethoprim
70-dhfr(7)753	*dfrA7 (dhfrVII)*	Dihydrofolate reductase	Trimethoprim
RB_dhfrVIII	*dfrA8 (dhfrVIII)*	Dihydrofolate reductase	Trimethoprim
RB_dhfrXII	*dfrA12 (dhfrXII)*	Dihydrofolate reductase	Trimethoprim
70-dhfr(13)929	*dfrA13 (dhfrXIII)*	Dihydrofolate reductase	Trimethoprim
70-dhfr(15)620	*dfrA15 (dhfrXV)*	Dihydrofolate reductase	Trimethoprim
CM_dhfrA20	*dfrA20*	Dihydrofolate reductase	Trimethoprim
RB_qnrA	*qnrA*	DNA gyrase/type IV topoisomerase-protecting protein	Quinolones
RB_qnrB1	*qnrB1*	DNA gyrase/type IV topoisomerase-protecting protein	Quinolones
RB_qnrB4	*qnrB4*	DNA gyrase/type IV topoisomerase-protecting protein	Quinolones

Table 1. (Continued)

Oligo	Gene[a]	Encoded protein/enzyme	AM preferred substrate[b, c, d]
RB_qnrS1	*qnrS1*	DNA gyrase/type IV topoisomerase-protecting protein	Quinolones
RB_qnrS2	*qnrS2*	DNA gyrase/type IV topoisomerase-protecting protein	Quinolones
F_ereA	*ereA*	Erythromycin esterase	Erythromycin
RB_ereA2	*ereA2*	Erythromycin esterase	Erythromycin
F_ereB	*ereB*	Erythromycin esterase	Erythromycin
F_mphA	*mphA*	Macrolide phosphorylase	Macrolides
F_arr3	*arr3*	Rifampin ADP-ribosyl transferase	Rifampin
CM_oqxA	*oqxA*	Olaquindox pump efflux	Olaquindox
CM_oqxB	*oqxB*	Olaquindox pump efflux	Olaquindox
F_qac	*qac*	Quaternary ammonium compounds resistance marker	Mobile element
70-int(1)1200	*intI1(1)*	Class 1 integron	Mobile element
70-int1(2)2368	*intI1(2)*	Class 1 integron	Mobile element
70-int1(3)	*intI1(3)*	Class 1 integron	Mobile element
RB_int2	*intI2*	Class 2 integron	Mobile element
RB_int2_TnsA	*intI2_tnsA*	Class 2 integron	Mobile element
RB_int3	*intI3*	Class 3 integron	Mobile element
RB_int3_repC	*intI3_repC*	Class 3 integron	Mobile element
F_tnpM	*tnpM*	Transposon Tn21 modulator protein	Mobile element

[a] Nomenclature according to van Hoek et al. [10] and Robert et al. [13]; in brackets, different name by which the gene is known.

[b] Source: ARDB, Antibiotic Resistance Genes Database, http://ardb.cbcb.umd.edu/browsegenes.html; CARD, The Comprehensive Antibiotic Resistance Database, http://arpcard.mcmaster.ca/; Wright and Thompson [50]; Bush and Jacoby [51]; Walther-Rasmussen and Hoiby [21].

[c] Abbreviation key: Kan, kanamycin; Amk, amikacin; Isep, isepamicin; Neo, neomycin; Paro, paromomycin; Rib, ribostamycin; Liv, lividomycin; But, butirosin; Gent, gentamicin; Dib, dibekacin; Net, netilmicin; Sis, sisomicin; Tob, tobramycin; Apr, apramycin; Pen, penicillin; Cephal I-II-III-IV, 1st-2nd-3rd-4th –generation cephalosporin; Cepham, cephamycin; CA, clavulanic acid.

[d] Bush-Jacoby-Medeiros, BJM Group 1: Pen, Cephal, Cepham not inhibited by CA; BJM group 2b: Pen, Cephal I-II; BJM group 2be: Cephal III-IV, monobactams; BJM group 2d: oxa- cloxacillin; BJM group 2df: oxa- cloxacillin, carbapenems; BJM group 2f: carbapenems, Cephal III, Cepham; BJM group 3a: carbapenemes not inhibited by CA.

DNA Labeling

Purified genomic DNA was quantified using a Nanodrop Spectrophotometer (Nanodrop Technologies, ThermoScientific, USA) and approximately 300 ng of DNA was subjected to fluorescent labeling using the Bioprime DNA labeling system (Invitrogen Life Technologies, Burlington, ON, Canada) as described previously [41]. Labeling efficiency and the percentage of dye incorporation was then determined by scanning the DNA sample in a NanoDrop spectrophotometer from 200 to 700 nm. Cy3 dye incorporation was calculated using a Web-based percent incorporation calculator (pangloss.com/seidel/Protocols/percent_inc.html).

Hybridization and Data Acquisition

Pre-hybridizations and hybridizations were performed following a protocol derived from Hamelin et al. [9]. For each hybridization, 500 ng of labeled DNA was dried under vacuum in a rotary desiccator without heating (Savant SpeedVac®, ArrayIt, USA). Dried labeled DNA was resuspended in hybridization buffer (DIG Easy Hyb Buffer, Roche Diagnostics, Laval, Quebec, Canada). Microarrays were pre-hybridized for at least one hour at 50°C with a pre-heated pre-hybridization buffer containing 5X SSC, 0.1% SDS and 1.0% BSA. After pre-hybridization, the microarrays were hybridized with a solution that consisted of 25 μL of hybridization buffer, 20 μL of Bakers Yeast tRNA (10 mg/mL) (Sigma Aldrich, St. Louis, MO, USA) and 20 μL of Sonicated Salmon Sperm DNA (10 mg/mL) (Sigma Aldrich), mixed together with the labeled DNA which had previously been denatured. Microarrays were hybridized overnight at 50°C in a SlideBooster (model SB800, Advalytix, Germany). After hybridization, stringency washes were performed with Advawash (Advalytix, Germany) using 1X SSC, 0.02% SDS preheated to 50°C. Microarray slides were scanned with a ScanArray Lite fluorescent microarray analysis system (Perkin-Elmer, Missasauga, Ontario, Canada) using with ScanArray Gx software (Perkin-Elmer, Foster City, CA, USA). Fluorescent spot intensities were quantified with QuantArray Version 3.0 (Packard Bioscience, Informer Technologies Inc., USA).

All microarrays were normalized using the same method. The mean value for each set of duplicate spotted oligonucleotides was divided by the correction factor taken from the negative controls spots. This value was then divided by the average of the empty spots to create a signal-to-noise ratio. Oligonucleotide spots with a signal-to-noise fluorescence ratio greater than the established threshold (2 in this case), were considered positive. These ratios were then converted into binary data where a value of 0 indicates a negative probe and a value of 1 indicates a positive probe.

Results

Commensal *E. coli* is commonly chosen as an indicator in the monitoring programs of microbiological resistance because of its ability to acquire plasmid-borne AMR and transfer it among Gram-negative bacteria. Moreover, monitoring of zoonotic microbial agents such as *Salmonella*, *Campylobacter,* and Shiga toxin-producing *E. coli* is currently included in these programs. Finally, *E. coli* can also be useful to monitor changes in bacteria possessing ESBLs. Genetic characterization of resistant strains is usually not included in monitoring data in the official programs since these data are not standardized. However, some programs do incorporate molecular typing results for selected isolates in peer-reviewed publications [31, 33, 35].

Detection of Single AMR Genes and Monitoring the Resistance to Antimicrobial Classes

The usual panel of antimicrobials tested in *E. coli* microbiological resistance monitoring programs includes gentamicin and streptomycin (aminoglycoside antimicrobial class), tetracycline, sulfonamides, trimethoprim, chloramphenicol, nalidixic acid and ciprofloxacin (quinolones), ampicillin (aminopenicillin), and cefotaxime (third generation cephalosporin). The frequency of phenotypic resistance to antimicrobials per bacterial species/ animal/year/country represents the type of data collected in monitoring programs. The dichotomized data of resistance (resistant/non-resistant strain to specific antibiotic per animal species/country) allow a trend analysis of AMR based on tables and graphs as tools for presenting the pattern of change over time [52]. Table 2 shows the frequency of positive AMR genes found in the two strain collections. Overall, the microarray detected 324 and 492 resistance genes in RC (all strains) and CC (40 out of 51 strains) collections, respectively. Between 2-30 and 1-41 AMR genes per strain were detected in RC and CC, respectively. In addition, the bulk of the detected genes (80.8% and 79.4%) belonged to the aminoglycoside, tetracycline, sulfonamide, trimethoprim, β-lactam, and phenicol resistance family of genes in RC and CC, respectively. In RC at least one AMR determinant for the above classes was detected in 25, 23, 20, 9, 15, and 4 strains, in that order. Similarly, in CC at least one AMR gene was detected in 34, 33, 23, 15, 31 and 14 strains. A lower number of AMR genes related to resistance to quinolones were detected both in RC and CC. The very high frequency of resistance to aminoglycosides, tetracyclines and sulfonamide-trimethoprim as noted by susceptibility testing in rabbit *E. coli* is well known in rabbitries of northern Italy and ranges from 72% to 87% among aminoglycosides (gentamicin, streptomycin, neomycin), 96 to 97% among tetracyclines (oxytetracycline, tetracycline, doxycycline) and 83% with sulfametossazole/trimethoprim [53]. The overlapping level of phenotypic and genotypic resistance to these classes of antimicrobials is in agreement with quantitative data on drug consumption in rabbits in northern Italy. More than 70% of the currently used antimicrobials belong to these above-mentioned antimicrobial classes. On one hand, this explains the phenomenon of the selective pressure of drug application on genetic resistance; on the other hand, this testifies to the mismatch between clinical and microbiological/molecular resistance. The use of antimicrobial mixtures or higher dosage and the lengthening of administration time can presumably overcome microbiological resistance, but simultaneously increase genetic resistance. This is an “iceberg model” where clinical resistance is the emergent part and microbiological/genetic resistance is the submerged. The selective pressure at the base of the iceberg model, suitable for aminoglycosides, tetracycline, and trimethoprim-sulphonamide, does not explain the genetic resistance to β-lactams and phenicols that are rarely used in rabbits. Genetic resistance to the latter is better associated with the mechanism of co-resistance and multi-resistance to antibiotics (see below).

The most frequent aminoglycoside resistance genes detected among RC were *aadA1* (in 24 out of 26 strains), *aac(3)-IV* (20 strains), *strB* and *strA* (21 and 16 strains). In particular, the *aac(3)-IV* gene in *E. coli* (and also in *Salmonella*, *Pseudomonas,* and *Campylobacter*), exhibits a broad resistance to numerous aminoglycosides, including apramycin whose use in rabbit is common [54, 55], while *aadA1*, *strA,* and *strB* genes confer resistance to streptomycin in many species of pathogenic Gram-negative bacteria. In addition, *aadA1,* which uses spectinomycin as a substrate, is associated with mobile elements (see below).

Twenty-one out of 26 isolates in RC were positive for tetracycline efflux pump *tet30* and *tetA* resistance types as well for *tetR*, which represses *tetA* in the absence of substrate [56].

The absence of harmonized data on phenotypic resistance and drug consumption for cattle in Iran precludes a similar assessment of our results with the rabbit collection. In this case, a more applicable model to assess the predictive value of data originating from genetic resistance would be a more generic comparative analysis aimed at establishing how much the data originating from resistant strain genotyping in a given field and region of investigation (cattle and Tehran in our case) corresponds to those with humans. The close relationship, and even co-habitation, between man and animal, in developing countries, justifies this approach.

Table 2. Distribution of genes detected in examined strains

Encoded enzyme/protein	Rabbit collection (n=26)		Cattle collection (n=51)	
	Genes detected[a]	Frequency[b]	Genes detected[a]	Frequency[b]
Aminoglycoside acetyltransferase	*aac(3)-IV, aacC2, aadE*	24 (20)	*aac(3)-VI, aacC2, aacA4, aadE, aac(3)-IV*	34 (23)
Aminoglycoside adenyltransferase	*aadA1, aadB*	26 (24)	*aadA1, aadA14, aadA7, aadB*	20 (17)
Aminoglycoside phosphotransferase	*aphA(1), aph(3')-IV*	7 (7)	*aphA(1), aphA(2), aph(3')-IV*	26 (23)
Streptomycin phosphotransferase	*strB, strA*	37 (21)	*strB, strA*	31 (18)
Tetracycline pump efflux/ribosomal protecting/inactivating protein	*tet30, tet(R), tet(A), tet(E), tet(C), tetD), tet(B), tet(H), tet(M)*	75 (23)	*tet(R), tet(A), tet30, tet(D), tet(B), tet(C) , tet(Y), tet(Q), tet(E), tet(M), tet(G), tet(X)*	110 (33)
Class A β-lactamase	bla_{TEM}, bla_{SHV-86}, bla_{SHV-37}, bla_{KPC-3}, $bla_{CTX-M-14}$,$bla_{CTX-M-1}$	13 (8)	bla_{TEM}, $bla_{CTX-M-14}$, $bla_{CTX-M-2}$, bla_{SHV-37}, bla_{KPC-3}, $bla_{CTX-M-1}$, bla_{SHV-86}, bla_{SME-1}	58 (27)
Class B β-lactamase	bla_{IMP-2}, bla_{VIM-2}	6 (5)	bla_{VIM-2}, bla_{IMP-2}	4 (3)
Class C β-lactamase	bla_{CMY-2}, *ampC, ampR*, bla_{FOX-2}	12 (9)	bla_{CMY-2}, *ampC*, bla_{FOX-2}, *ampR*	33 (24)
Class D β-lactamase	bla_{OXA-2}, bla_{OXA-26}, bla_{OXA-9}	7 (5)	bla_{OXA-61}, bla_{OXA-2b}, bla_{OXA-9}	4 (3)
Chloramphenicol acetyltransferase	*catII, catIII, catBe, cat; cmlA1*	5 (4)	*cmlA1, catI, catII, catIII, cat*	20 (14)
Chloramphenicol/florfen icol exporter	*floR*	2 (2)	*floR*	2 (2)
Sulphonamide dihydropteroate	*sul2, sul1, sul3*	28 (20)	*sul2, sul3, sul1*	26 (23)
Dihydrofolate reductase	*dfrA1, dfrA7, dfrA13, dfrA5, dfrA9, dfrA8, dfrA12*	19 (9)	*dfrA5, dfrA8, dfrA1, dfrA15, dfrA9, dfrA7, dfrA13, dfrA12*	23 (15)
Erythromycin esterase	*ereB, ereA, ereA2*	7 (5)	*ereA, mphA, ereA2, ereB*	7 (5)
Rifampin transferase			*arr3*	2 (2)
Fluoroquinolone resistance protein	*qnrB4, qnrB1, qnrA*	4 (3)	*qnrS1, qnrB1, qnrB4, qnrA, qnrS2*	15 (11)

Table 2. (Continued)

Encoded enzyme/protein	Rabbit collection (n=26)		Cattle collection (n=51)	
	Genes detected[a]	Frequency[b]	Genes detected[a]	Frequency[b]
Olaquindox resistance determinant	*oqxB, oqxA*	3 (3)	*oqxB, oqxA*	6 (6)
Quaternary ammonium resistance protein	*qac*	13 (13)	*qac*	4 (4)
Integrase	*intI1(2), intI1(3), intI1(1)*	28 (14)	*intI1(1), intI1(3), intI1(2), intI2_tnsA, intI2, intI3_repC, intI3*	51 (23)
Transposase	*tnpM*	8 (8)	*tnpM*	16 (16)

[a] Listed in the order of decreasing frequency; [b] Occurrence of detection (n° of involved strains).

Several studies have focused on the nosocomial emergence of resistant bacteria in Tehran (Iran). Khalili et al. [57] reported resistance data through susceptibility testing of 238 *E. coli* strains from the bloodstream and urinary tracts of hospitalized patients. Resistance levels among aminoglycosides were variable ranging from 47.5 to 66.0% (gentamicin, amikacin) up to 96.6% (tobramycin) and a high or very high resistance was found against ciprofloxacin (62.2%), penicillins (piperacillin, 99.6%) and cephalosporins (cefalozin, 91.2%; cefuroxime, 99.2%; ceftazidime, 68.5%; ceftriaxone, 73.1%). Ghadiri et al. [58], in a similar study, reported different data but similar proportions between antimicrobial classes, with a lower resistance level for aminoglycosides (amikacin, gentamicin ranging from 2.6 to 3.7%), penicillins (penicillin, ampicillin 18.0 to 24.3%), cephalosporins (ceftazidime, cephalotin, 7.7 to 8.4%) and quinolones (ciprofloxacin 22.9%).

Our results of uniform genetic resistance to aminoglycosides and β-lactams/cephalosporins ranging from 60.8 to 66.7%, including a relatively low level of genetic resistance to quinolones (21.6%), do not match with the above profiles in humans. The prevalent aminoglycoside resistance genes in CC were *aphA1*, *aac(3)-VI*, *strB* and *aadA1* (in 22, 21, 18, and 14 out of 51 strains, respectively). In available nosocomial data on aminoglycoside-related resistance genes in *E. coli* isolated from the urinary tract (Lorestan, West Iran), the *aph(3')-Ia* (*aphA1*) determinant was not prevalent and in contrast, the prevalent *aac(3)-IIa* (*aacC2*) gene does not match our results (six strains in CC) [59].

Similarly, the prevalent patterns of genetic resistance to β-lactams in CC were bla_{TEM}, CTX type β-lactamases, $bla_{CMY\text{-}2}$ and *ampC* type (in 24, 21, 16 and 12 strains) are not comparable to that of the available nosocomial data in Kerman (South-East Iran) where the TEM (43.5%) and SHV (34.8%) types are the most prevalent and CTX and *ampC* types were detected in 15.9 and 9.6% of *E. coli* from the urinary tract [60]. The low frequency of bla_{SHV} genes in Iranian cattle was confirmed by Momtaz et al. [61] in an investigation on *E. coli* from mastitic milk.

Multi-Resistance Gene Profiles and Investigating Genetic Multi-Resistance

To assess multi-resistance at the molecular level, isolates exhibiting three or more resistance genes families may be referred to as putative multi-resistant [62]. This is consistent with the investigation of antimicrobial multi-resistance in monitoring programs based on phenotypic resistance [35]. Furthermore, in order to focus on multi-resistance patterns involving the critical important antimicrobials (CIA) for human health, animal-derived *E. coli* isolates exhibiting resistance genes to third-generation cephalosporins and fluoroquinolones, are important monitoring outputs.

Figures 1 and 2 show the patterns of AMR-related family genes in positive strains of RC and CC respectively. In an attempt to exclude the resistance-related genes whose contribution to the intrinsic or acquired multi-resistance is uncertain [62, 63], diagrams of multi-resistance were constructed with exclusion of genes encoding drug exporters through efflux pumps imparting resistance to tetracycline and phenicols.

In RC, multi-resistance gene profiles involving at least three antimicrobial classes were found in 17 of 26 strains, of which 12 carried at least one resistance gene to β-lactams. Five strains exhibited broad multi-resistance to five or more antimicrobial classes comprising β-lactams (3 strains), or co-resistance to β-lactams and quinolones (2 strains).

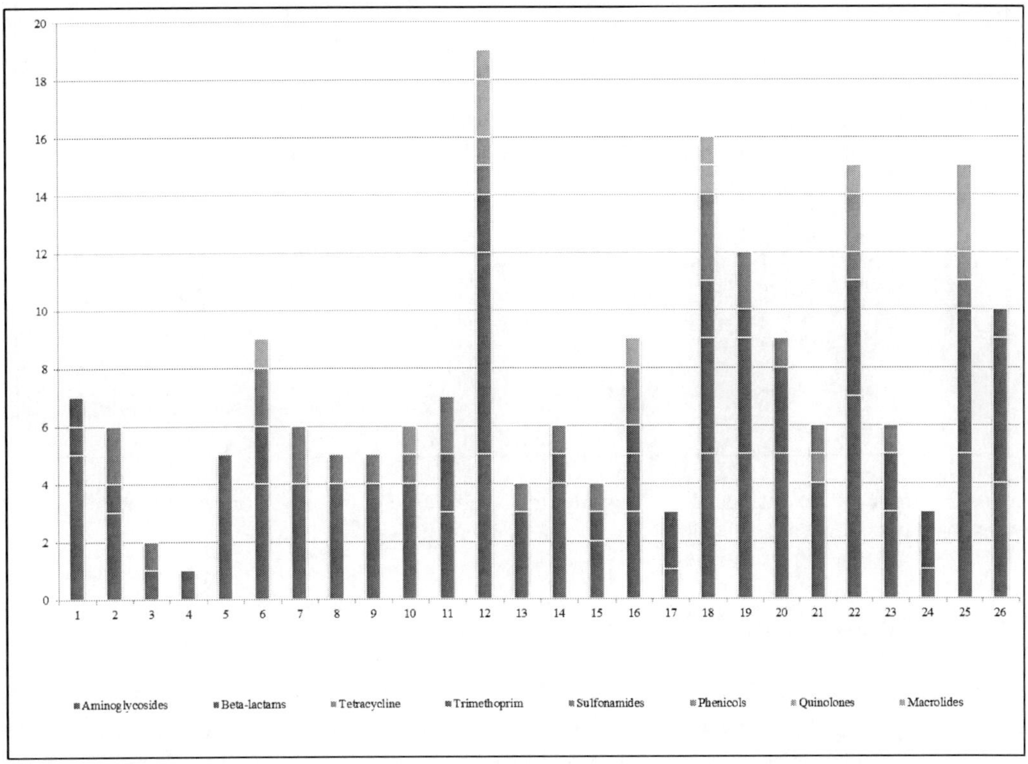

Figure 1. Number of positive genes per antimicrobial class (except protein exporter/pump efflux encoding genes) from rabbit collection.

All strains carrying resistance genes to β-lactams harbored several recognized or putative gene cassettes (see below), indicating that the resistance to β-lactams in rabbit *E. coli* strains, in the absence of selective pressure due to drug consumption, was likely due to HGT through conjugative mechanisms. In particular, the circulation of multi-resistant strains harboring β-lactamase-encoding genes, such as $bla_{CMY\text{-}2}$, $bla_{IMP\text{-}2}$, or $bla_{KPC\text{-}3}$, is an alarming observation indicating a potential reservoir of ESBL and carbapenemase-producing *E. coli*. The low level of quinolone resistance is consistent with the limited use of quinolones in rabbit production and the available data on phenotypic resistance in rabbit *E. coli* in northern Italy [53].

Conversely, in CC at least three antimicrobials were found in 26 strains (out of 51), of which 23 carried at least one resistance gene to β-lactams. In seven strains, five or more antimicrobial classes comprising β-lactam and quinolone resistance determinants were detected. The above mentioned seven strains consistently showed the pattern *aac3-VI*, *bla*$_{TEM}$, *bla*$_{CTX}$ type, *sul*, and *qnr* family genes associated with class 1 integrons and the transposon determinant *tnpM* (see below).

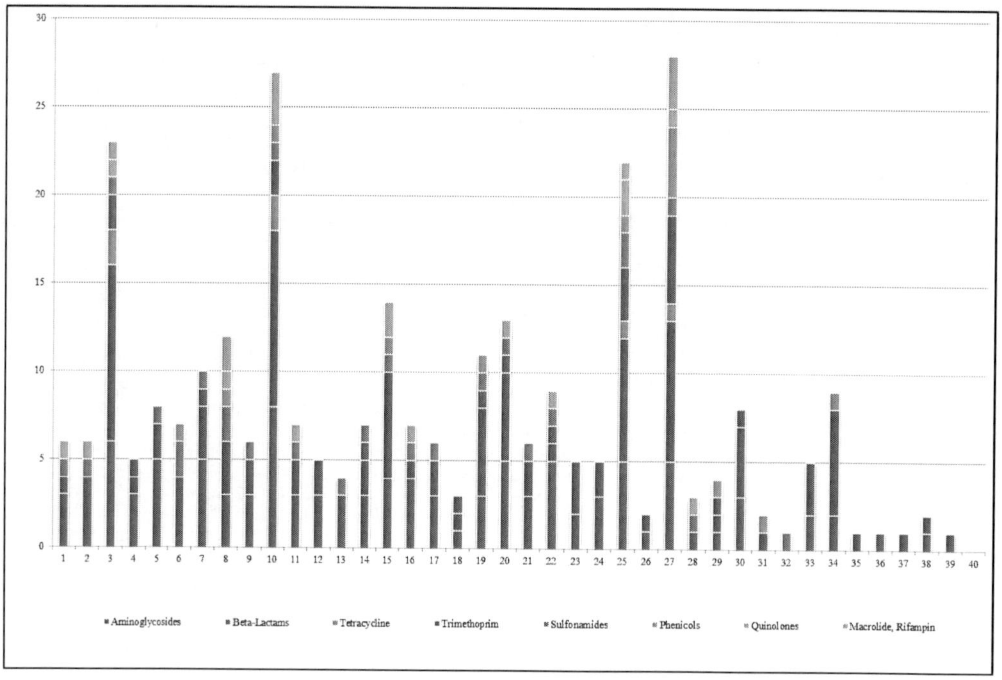

Figure 2. Number of positive genes per antimicrobial class (except protein exporter/pump efflux encoding genes) from cattle collection.

Mobile Genetic Element Profiles

More than 500 cassette arrays and related integrase encoding genes in *E. coli* were deposited in Integrall©, an integron database available on http://integrall.bio.ua.pt [64]. The class 1 integron is the most prevalent integrase recognized in *E. coli* and the related patterns of cassette arrays might be a good indicator in assessing AMR gene profiles with regard to their level of integration. In particular, *aadA1*, *dfrA1*, *dfrA17*, *aadA2*, *aadA5*, *qac*, *sul1*, *dfrA12*, *aacA4*, *aadB*, and *cmlA1* are a putative gene cassette in combination of two or more, and *aadA1-dfrA1*, *aadA5-dfrA17*, *sul1-qac*, *aadA1-sul1-qac*, *cmlA1- aacA4*, *aadA1-bla*$_{OXA}$, *and cmlA1-aacA4* are the most frequent array in class 1 integron profiles. Furthermore, detection of the *tnpM* determinant of plasmid-borne transposon Tn21 can provide indicative information with regard to transferable resistance genes in *Enterobacteriaceae* [29].

Table 3 and 4 list the integron-associated gene profiles in RC and CC respectively. In RC only class 1 integron genes were detected in 14 of 26 strains with six of them being *tnpM* positive. All 14 integron-harboring strains possessed the *aadA1* gene cassette and about 50% of them possessed the cassette array *aadA1-aac(3')-IV-sul1-qac*, thus indicating a high level of integration and potential transferability of resistance genes. The distribution of cephalosporinase and carbapenemase-encoding genes among class 1 integron-harboring strains does not correlate with the prevalent cassette array and detection of *tnpM*, indicating a slow or recent evolution of this multidrug genetic character. In a recent study on avian *E. coli* in northern Italy, the integron and gene cassette profiles, detected by PCR and sequencing, differ from our results presented in this chapter [65]. The presence of class 2 integrons in about 9% of strains and the *dfhrA1-aadA1* cassette array were predominately found, thereby demonstrating species-specific niches of AMR genetic evolution in *E. coli* in different chain productions.

Conversely, in the CC, class 1, 2, and 3 integrons were detected in 23 strains (out of 51) but only class 1 integron was strongly associated with the *tnpM* gene in 13 strains (about 25%). However, the latter harbored more than 50% of all positive genes detected in the strain collection.

Table 3. Mobile genetic elements and associated multi-resistance gene profiles from rabbit collection

Integron variable	*tnpM*	Multiresistence-associated gene profiles
intI1(2)	no	*aphA1, aacC2, aac(3)-IV, aadA1, strB, tet30,* bla_{TEM}, *dfrA1, qac*
intI1(2)	no	*aphA1, aac(3)-IV, aadA1, strA, strB, qac*
intI1(2), intI1(3)	no	*aacC2, aac(3)-IV, aadA1, tet(R),* bla_{TEM}, *sul1, sul2*
intI1(2), intI1(3)	no	*aac(3)-IV, aadA1, strA, strB, tet30, tet(R), tet(A),* bla_{TEM}, bla_{CMY-2}, bla_{KPC-3}, bla_{VIM-1}, bla_{IMP-2}, *dfrA5*
intI1(2), intI1(3)	no	*aadA1, strA, strB, tet30, tet(R), tet(A),* bla_{TEM}, bla_{OXA26}, *sul1, sul2, dfrA7, ereB, qac*
intI1(2), intI1(3)	no	*aadA1, tet30, tet(R), tet(A),* bla_{TEM}, bla_{CMY-2}
intI1(2), intI1(3)	no	*aadA1 strA, strB, tet30, tet(R), tet(A), blaCMY-2, blaIMP-2, sul1, sul2, qac*
intI1(1), intI1(2), intI1(3)	no	*strB, sul2, qac*
intI1(2), intI1(3)	yes	*aac(3)-IV, aadA1, strA, strB, tet30, tet(R), tet(A), sul1, sul2, qac*
intI1(2), intI1(3)	yes	*aphA1, aac(3)-IV, aadA1, strA, strB, tet30, tet(R), tet(A), tet(C), tet(D), tet(E),* bla_{TEM}, *ampC,* bla_{SHV-86}, bla_{CMY-2}, $bla_{CTX-M-14}$, bla_{KPC-3}, *catII, floR, sul1, dfrA1, dfrA8, ereA, qnrB1, qnrB4, qac*
intI1(2), intI1(3)	yes	*aphA1, aac(3)-IV, aadE, aadA1, aadB, strB, tet30, tet(R), tet(A),* bla_{CMY-2}, bla_{OXA-2}, bla_{OXA-26}, bla_{IMP-2}, *sul1, sul2, dfrA1, qac*
intI1(2), intI1(3)	yes	*aphA1, aac(3)-IV, aadA1, strA, strB, tet30, tet(R), tet(A), tet(B), cat, sul2, dfrA1, dfrA5, dfrA7, dhfrX, dfrA13, ereA, ereA2, ereB, oqxA, qac*
intI1(2), intI1(3)	yes	*aac(3)-IV, aadA1, strB, tet30, tet(R), tet(A),* bla_{CMY-2}, bla_{IMP-2}, *sul2*
intI1(1), intI1(2), intI1(3)	yes	*aac(3)-IV, aadA1, strA, strB, tet30, tet(R), tet(A), sul1, sul2, dfrA1, dfrA13,qnrA, qac*

Table 4. Mobile genetic elements and associated multi-resistance gene profiles from cattle collection

Integron (variable)	*tnpM*	Multiresistence-associated gene profiles
intI1(2)	no	*aph3, aph6, aphA1; tetB; sul2*
intI1(1)	no	*aac3-VI, aadA1, aphA1; tetR;* bla_{CMY-2}; bla_{SHV-37}; *cmlA1; sul3; qnrS1*
IntI1(1)	no	*aphA1;* bla_{TEM}; *dfrA5*
intI1(1), intI1(3)	no	*aac3-VI, aadA1, aphA1; tet30, tetA, tetB, tetR;* bla_{TEM}; *dfrA8*
intI1(1), intI1(3)	no	*aadA1; sul3; cmlA1; qnrS1*
intI1(1), intI1(2), intI1(3)	no	*aacC2, aphA1, aph6; tetB;* bla_{TEM}, $bla_{CTX-M-12}$, bla_{CMY-2}, bla_{OXA-9}, bla_{OXA-61}; *catI; cmlA1; sul2; dfrA13; qac*
intI2, intI2_tnsA	no	*aadA1; tetB; dfrA1*
intI2, intI2_tnsA	no	*aadA1; tetB;* bla_{TEM}; *catI; dfrA1*
intI3_repC	no	*aac3-VI;* bla_{CMY-2}
intI3_repC	no	*aphA1, aph3, aph6; tet30, tetA, tetR;* bla_{TEM}, bla_{CMY-2}; *sul2*
intI1(1), intI1(3)	yes	*aadA1, aph6; tetA, tetB, tetR; cmlA1; sul3; qnrS1*
intI1(1), intI1(3)	yes	*aacC2, aadA1, aphA1, aph6; tetB; cmlA1; catI; sul2, sul3; qac*
intI1(1), intI1(3)	yes	*aac3-VI, aadA1, aphA1, aph3, aph6; tet30, tetA, tetC, tetD, tetR;* $bla_{CTX-M-14}$, bla_{TEM}, *ampC; sul2; dfrA5; oqxB*
intI1(1), intI1(3)	yes	*aac3-VI, aac6-Ib, aacC2, aadA1, aphA1; tet30, tetA, tetB, tetC, tetD, tetR, tetY;* bla_{TEM}; $bla_{CTX-M-2}$, $bla_{CTX-M-14}$, bla_{VIM-2}, *ampC; cmlA1; sul3; dfrA8; qnrS1; oqxB*
intI1(1), intI1(3)	yes	*aac3-VI, aadA1, aphA1, aph3, aph6; tet30, tetA, tetR;* bla_{TEM}; *sul2; dfrA5; mphA*
intI1(1), intI1(3)	yes	*aphA1, aph3, aph6; tet30, tetA, teetR;* bla_{TEM}; bla_{CMY-2}; *sul2; dfrA5*
intI1(1), intI1(3)	yes	*aac3-VI, aadA14, aadB, aph3, aph6; tetD, tetQ, tetR;* bla_{TEM}, $bla_{CTX-M-1}$, $bla_{CTX-M-2}$, bla_{KPC-3}, bla_{VIM-2}, *ampC,* bla_{FOX-2}, bla_{OXA-2B}; *cat, catI, catII, catIII, floR; sul2; dfrA5, dfrA7, dfrA8, dfrA9, dfrA15; ereA, ereA2, ereB; qnrA*
intI1(1), intI1(3)	yes	*aac3-VI, aadA1, aphA1; tet30, tetA, tetR;* bla_{CMY-2}; *cmlA1; sul3; qnrS1*
intI1(1), intI1(2), intI1(3)	yes	*aac3-VI, aphA1, aph6; tetB;* bla_{TEM}, $bla_{CTX-M-14}$, bla_{CMY-2}; *catI; mphA; sul1, sul3; arr3; qnrB1; qac*
intI1(1), intI1(2), intI1(3), intI2, intI2_tnsA	yes	*aac3-VI, aadA1, aphA1, aph3, aph6; tet30, tetA, tetB, tetC, tetD, tetE, tetQ, tetR;* bla_{TEM}, bla_{SHV-37}, $bla_{CTX-M-1}$, $bla_{CTX-M-2}$, $bla_{CTX-M-14}$, bla_{KPC-3}, *ampC; catII; sul1, sul2; dfrA1, dfrA8, dfrA12; arr3; qnrA, qnrS2; oqxB; qac*
intI1(1), intI1(2), intI3	yes	*aac3-VI, aac6, aac6-Ib, aacC2; aadA1, aadA7, aadA14, aphA1; tet 30, tetA, tetB, tetC, tetD, tetE, tetM, tetQ, tetR, tetY;* bla_{TEM}, bla_{SHV-37}, $bla_{CTX-M-1}$, $bla_{CTX-M-2}$, $bla_{CTX-M-14}$, bla_{KPC-3}, *ampC,* bla_{CMY-2}, bla_{FOX-2}; *catII; cmlA1, floR; sul3; dfrA8, dfrA15; qnrB1, qnrB4, qnrS1*
Integron (variable)	tnpM	*Multiresistence-associated gene profiles*
intI1(2), intI2, intI2_tnsA, intI3_repC	yes	*aac3-VI, aadA1, aph(3')-IV, aphA1, aphA2, aph6; tet30, tetA, tetC, tetD, tetG, tetM, tetR, tetX;* bla_{TEM}, bla_{SHV-86}, $bla_{CTX-M-2}$, bla_{KPC-3}, bla_{SME-1}, bla_{IMP-2}, bla_{VIM-2}, *ampC,* bla_{CMY-2}, bla_{FOX-2}; *sul2; dfrA1, dfrA9; ereA; qnrB4; oqxB*
intI1(1), intI1(2), intI2_tnsA, intI3_repC	yes	*aac3-VI, aadA14, aphA1, aphA2; tet30, tetA, tetC, tetD, tetR;* bla_{TEM}, $bla_{CTX-M-14}$, bla_{SHV-37}, bla_{SHV-86}, *ampC,* bla_{CMY-2}; *catIII; sul3; qnrB1, qnrB4; oqxB*

The above 13 strains were composed of genes encoding for resistance to CIA such as class A and/or C *bla*$_{TEM}$, *bla*$_{CTX}$, ampC, bla$_{CMY}$ (11 strains), partly associated with carbapenemase *bla*$_{KPC}$ or bla$_{VIM}$ types (5 strains), as well as resistance determinants to quinolone in 9 strains. Despite the concentration of positive AMR genes, few putative gene cassettes were detected. Among integron-harboring strains, only a cluster of seven strains contained the recognized *aadA1-cmlA1* cassette array, but not always associated with the *tnpM* determinant. The broad presence of integron and transposase encoding genes and a few gene cassettes or cassette arrays are indicative of a significant potential of translocation but a poor level of genetic integration. Najibi et al. [66] described gene cassette profiles in *E. coli* from diarrheagenic human strains in Iran where the *dfr-aadA* array was prevalent, confirming a certain phylogenetic distance between strains examined in this study and strains of human origin.

Using DNA Microarray to Analyze Genetic AMR within Integrated Monitoring Programs

The application of microarray technology in detecting Gram-negative AMR genes has been expanding in recent years and validation or research data are available for comparative studies or epidemiological findings, in both the human and animal fields. DNA microarray-based specific tools were used to characterize ESBL genotypes [67, 68] or resistance to aminoglycosides [69, 70]. Card et al. [71] developed a microarray for detecting resistance genes towards 75 antibiotics and tested 132 strains of Gram-negative pathogenic bacteria, underscoring the use of microarray technology in clinical diagnostics. Glenn et al. [72] have analyzed the multi-resistance of 32 avian *E. coli* strains by microarrays as well as class 1 integron-related genes by PCR, focusing on the potential risk of resistance gene transfer associated with integron presence. A universal DNA microarray based on 775 oligonucleotides identified in GenBank was developed by Frye et al. [73] for detecting AMR determinants in Gram-negative and -positive bacteria, which provided a clear demonstration of the advantages that microarrays could bring to epidemiological studies. Similar to our approach of integrating genotyping and phenotyping data within monitoring programs of AMR, Castanheira et al. [74] and Sheng et al. [75] reported on the prevalence of nosocomial EBSL genes among *Enterobacteriaceae* using either a commercial microarray or multiplex PCR, in the USA (SENTRY program) or Asia-Pacific region (SMART program) respectively. The integration of phenotyping and genotyping, by using a microarray approach, is very applicable to multi-resistant and mobile genetic element-harbouring bacteria like the Gram-negatives, since AMR in Gram-negative bacteria is characterized by fragmentation of multidrug-resistance phenotypic patterns [39]. Allowing the clustering of AMR types with their potential mobilization, genotyping may have a predictive value useful to refine monitoring programs and possibly introduce elements of surveillance (observations for the purpose of control) of the AMR phenomenon. The DNA microarray technology may be an appropriate technology for inclusion in AMR monitoring programs because of its adaptability to and ease of data collection for any particular monitoring program. Furthermore, according to the action plans of major international institutions on countering AMR, there will be several activities in animal productions such as the reduction/prudential use of antibiotics, the strengthening of biosafety measures and use of non-antibiotic antimicrobials. Consequently, it will be necessary to assess the effectiveness of these actions and measure the change in the level of resistance in indicator bacteria. Microarray technology can play an appropriate role in conducting genotyping studies on indicator bacteria, before and during the above activities. Similarly, the proposed tool for genotyping bacterial populations could be used in impact assessments on AMR levels in antibiotic-free animal production systems through comparative studies with traditional systems. In general, data coming from the application of microarrays in detecting AMR genes are sufficiently advanced as they relate to the number of identifiable AMR genes, methodology and AMR probe validation. However, the standardization process requires additional phases, related to the objectives laid down within specific monitoring programs, to improve the specificity of the method relative to the species and geographical area under investigation. Furthermore:

1. methods should be aimed at the detection of gene cassettes and gene transfer determinants as an added value to supplement phenotypic data
2. microarray designs should be continuously updated from sequencing data
3. bioinformatics tools to improve statistical analysis should be developed.

Conclusion

In this chapter, we described a methodology for interpreting genetic data underlying AMR in *E. coli* generated by a DNA microarray, involving a major integration between genotyping and phenotyping AMR in monitoring programs. This methodology has highlighted how data interpretations may be different for rabbit-derived *E. coli* strains coming from an European region with intensive rabbit production or for cattle-derived *E. coli* strains from a densely populated area in a developing country.

In the first case, available data on drug consumption and phenotypic AMR in the field of investigation allowed the discrimination of AMR presumably due to selective pressure, from genetic resistance to antimicrobials not used in rabbits, such as β-lactams, indicative of horizontal gene transfer among bacteria. In the second case, comparative analysis of multi-resistance and mobile genetic element profiles between our results and available data on nosocomial AMR in Iran allowed us to determine that AMR types in cattle poorly matched those with human AMR profiles. Finally, analysis of integron-inserted gene cassette profiles and transposon determinants such as *tnpM* allowed us to speculate about differences on the level of integration and potential transferability of resistance genes in both strain collections.

Acknowledgments

Our thanks go to Fabrizio Agnoletti, Istituto Zooprofilattico Sperimentale delle Venezie (Treviso, Italy) and Taghi Zahraei Salehi, Faculty of Veterinary Medicine, University of Tehran (Iran) for supplying the strain collections, and to our colleague Anna Abass for her good laboratory technical support.

References

[1] European Council. Council conclusions of 22 June 2012 on the impact of antimicrobial resistance in the human health sector and in the veterinary sector – a 'One Health' perspective. *Official Journal of the European Union* 2012;55,C211;IV:2-5.

[2] World Organization for Animal Health (OIE). Resolution n. 27. The "One Health" approach to address health risks at the animal-human ecosystem interface. Resolutions adopted during the 80th OIE General Session 2012; Paris, 20-25 May 2012. Available at: http://www.oie.int/about-us/key-texts/resolutions-and-recommendations/resolutions-adopted-by-the-oie-international-committee/. Accessed on 9 August 2013.

[3] Levy SB, Fitzgerald GB, Macone AB. Changes in intestinal flora of farm personnel after introduction of a tetracycline-supplemented feed on a farm. *N Engl J Med* 1976;295:583-588.

[4] Pappas G. An animal farm called extended-spectrum beta-lactamase: antimicrobial resistance as a zoonosis. *Clin Microbiol Infect* 2011;17:797-798.

[5] Acar JF, Moulin G. Antimicrobial resistance: a complex issue. In: Acar JF, Moulin G, eds. Antimicrobial resistance in animal and public health. Paris: World Organization for Animal Health. *Rev Sci Tech OIE* 2012;31:23-28.

[6] Accogli M, Fortini D, Giuffrè M, Graziani C, Dolejska A, Carattoli A et al. IncI 1 plasmids associated with the spread of CMY-2, CTX-M-1 and SHV-12 in *Escherichia coli* of animal and human origin. *Clin Microbiol Infect* 2013;19:238-240.

[7] Saenz Y, Brinas L, Dominguez E, Ruiz J, Zarazaga M, Vila J et al. Mechanisms of resistance in multiple-antibiotic-resistant *Escherichia coli* strains of human, animal and food origins. *Antimicrob Agents Chemother* 2004;48:3996-4001.

[8] Maynard C, Fairbrother JM, Bekal S, Sanschagrin F, Levesque RC, Brousseau R et al. Antimicrobial resistance genes in enterotoxigenic *Escherichia coli* O149:K91 isolates obtained over a 23-year period from pig. *Antimicrob Agents Chemother* 2003;47:3214-3221.

[9] Hamelin K, Bruant G, El-Shaarawi A, Hill S, Edge TA, Bekal S et al. A virulence and antimicrobial resistance DNA microarray detects a high frequency of virulence genes in *Escherichia coli* isolates from great lakes recreational waters. *Appl Environ Microbiol* 2006;72:4200-4206.

[10] van Hoek AHAM, Mevius D, Guerra B, Mullany P, Roberts AP, Aarts HJM. Acquired antibiotic resistance genes: an overview. *Front Microbiol Antimicrob Resist Chemother* 2011;2:1-27.

[11] Lambert T. Antibiotics that affect the ribosome. In: Acar JF, Moulin G, eds. Antimicrobial resistance in animal and public health. Paris: World Organization for Animal Health. *Rev Sci Tech OIE* 2012;31:57-63.

[12] Cambau E, Guillard T. Antimicrobials that affect the synthesis and conformation of nucleic acids. In: Acar JF, Moulin G, eds. Antimicrobial resistance in animal and public health. Paris: World Organization for Animal Health. *Rev Sci Tech* OIE 2012;31:77–87.

[13] Roberts MC, Schwarz S, Aarts HJM. Erratum: acquired antibiotic resistance genes: an overview. *Front Microbiol Antimicrob Resist Chemother* 2012;3:1-17.

[14] Leverstein-van Hall MA, Dierikx CM, Cohen Stuart J, Voets GM, van den Munckhof MP, van Essen-Zandbergen A et al. Dutch patients, retail chicken meat and poultry

share the same ESBL genes, plasmids and strains. *Clin Microbiol Infect* 2011;17:873-880.

[15] Bush K. Antimicrobial agents targeting bacterial cell walls and cell membranes. In: Acar JF, Moulin G, eds. Antimicrobial resistance in animal and public health. Paris: World Organization for Animal Health. *Rev Sci Tech OIE* 2012;31:43-56.

[16] Platteel TN, Leverstein-van Hall MA, Cohen Stuart JW, Voets GM, Bonten MJM. Differences in the antibiotic susceptibility of human *Escherichia coli* with poultry-associated and non-poultry-associated extended-spectrum beta-lactamases. *Eur J Clin Microbiol Infect Dis* 2013:32:1091-1095.

[17] Ambler RP. The structure of beta-lactamases. *Philos Trans R Soc Lond B Biol Sci* 1980;289:321-331.

[18] Bauernfeind A, Wagner S, Jungwirth R, Schneider I, Meyer D. A novel class C β-lactamase (FOX-2) in *Escherichia coli* conferring resistance to cephamycins. *Antimicrob Agents Chemother* 1997;41:2041-2046.

[19] Paterson DL, Bonomo RA. Extended-spectrum beta lactamases: a clinical update. *Clin Microbiol Rev* 2005;18:657-686.

[20] Zhanel GG, Johanson C, Embil JM, Noreddin A, Gin A, Vercaigne L et al. Ertapenem: review of a new carbapenem. *Expert Rev. Anti Infect. Ther.* 2005;3:23-39.

[21] Walther-Rasmussen J, Hoiby N. OXA-type carbapenemases. *J Antimicrob Chemother* 2006;57:373–383.

[22] Drawz SM, Bonomo RA. Three decades of β-lactamase inhibitors. *Clin Microbiol Rev* 2010;23:160–201.

[23] Hall RM, Collis CM. Antibiotic resistance in gram-negative bacteria: the role of gene cassettes and integrons. *Drug Resist Update* 1998;1:109-119.

[24] Fluit AC, Schmitz FJ. Resistance integrons and super-integrons. *Clin Microbiol Infect* 2004;10:272-288.

[25] Mazel D. Integrons: agents of bacterial evolution. *Nat Rev Microbiol* 2006;4:608-620.

[26] Partridge SR, Tsafnat G, Coiera E, Iredell JR. Gene cassettes and cassette arrays in mobile resistance integron. *FEMS Microbiol Rev* 2009;33:757-784.

[27] Gillings MR, Xuejun D, Hardwick SA, Holley MP, Stokes HW. Gene cassettes encoding resistance to quaternary ammonium compounds: a role in the origin of clinical class 1 integrons?. *ISME J* 2009;3:209-215.

[28] Francia MV, Lobo JMG. Gene integration in the *Escherichia coli* chromosome mediated by Tn21 integrase (Int21). *J Bacteriol* 1996;178:894-898.

[29] Liebert CA, Hall RM, Summers AO. Transposon Tn21, flagship of the floating genome. *Microbiol Mol Biol Rev* 1999;63:507-522.

[30] Vizvaryova M, Valkova D. Transposons. The useful genetic tools. *Biologia Bratisl* 2004;59:309-318.

[31] European Food Safety Authority (EFSA). Technical specifications for the analysis and reporting of data on antimicrobial resistance in the European Union. *EFSA J.* 2012;10:2587.

[32] European Food Safety Authority (EFSA). Technical specifications on the harmonised monitoring and reporting of antimicrobial resistance in *Salmonella*, *Campylobacter* and indicator *Escherichia coli* and *Enterococcus* spp. bacteria transmitted through food. *EFSA J.* 2012;10:2742.

[33] Silley P, Simjee S, Schwarz S. Surveillance and monitoring of antimicrobial resistance and antibiotic consumption in humans and animals. In: Acar JF, Moulin G, eds. Antimicrobial resistance in animal and public health. Paris: World Organization for Animal Health. *Rev. Sci. Tech. OIE* 2012;31:105-120.

[34] Grave K, Torren J, Muller A, Mackay D. European surveillance of veterinary antimicrobial consumption (ESVAC) project. Results and further need for harmonisation with ESAC-net. ECDC 2nd joint meeting of the antimicrobial resistance and healthcare-associated infections (ARHAI) networks [Session 19]; 2012 November 26-28; Berlin, Germany. Available at: http://www.ecdc.europa.eu/en/press/events/Lists/Events/ECDC_DispForm.aspx?List=43564830-6b8a-442f-84e7-2495fa49489b&ID=230. Accessed on 9 August 2013.

[35] European Food Safety Authority (EFSA); European Centre for Disease Prevention and Control (ECDC). The European Union Summary Report on antimicrobial resistance in zoonotic and indicator bacteria from humans, animals and food in 2011. *EFSA J.* 2013;11:3196.

[36] Agnoletti F. Update on rabbit enteric diseases: despite improved diagnostic capacity, where does disease control and prevention stand? Proceedings of 10th World Rabbit Congress; 2012 Sept 3-6; Sharm El-Sheikh, Egypt. World Rabbit Science Association Proceedings; 2012:1113-1127. Available at: http://world-rabbit-science.com/WRSA-Proceedings/Congress-2012-Egypt/Papers/07-Pathology/P00-Agnoletti.pdf. Accessed on 9 August 2013.

[37] Mardani M. Trends in antimicrobial resistance. *Iran J Clin Infect Dis* 2006;1:109-111.

[38] Pourakbani B, Sadr A, Ashtiani MTH, Mamishi S, Dehghani M, Mahmoudi S et al. Five-year evaluation of the antimicrobial susceptibility patterns of bacteria causing bloodstream infections in Iran. *J Infect Dev Ctries* 2012;6:120-125.

[39] O'Brien TF, Stelling J. Integrated multilevel surveillance of the world's infecting microbes and their resistance to antimicrobial agents. *Clin Microbiol Rev* 2011;24:281-295.

[40] Fang H, Xu J, Ding D, Jackson SA, Patel IR, Frye JG et al. An FDA bioinformatics tool for microbial genomics research on molecular characterization of bacterial foodborne pathogens using microarrays. *BMC Bioinformatics* 2010;11:2-11.

[41] Bruant G, Maynard C, Bekal S, Gaucher I, Masson L, Brousseau R et al. Development and validation of an oligonucleotide microarray for detection of multiple virulence and antimicrobial resistance genes in *Escherichia coli. Appl Environ Microbiol* 2006;72:3780-3784.

[42] Frigon D, Biswal B, Mazza A, Masson L, Gehr R. Biological and physicochemical wastewater treatment processes reduce the virulence risk of *Escherichia coli. Appl Environ Microbiol* 2013;79:835-844.

[43] Ramírez Castillo FY, Avelar González FJ, Garneau P, Márquez Díaz F, Guerrero Barrera AL, Harel J. Presence of multi-drug resistant pathogenic *Escherichia coli* in the San Pedro River located in the State of Aguascalientes, Mexico. *Front Microbiol* 2013;4:147.

[44] Tonelli A, Badagliacca P, Bruant G, Letizia A, Di Provvido A, Harel J et al. Genetic characterization of rabbit *Escherichia coli* strains with the use of microarray technology. Proceedings of 9th World Rabbit Congress; 2008 June 10-13; Verona, Italy. World Rabbit Science Association Proceedings 2008:1097-101. Available at:

http://world-rabbit-science.com/WRSA-Proceedings/Congress-2008-Verona/Papers/P-Tonelli.pdf. Accessed on 27 August 2013.

[45] Bonnet C, Diarrassouba F, Brousseau R, Masson L, Diarra MS. Pathotype and antibiotic resistance gene distribution of *Escherichia coli* isolates from broiler chickens raised on antimicrobial supplemented diets. *Appl Environ Microb* 2009;75:6955-6962.

[46] Baines D, Erb S, Turkington K, Kudau G, Juba J, Masson L et al. Mouldy feed, mycotoxins and Shiga toxin-producing *Escherichia coli* colonization associated with Jejunal Hemorrhage Syndrome in beef cattle. *BMC Vet Res* 2011;7:24.

[47] Jakobsen L, Garneau P, Kurbasic A, Bruant G, Stegger M, Harel J et al. Microarray-based detection of extended virulence and antimicrobial resistance gene profiles in phylogroup B2 *Escherichia coli* of human, meat and animal origin. *J Med Microbiol* 2011 Oct;60:1502-1511.

[48] Badagliacca P, Agnoletti F, Tonelli A, Drigo I, Merildi V, Lopes F et al. Virulence gene profiles of rabbit enteropathogenic *Escherichia coli* strains isolated in northern Italy. *Vet. Ital.* Submitted 2013.

[49] Staji H, Badagliacca P, Salehi TZ, Tamai IA, Iorio M, Mazza A et al. Pathotyping of diarrheagenic cattle *Escherichia coli* strains isolated in the Province of Tehran, Iran. *Vet Ital* Submitted 2013.

[50] Wright GD, Thompson PR. Aminoglycoside phosphotransferases: proteins, structure and mechanism. *Front Biosci* 1999;4:d9-21.

[51] Bush K, Jacoby GA. Updated functional classification of β-lactamases. *Antimicrob Agents Chemother* 2010;54:969-976.

[52] Aerts M, Faes C, Nysen R. Development of statistical methods for the evaluation of antimicrobial resistance data in bacterial isolates from animals and food. Scientific report submitted to EFSA by I-BioStat, Interuniversity Institute for Biostatistics and Statistical Bioinformatics, Hasselt University and Katholieke Universiteit. Leuven, Belgium. [Question N° EFSA-Q-2010-00111. Accepted for Publication on 22 August 2011]. Available at: http://www.efsa.europa.eu/en/supporting/ doc/186e.pdf. Accessed on 9 August 2013.

[53] Pisoni AM, Poloni R, Tosi R, Gallazzi D, Grilli G. Osservazioni sulla sensibilità agli antibiotici di alcuni patogeni isolati da conigli da carne [Isolation and antimicrobial susceptibility of bacterial pathogens of the rabbit]. Atti delle Giornate di Coniglicoltura ASIC [Proceedings Annual ASIC Meeting]; 2007 Sept 26-27; Forlì, Italy. Associazione Scientifica Italiana di Coniglicoltura [Italian Rabbit Science Association]. ASIC Meetings 2007:79-81. Available at: http://www.asic-wrsa.it/documenti/giornate 2007/12_Pisoni.pdf. [Italian]. Accessed on 27 August 2013.

[54] Zhang XY, Ding LJ, Fan MZ. Resistance patterns and detection of *aac(3)-IV* gene in apramycin-resistant *Escherichia coli* isolated from farm animals and farm workers in northeastern of China. *Res Vet Sci* 2009;87:449-454.

[55] Magalhaes MLB, Blanchard JS. The kinetic mechanism of *aac(3)-IV* aminoglycoside acetyltransferase from *Escherichia coli. Biochemistry* 2005;44:16275-16283.

[56] Ramos JL, Martinez-Bueno M, Molina-Henares AJ, Teràn W, Watanabe K, Zhang X et al. The *tetR* family of transcriptional repressors. *Microbiol Mol Biol Rev* 2005;69:326-356.

[57] Khalili H, Soltani R, Afhami S, Dashti-Khavidaki S, Alijani B. Antimicrobial resistance pattern of Gram-negative bacteria of nosocomial origin at a teaching hospital in Islamic Republic of Iran. *East Mediterr Health J* 2012;18:172-177.

[58] Ghadiri H, Vaez H, Khosravi S, Soleymani E. The antibiotic resistance profiles of bacterial strains isolated from patients with hospital-acquired bloodstream and urinary infections. *Crit Care Res Pract* 2012;Article ID 890797:1-6.

[59] Mofrad SM, Shakib P, Goudarzi G, Noroozi J. Redundancy of encoding genes of aminoglycoside modifying enzymes on *Escherichia coli* isolated from patients with urinary tracts infections: case study. *Life Sci J* 2013;10:829-834.

[60] Kalantar D, Mansouri S. Emergence of multiple β-lactamases produced by *Escherichia coli* clinical isolates from hospitalized patient in Kerman, Iran. *Jundishapur J Microbiol* 2010;3:137-145.

[61] Momtaz H, Dehkordi FS, Taktaz T. Shiga toxin-producing *Escherichia coli* isolated from bovine mastitic milk: serogroups, virulence factors and antibiotic resistance properties. *Scientific World J* 2012;article ID 618709:1-9.

[62] Schwarz S, Silley P, Simjee S, Woodford N, van Duijkeren E, Johnson AP et al. Assessing the antimicrobial susceptibility of bacteria obtained from animals. *J Antimicrob Chemother* 2010;65:601-604.

[63] Poole K. Efflux-mediated antimicrobial resistance. *J Antimicrob Chemother* 2005;56:20-51.

[64] Moura A, Soares M, Pereira C, Leitao N, Henriques I, Correia A. Integrall: a database and search engine for integrons, integrases and gene cassettes. *Bioinformatics* 2009;25:1096-1098.

[65] Dotto G, Giacomelli M, Giovinardi D, Grilli G, Piccirillo A. Antibiotico-resistenza e integroni di classe 1 e 2 in ceppi di *Escherichia coli* isolati da specie avicole commerciali [Antimicrobial resistance and class 1 and 2 integrons in *Escherichia coli* strains isolated from commercial poultry species]. Atti del LII Congresso della Società Italiana di Patologia Aviare [Proceedings 52th SIPA Annual Congress]; 2013 Apr 11-12, Forlì, Italy. Società Italiana di Patologia Aviare, Atti dei Convegni [Italian Avian Pathology Association, SIPA Meetings] 2013:156-164.

[66] Najibi S, Bakhshi B, Fallahzad S, Pourshafie MR, Katouli M, Sattari M et al. Distribution of class 1 integrons among enteropathogenic *Escherichia coli*. *Can J Microbiol* 2012;58:637-643.

[67] Leinberger DM, Grimm V, Rubtsova M, Weile J, Schroppel K, Wichelhaus C et al. Integrated detection of extended-spectrum-beta-lactam resistance by DNA microarray-based genotyping of TEM, SHV and CTX-M genes. *J Clin Microbiol* 2010;48:460-471.

[68] Bogaerts P, Hujer AM, Naas T, Rezende de Castro R, Endimiani A, Nordmann P et al. Multicenter evaluation of a new DNA microarray for rapid detection on clinically relevant *bla* genes from β-lactam-resistant Gram-negative bacteria. *Antimicrob Agents Chemother* 2011;55:4457-4460.

[69] Davis MA, Baker KNK, Orfe LH, Shah DH, Besser TE, Call DR. Discovery of gene conferring multiple-aminoglycoside resistance in *Escherichia coli*. *Antimicrob Agents Chemother* 2010;54:2666-2669.

[70] Szmolka A, Anjum MF, Ragione RM, la Kaszanyitzky EJ, Nagy B. Microarray based comparative genotyping of gentamicin resistant *Escherichia coli* strains from food animals and humans. *Vet Microbiol* 2012;156:110-118.

[71] Card R, Zhang J, Das P, Cook C, Woodford N, Anjum MF. Evaluation of an expanded microarray for detecting antibiotic resistance genes in a broad range of Gram-negative bacterial pathogens. *Antimicrob. Agents Ch.* 2013;57:458-465.

[72] Glenn LM, Englen MD, Lindsey RL, Frank JF, Turpin JE, Berrang ME et al. Analysis of antimicrobial resistance genes detected in multiple-drug-resistant *Escherichia coli* isolates from boiler chicken carcasses. *Microb Drug Resist* 2012;18:453-463.

[73] Frye JG, Lindsey RL, Rondeau G, Porwollik S, Long F, McClelland M et al. Development of a DNA microarray to detect antimicrobial resistance genes identified in the National Center for Biotechnology Information Database. *Microb Drug Resist* 2010;16:9-19.

[74] Castanheira M, Farrell SE, Deshpande LM, Mendes RE, Jones RN. Prevalence of β-lactamase-encoding genes amon *Enterobacteriaceae* bacteremia isolates collected in 26 U.S. hospitals: report from the SENTRY antimicrobial surveillance program 2010. *Antimicrob Agents Chemother* 2013;57:3012-3020.

[75] Sheng WH, Badal RE, Hsueh PR. Distribution of extended-spectrum β-lactamases, ampC β-lactamases, and carbapenemases among *Enterobacteriaceae* isolates causing intra-abdominal infections in the Asia-Pacific region: results of the study for monitoring antimicrobial resistance trends. *Antimicrob Agents Chemother* 2013;57:2981-2988.

Reviewed by: Moussa S. Diarra; Affiliation: Pacific Agri-Food Research Centre, Agriculture and Agri-Food Canada, P.O. Box 1000, 6947 Highway 7, Agassiz, BC, Canada V0M 1A0. Phone: (604) 796-2221. Fax: (604) 796-0359. E-mail: moussa.diarra@agr.gc.ca; web site: http://www.agr.gc.ca/eng/science-and-innovation/research-centres/british-columbia/pacific-agri-food-research-centre-agassiz/?id=1351005326819.

In: Microarrays: Principles, Applications and Technologies
Editor: James V. Rogers
ISBN: 978-1-62948-669-7

Chapter 11

Phenotypic Anchoring of Liver Injury in Rats by Gene Expression Profiling

Minjun Chen[1,*], Reagan Kelly[1], Lillian Tong[1], Jürgen Borlak[2] and Weida Tong[1]
[1]Division of Bioinformatics and Biostatistics,
FDA's National Center for Toxicological Research, Jefferson, Arkansas, US
[2]Center of Pharmacology and Toxicology,
Hannover Medical School, Hannover, Germany

Abstract

Toxicogenomics offers the possibility to simultaneously monitor the expression of thousands of genes to detect early as well as transient changes that allude to the pathogenesis of toxicity. Specifically, phenotypic anchoring has been proposed to improve understanding of toxicogenomics data by linking the pattern of gene expression to adverse reactions. Here, we attempted to anchor gene expression profiling to liver injury in rats as measured by serum alanine aminotransferase (ALT), a widely used biomarker for detecting signals of hepatotoxicity. The aim of the chapter was to identify a gene expression-based biomarker that had a better specificity and sensitivity than the traditional ALT. To that end, a multivariate model was developed based on gene expression profiling from rat livers treated by a compendium of eight hepatotoxins, and the derived biomarker was further verified by an independent dataset from rats treated by acetaminophen. The results demonstrated that the model prediction not only correlated better with hepatocellular necrosis than with the measured ALT, but could also identify signals of liver injury earlier than the measured serum ALT. The regulated biomarker gene set was related to mitochondrial dysfunction, cholestasis and bile acids synthesis, drug metabolism, and oxidative stress. Taken collectively, gene expression based biomarkers have the potential to improve the detection of liver injury in rats, and can be potentially extended to humans.

* for correspondence Email: minjun.chen@fda.hhs.gov.

Disclaimer: The views presented in this article do not necessarily reflect current or future opinion or policy of the US Food and Drug Administration. Any mention of commercial products is for clarification and not intended as an endorsement.

Introduction

Toxicogenomics has enjoyed tremendous attention since its inception, and was expected to change traditional approaches for assessing toxicity [1, 2]. Taking advantage of the microarray methodology, tens of thousands of individually-expressed genes could be simultaneously monitored, permitting the assessment of significant alterations in gene expression profiles induced by different compounds. After decades of development, enormous amounts of microarray data have been accumulated in the public domain that offers an unprecedented opportunity to convert these data into knowledge that will enhance toxicity prediction and mechanism understanding. Phenotypic anchoring is one of these approaches proposed to improve understanding the toxicogenomics data by linking the pattern of gene expressions to adverse reactions [3-5].

In this regard, drug-induced liver injury (DILI) is a leading cause for failures in drug development at clinical phases of investigation and for post-marketing regulation events like drug withdrawals, black box warnings, and Dear Doctor Letters [6-9]. DILI accounts for more than half of the acute liver failure cases in the United States [10]. Application of biomarkers is an appropriate solution to reduce the risks caused by DILI. However, the current biomarkers of hepatic injury are insufficient to satisfy the requirements from the clinic, industry, and regulatory agencies [11]. One of the most commonly used serum biomarkers for detecting liver injury is serum alanine aminotransferase (ALT), which has been widely used in preclinical studies and clinical trials for monitoring potential drug-related hepatotoxicity. For example, it is widely accepted that liver injury could be defined as an ALT elevation of more than three times of the upper limit of the normal (ULN) range [12]. Nonetheless, ALT lacks the specificity to predict DILI, especially severe DILI [13-15]. This is largely due to the fact that, besides the liver, ALT is also detected in diseases of the kidney, heart, muscle, and other tissues. For instance, Nathwani et al. [16] reviewed the records of 16 patients who had muscle necrosis without evidence of liver disease, and confirmed that muscle injury can also be a cause of elevated ALT. In this chapter, we investigated whether changes in the expression of a set of genes was associated with liver injury at an early stage and thus could offer potential to be of diagnostic value. For this purpose, we linked the gene expression profile to ALT measurement and the phenotype of liver injury, and expected that the predictive model based on the genes from diverse biological pathways would be more robust than ALT measurements alone. To test this hypothesis, we utilized a published dataset of rats treated with eight hepatotoxins to develop a toxicogenomics model, and applied this to another published dataset of rats treated with acetaminophen (APAP) as an independent verification. The phenotypic anchoring model has shown the potential to perform better than the measurement of ALT to detect liver injury in rats.

Materials and Methods

Datasets

The gene expression data and associated histopathological and clinical chemistry data were obtained from a National Institute of Environmental Health Sciences (NIEHS) study as detailed previously [17]. Eight compounds known to cause liver necrosis in rats were used in the study, including 1,2-dichlorobenzene, 1,4-dichlorobenzene, bromobenzene, diquat dibromide, galactosamine, monocrotaline, n-nitrosomorpholine, and thioacetamide. For each compound, four to six male Fischer 344 rats were administrated with four dosages (i.e., vehicle control, low, middle, and high) of the hepatotoxins at an age of 12-weeks. Six compounds were given by oral gavage, while diquat dibromide and galactosamine were administered by intraperitoneal injection. Animals were sacrificed at 6, 24, or 48 hours after the single dose treatment. Liver samples were isolated for histopathological examination and gene expression profiling using Affymetrix Rat 230 2.0 array. Blood samples were collected for clinical chemistry and hematology tests. A brief summary of the samples is displayed in Table 1. The second dataset was generated in rats treated with APAP [18, 19]. Similarly, male Fischer 344 rats were administrated APAP with 4 different doses by oral gavage at 12 weeks old with their vehicle control groups in both light and dark conditions as indicated in Table 1. There were 4 animals in each dose/time/light condition group and 3 animals in the control groups. All animals were sacrificed at 6, 18, 24, or 48 hours after treatment. Liver samples were collected for histopathological examination and gene expression profiling using Agilent Rat Oligo Microarray (22K). Blood samples were collected for clinical chemistry and hematology tests. A brief summary of the experimental conditions for the NIEHS and APAP studies is displayed in Table 1.

Table 1. Brief description of samples for the training and test sets in the NIEHS study and independent validation set from the APAP study

Data set	Route	Compound	Dose (mg/kg)	Time points (hour)	# of samples
Training set from NIEHS study	oral gavage	1, 2-dichlorobenzene	0/15/150/1500	6/24/48	46
	oral gavage	1, 4-dichlorobenzene	0/15/150/1500	6/24/48	48
	oral gavage	bromobenzene	0/25/75/250	6/24/48	48
	oral gavage	monocrotaline	0/10/50/300	6/24/48	44
	oral gavage	n-nitrosomorpholine	0/10/50/300	6/24/48	48
Test set from NIEHS study	oral gavage	thioacetamide	0/15/50/150	6/24/48	46
	intraperitoneal injection	diquat dibromide	0/25/100/400	6/24/48	48
	intraperitoneal injection	galactosamine	0/5/10/25	6/24/48	90
Independent validation from APAP study	oral gavage	acetaminophen	0/50/150/1500/2000	6/18/24/48	152

Data Analysis

Training, Test, and Validation Sets

The NIEHS dataset was divided into the training and test sets using a split-sample method according to the experimental sequence; specifically, 234 samples from five compounds (1,2- dichlorobenzene, 1,4-dichlorobenzene, bromobenzene, monocrotaline, and N-nitrosomorpholine) administrated by oral gavage were set into the training set for model development, and the remaining 184 samples treated by three compounds (diquat dibromide, galactosamine, and thioacetamide) were used as the test set to assess the performance of the developed model. Two compounds (diquat dibromide, galactosamine) administered through intraperitoneal injection were set into the validation set to confirm whether the multi-gene model could identify false positives of elevated ALT possibly caused by muscle injury. One hundred fifty two samples from the APAP study were used as the independent validation set.

Data Preprocessing

A RefSeq-based approach was employed for probe mapping between the Affymetrix Rat 230 2.0 microarray and Agilent Rat Oligo microarray (22K) as detailed in [17]. Specifically, the probe IDs of Agilent platform and the probe-set IDs of Affymetrix platform were mapped to the RefSeq databases using the ArrayTrack Library [20] dated December 2012 that mirrored the RefSeq database in the NCBI, resulting in 4,860 shared RefSeq IDs identified using a Venn diagram approach. As suggested in [21], a median expression value was calculated if multiple probes (or probe-sets) IDs corresponded to the same RefSeq ID.

The intensity data instead of the ratio between two channels was used for the Agilent arrays analysis, since the intensity value from a single channel of a two-color array like the Agilent Rat Oligo Microarray has been demonstrated to have similar sensitivity/specificity as the two-color ratio [22]. Here, we only used the intensity data from the Agilent arrays [21] in order to facilitate transferring the model developed to the Affymetrix platform. Furthermore, the expression value of each gene in both Affymetrix and Agilent arrays was centralized by subtracting its average across samples in the training/validation sets independently to reduce the batch effects between two platforms.

Only the genes with average expression intensity over 1,000 in the control or treated groups were kept in order to reduce background noise as suggested previously [22]. Considering that informative genes are associated with high variation across the control and treated samples [23], we employed a variation-based approach for gene pre-selection. Specifically, we ranked genes based on their relative variation levels (i.e., variation/average) across the studied samples in the training set, and kept the 200 top-ranked genes for further analysis.

Partial Least-squares Regression (PLS) and Model Development

PLS is a widely-used multivariate algorithm to determine the quantitative relationship between dependent variables and independent variables [24]. Compared with other methods, the PLS method is considered to be more robust since its parameters are not as sensitive to the addition of new calibration samples [24]. Given n samples, suppose $\mathbf{y}$ $(n \times 1)$ is the dependent variable (e.g., ALT measurement), and $\mathbf{X} = (\mathbf{x}_1, \cdots\cdots, \mathbf{x}_p)$ is a $n \times p$ covariate matrix, in which each column represents a gene. In the classical multiple linear regression, a direct relationship between $\mathbf{y}$ and is established as following:

$$\mathbf{y} = \mathbf{Xb} \qquad \text{(Eq. 1)}$$

where, $\mathbf{b}$ is the coefficient vector $(p \times 1)$ to be estimated. Unlike other methods, PLS constructs new latent variables $\mathbf{t}_1, \cdots, \mathbf{t}_s$, which are a combination of the original covariates $\mathbf{x}_i$, and these latent variables are used to connect the dependent and covariate variables as following:

$$\mathbf{X} = \mathbf{t}_1\mathbf{p}_1^{'} + \cdots \mathbf{t}_s\mathbf{p}_s^{'} + \mathbf{E}_\mathbf{x} \qquad \text{(Eq. 2)}$$

and

$$\mathbf{y} = \mathbf{t}_1 q_1 + \cdots \mathbf{t}_s q_s + \mathbf{e}_\mathbf{y} \qquad \text{(Eq. 3)}$$

The superscript represents the matrix/vector transform, $\mathbf{p}_i$ is the weight vector of the PLS components, and q_i is the coefficient of the latent variables. is the dimension of the latent variables. $\mathbf{E}_\mathbf{x}$ and $\mathbf{e}_\mathbf{y}$ are the residual items in the equation. The PLS algorithm was implemented using a PLS package (version 2.3-0) [25].

Selection of the optimal number of latent variables s (called PLS components thereafter) was a balance of model complexity and over-fitting possibility. Mostly, a cross-validation approach was applied, and PRESS (prediction error sum of squares) was used as a measure of the fitting of a PLS model prediction to the dependent data. The optimal number of PLS components was selected when it yielded the minimum PRESS.

Gene Selection for Modeling

The coefficients of the first PLS weight vector has been demonstrated to be valid for ranking independent variables as equivalent to the F-statistic, a feature ranking method widely used in microarray data analysis [26]. We applied a leave-one compound-out cross-validation strategy in order to alleviate dataset dependence [27]. Basically, the training set was divided into five parts, and each part associated with a specific compound was set aside sequentially, and the remained samples were used to calculate the PLS weight vectors. After all compounds were set left out once, the PLS weight vectors were put together to rank genes by calculating the average for each coefficient. Considering the general requirement of the sample sizes being 3~5 times more than the variables to avoid over-fitting [28], we selected the top 40 genes for further model development due to the limited number of samples (i.e., n=234) in the training set.

Receiver Operating Characteristic (ROC) Curve

ROC curves have been widely used to compare the diagnostic performance of different tests [29]. The ROC curve is a graphical plot of the false-positive rate and the true-positive rate for a binary classifier system as its discrimination threshold is varied. A case is called test-positive (e.g., predicted to cause necrosis) if the threshold is above a defined value, and all of the cases together give the false-positive rates and true-positive rates. The total area under the ROC curve (AUC) is a measure of the test's capability to correctly predict the expected outcome. An AUC of 1 represents a perfect test, while 0.5 or less represents a test of no better than random chance. The ROC analysis was implemented using a ROCR package (version 1.0-4) [30].

Results

The training set from the NIEHS study was used to identify the genes relevant to liver injury. The biological significance of the selected genes was confirmed by pathway analysis, network analysis, and gene ontology analysis. Furthermore, a predictive model was developed using these selected genes and the PLS algorithm, and was subsequently assessed by the NIEHS test set and the APAP validation set. Finally, we assessed the robustness of the model prediction using histological findings and the false positive tests from the measured ALT. The workflow is detailed in Figure 1.

Selection of the Informative Genes Relevant to Liver Injury

There were 40 genes selected based on absolute values of the PLS weight coefficients in Table 2. Notably, many of the repressed genes are correlated with ALT activity and thus, liver injury. Obviously, the mode of actions of different hepatotoxins was different, and therefore, it is reasonable that many genes were identified as well correlated with the presence of liver toxicity.

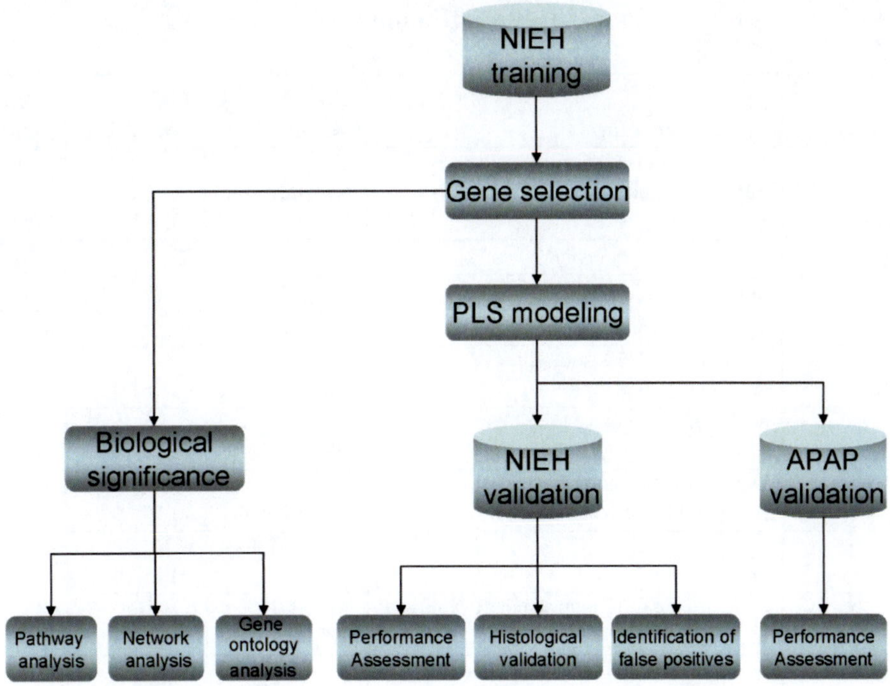

Figure 1. The workflow of model development and evaluation.

Table 2. Top 40 genes relevant to liver injury determined by the PLS weight coefficient using a leave-one compound-out cross-validation strategy

Gene Symbol	Entrez Gene Name	Ref Seq	PLS Coefficient	Correlation with ALT
RGN	regucalcin (senescence marker protein-30)	NM_031546	-0.26	0.77
SLCO1B2	solute carrier organic anion transporter family, member 1B3	NM_031650	-0.21	0.81
SLC10A1	solute carrier family 10 (sodium/bile acid cotransporter family), member 1	NM_017047	-0.21	0.80
RGD1307603_ PREDICTED	cDNA sequence BC021614	XM_215189	-0.20	0.77
SLC27A5	solute carrier family 27 (fatty acid transporter), member 5	NM_024143	-0.20	0.82
CES3	carboxylesterase 1 (monocyte/macrophage serine esterase 1)	NM_133295	-0.20	0.77
ABAT	4-aminobutyrate aminotransferase	NM_031003	-0.19	0.84
CYP2C	cytochrome P450, family 2, subfamily C, polypeptide 9	NM_019184	-0.19	0.80
OAT	ornithine aminotransferase (gyrate atrophy)	NM_022521	-0.19	0.63

Table 2. (Continued)

Gene Symbol	Entrez Gene Name	Ref Seq	PLS Coefficient	Correlation with ALT
SERPINA6	serpin peptidase inhibitor, clade A (alpha-1 antiproteinase, antitrypsin), member 6	NM_0010096 63	-0.18	0.77
NDRG2	NDRG family member 2	NM_133583	-0.17	0.81
DPYS	dihydropyrimidinase	NM_031705	-0.17	0.84
ALDOA	aldolase A, fructose-bisphosphate	NM_012495	0.16	-0.72
CYP2A2	cytochrome P450, subfamily 2A, polypeptide 1	NM_012693	-0.16	0.86
ACSL1	acyl-CoA synthetase long-chain family member 1	NM_012820	-0.16	0.84
KMO	kynurenine 3-monooxygenase (kynurenine 3-hydroxylase)	NM_021593	-0.16	0.84
PON3	paraoxonase 3	NM_0010040 86	-0.15	0.78
DHRS3	dehydrogenase/reductase (SDR family) member 3	NM_0010371 99	-0.15	0.76
BAAT	bile acid Coenzyme A: amino acid N-acyltransferase (glycine N-choloyltransferase)	NM_017300	-0.14	0.79
CDO1	cysteine dioxygenase, type I	NM_052809	-0.14	0.81
PGCP	plasma glutamate carboxypeptidase	NM_031640	-0.14	0.81
FTCD	formiminotransferase cyclodeaminase	NM_053567	-0.14	0.83
LIPC	lipase, hepatic	NM_012597	-0.13	0.80
HSD3B	hydroxy-delta-5-steroid dehydrogenase, 3 beta- and steroid delta-isomerase 5	NM_012584	-0.13	0.77
HADH	hydroxyacyl-Coenzyme A dehydrogenase	NM_057186	-0.13	0.82
CFL1	cofilin 1 (non-muscle)	NM_017147	0.12	-0.84
HMGCS2	3-hydroxy-3-methylglutaryl-Coenzyme A synthase 2 (mitochondrial)	NM_173094	-0.12	0.77
GNMT	glycine N-methyltransferase	NM_017084	-0.12	0.73
CYP27A1	cytochrome P450, family 27, subfamily A, polypeptide 1	NM_178847	-0.12	0.82
SDC2	syndecan 2	NM_013082	-0.12	0.80
MAOB	monoamine oxidase B	NM_013198	-0.12	0.79
PXMP2	peroxisomal membrane protein 2, 22kDa	NM_031587	-0.12	0.80
SELENBP1	selenium binding protein 1	NM_080892	-0.12	0.77
NNT	nicotinamide nucleotide transhydrogenase	NM_0010131 57	-0.12	0.65
ST3GAL6	ST3 beta-galactoside alpha-2,3-sialyltransferase 6	NM_207602	-0.12	0.71
AQP9	aquaporin 9	NM_022960	-0.11	0.78
F10	coagulation factor X	NM_017143	-0.11	0.82

Gene Symbol	Entrez Gene Name	Ref Seq	PLS Coefficient	Correlation with ALT
RGD1305679	chromosome 6 open reading frame 105	NM_0010141 44	-0.11	0.85
CYP2C23	cytochrome P450, family 2, subfamily c, polypeptide 44	NM_031839	-0.10	0.78
SDCBP	syndecan binding protein (syntenin)	NM_031986	0.10	-0.79

To interpret the biological significances, all of the selected 40 genes were mapped in the Ingenuity Pathway Analysis (IPA) knowledge base (www.ingenuity.com) [31], and twelve significant canonical functional pathways were identified and depicted in Figure 2, including fatty acid metabolism, tryptophan metabolism, butanoate metabolism, bile acids biosynthesis, and hepatic cholestasis. Moreover, these selected rat genes were converted into the human genes using the ID converter in ArrayTrack [20] and then imported into our in-house developed network analysis tool, called atBioNet [32], as the seed genes to identify biological meaningful modules (clusters). Figure 3A shows the overall network formed by the 40 selected genes, and the network clustering algorithm SCAN was employed to identify the biological modules inside the network [32]. Only two modules containing at least 3 seed genes were kept as shown in Figure 3-B&C. The genes involved in these two modules were mapped to KEGG pathways as listed in Table 3. In the first module, the significantly-enriched pathways, such as synthesis and degradation of ketone bodies, butanoate metabolism, and valine, leucine, and isoleucine degradation are all related to mitochondrial dysfunction.

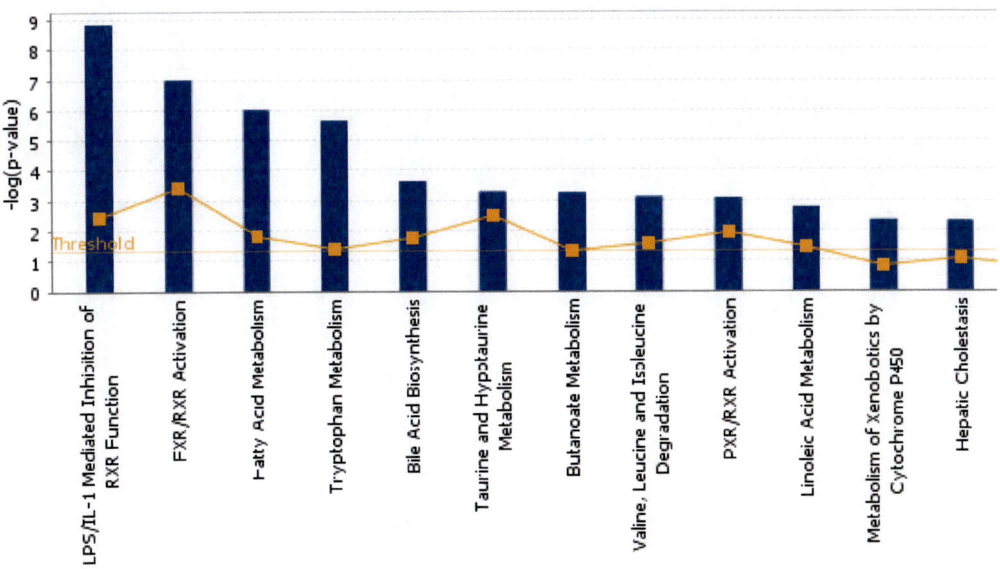

Figure 2. The most significant canonical pathways analysis identified from Ingenuity Pathway Analysis tool.

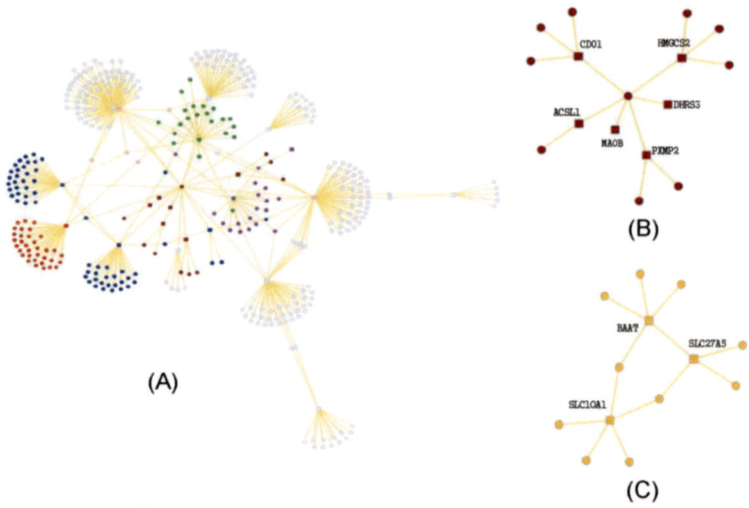

Figure 3. (A) The extracted protein-protein interaction network using the selected 40 genes as seeds. A square represents the seed gene and a dot represents the added genes; (B-C) the two selected modules (cluster) identified by atBionet with at least 3 seed genes.

Figure 4. The top gene ontology path in the molecular component identified by GOFFA, a tool developed by the US FDA.

Meanwhile, the pathways involved in the second module like bile secretion, primary bile acids biosynthesis, and ABC transporters were related to cholestasis. Finally, the gene ontology analysis was further used to confirm the findings through the network analysis, and the most significant gene ontology path in the molecular component identified by GOFFA (a gene ontology tool developed by the FDA [33]) were specifically linked to the mitochondrial inner membrane as shown in Figure 4. Clearly, these selected genes are connected with mitochondrial dysfunction and cholestasis, two important mechanisms involved in liver injury induced by chemicals.

Table 3. Pathway analysis of two modules with at least three seed genes identified by atBioNet

Module	Map Title in KEGG	# of genes mapped in the pathway	Fisher p value
Mitochondrial dysfunction	Synthesis and degradation of ketone bodies (hsa00072)	3	0.00001
	Peroxisome (hsa04146)	4	0.0003
	Butanoate metabolism (hsa00650)	3	0.00005
	Valine, leucine and isoleucine degradation (hsa00280)	3	0.00016
	Taurine and hypotaurine metabolism (hsa00430)	2	0.00026
	Metabolic pathways (hsa01100)	9	0.00051
Cholastasis	Bile secretion (hsa04976)	6	1.0E-8
	Primary bile acid biosynthesis	3	0.000002
	PPAR signaling pathway (hsa03320)	3	0.0002
	ABC transporters (hsa02010)	2	0.0022
	Peroxisome (hsa04146)	2	0.007

Development of a Multi-gene Model for Detecting Liver Injury

Two hundred thirty-four samples treated by five hepatotoxins in the NIEHS training set were used for model development. As mentioned above, a leave-one compound-out cross-validation was used to determine the number of PLS components. After checking PRESS values from cross-validation against the number of PLS components involved in the model, we selected three components for the final PLS model with the lowest PRESS values.

We employed another leave-one compound-out cross-validation to assess the prediction performance of the developed model. As indicated in Figure 5A, the Pearson correlation between the model prediction and the ALT measurement is as high as 0.92 for the training set, suggesting that the gene expression-based association correlated with the ALT measurement. The predictive performance of this statistical association was hardly affected when regulated genes were removed from computations, suggesting the robustness of the relationship. The developed PLS model was further assessed using the NIEHS test set, and the Pearson correlation between the model prediction and the ALT readout is 0.86 as shown in Figure 5B. The correlation of the model prediction and the ALT measurement is comparable between those obtained from cross-validation and the test set.

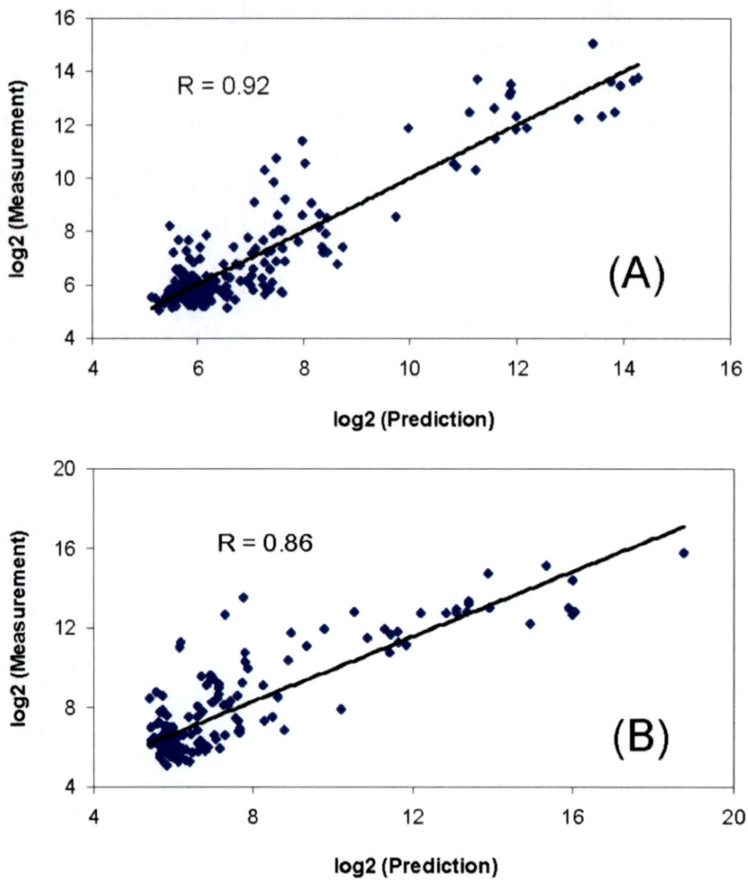

Figure 5. The correlation between the model prediction and the ALT readout. (A) cross-validation in the NIEHS training set; (B) external assessment in the NIEHS test set.

We further assessed the model prediction using the liver pathological information [34]. In the present study, the severity of hepatocyte necrosis for each animal was scored as "none", "minimal", "mild", "moderate", or "marked" by pathologists [17]. The necrosis scoring was converted to a binary variable depending on the presence of necrosis, and ROC curves were used to evaluate the prediction of necrosis presence detected by the measured ALT or the model prediction.

As shown in Figure 6, the model prediction has a higher AUC value than the ALT measurement (0.94 versus 0.88), suggesting that the developed model correlated well with liver injury. To confirm this finding, we further scrutinized the ALT measurement and model predictions of samples among the tested compounds and found that the model prediction identified liver injury earlier in some compounds. For example, the model prediction can identify significant elevation (i.e., > 2*ULN) 6 hours after treatment of n-nitrosomorpholine for both moderate and high doses, while the measured ALT failed until 24 hours after treatment as shown in Table 4.

Table 4. The ALT readout and model prediction in the animals treated by n-nitrosomorpholine at 6 and 24 hours after treatment. The bolded, italicized numbers indicate that the measured ALT or model prediction were two times greater than the ULN

Compound	Dose (mg/kg)	Duration (hours)	Model prediction	ALT measurement
	10	6	77	42
	10	6	68	36
	10	6	61	40
	10	6	71	37
	50	6	***192***	53
	50	6	***132***	55
	50	6	***152***	49
	50	6	***150***	76
	300	6	***164***	58
	300	6	***154***	50
	300	6	***171***	67
N-nitrosomorpholine	300	6	128	64
	10	24	70	56
	10	24	90	49
	10	24	74	52
	10	24	94	46
	50	24	***329***	***150***
	50	24	***343***	***361***
	50	24	***346***	***152***
	50	24	***321***	***173***
	300	24	***1794***	***1520***
	300	24	***2387***	***1260***
	300	24	***849***	***376***
	300	24	***1858***	***1400***

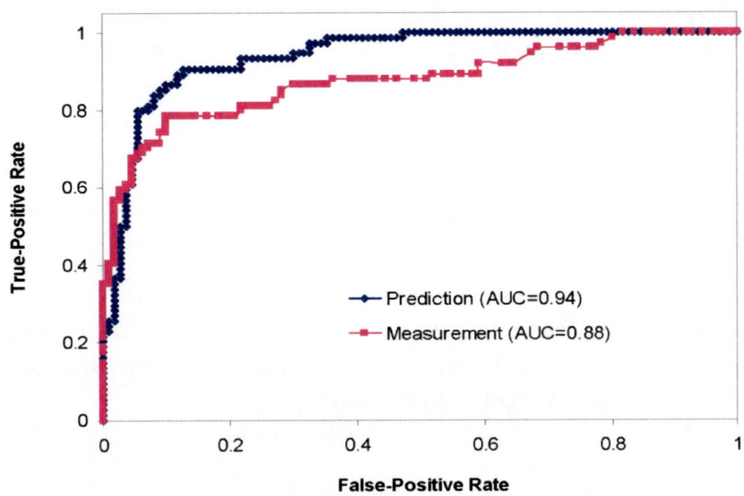

Figure 6. Receiver operating characteristic (ROC) curves of model prediction (blue) and the measured ALT (pink). AUC: area under the curve.

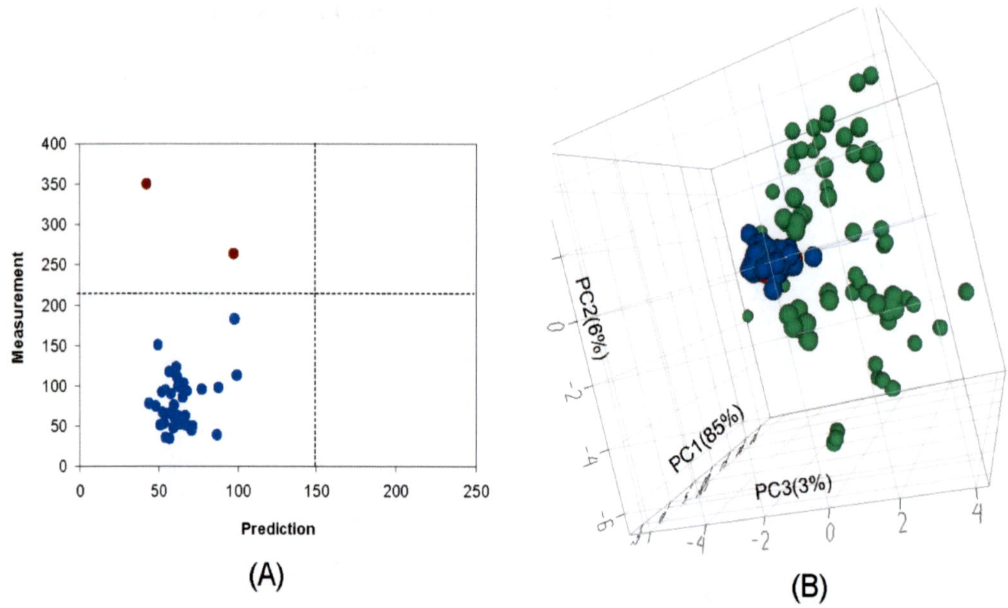

Figure 7. (A) The false positive tests based on the measured ALT. The dash line represents the threshold of >2*ULN (upper limits of normal). (B) Plot of the first 3 principle components (PCs) based on the 40 selected genes, and the total variances explained by the first 3 PCs exceed 90%. Red: false positives tests of the measured ALT in the controls; Blue: true negatives tests of the measured ALT in the controls; Green: true positives tests of the measured ALT in the treated groups.

As mentioned before, muscle injury can cause abnormal ALT elevation (i.e., false positives of liver injury). After examining 36 control samples with vehicle treatment administered by intraperitoneal injection in the NIEHS test set, at least two animals showed abnormal ALT elevation (i.e., > 2*ULN; Figure 7A). However, the histological examination did not identify any liver-related injury, suggesting possible false positive tests from the ALT measurement. In contrast, none of the model predictions suggested hepatic injury in these two animals, and the principle component analysis of the gene expression profiles also indicated that these two animals did not show any significant changes as compared to the other controls (see Figure 7B). Taken together, these results suggested that the multi-gene model might have better specificity than the measured ALT.

Independent Validation by the APAP Dataset

We applied the developed multi-gene model to the APAP dataset with 152 samples as an independent validation. As shown in Figure 8, the correlation between the model prediction and the ALT measurement was high and the Pearson correlation is 0.89, further demonstrating the robustness of the model we developed.

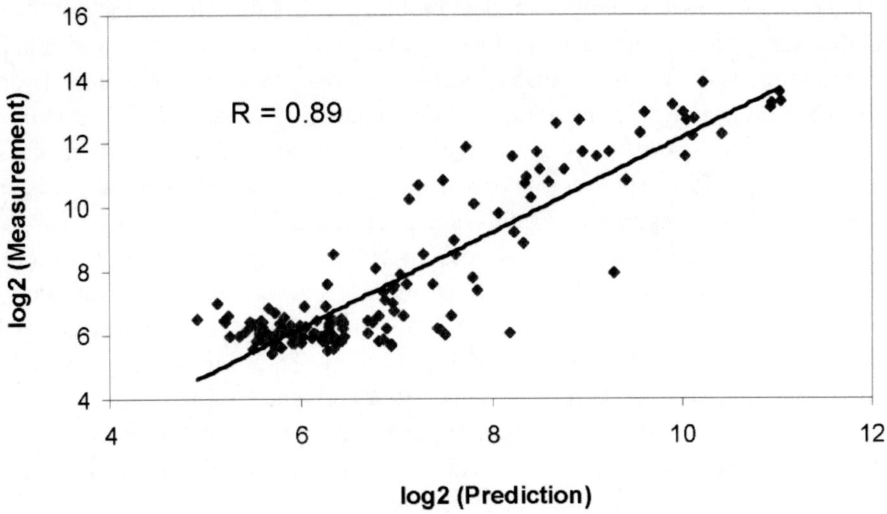

Figure 8. The correlation between model prediction and the ALT readout in the APAP dataset.

Conclusion

Phenotypic anchoring is an important approach to extract knowledge from the huge amount of gene expression data associated with toxicogenomics studies. In this chapter, we have developed a multi-gene predictor to anchor the ALT activity in rats, a commonly used serum biomarker of hepatotoxicity. Published gene expression data from the rats treated by a compendium of eight hepatotoxins were used to train and test the proposed model, and another APAP study was used as an independent validation. Our study indicated that the developed multi-gene model has successfully anchored the phenotype of liver injury, and the results seem to be robust, as shown by through the external validation of the NIEHS dataset and independent validation of the APAP dataset. Recently, our group reported that toxicogenomics carried out in rats did not perform better then serum biomarkers to predict drug-induced liver injury in humans [35]. However, when toxicogenomics is applied for the prediction of liver injury within the same species, it performed better than ALT measurements alone, suggesting an important advantage in preclinical testing strategies.

The correlation between the ALT elevation and model prediction was used to check the validity of phenotype anchoring of the proposed model. The Pearson correlations are 0.92 for the cross-validation in the training set, 0.86 for the NIEHS test set, and 0.89 for the APAP dataset as an independent validation, suggesting the robustness of the multi-gene model as compared to the ALT measurement. It is interesting to note that more samples fell above the linear trend line in both the NIEHS and APAP studies, indicating that the readout of ALT was overall higher than the model prediction. The validity of the model prediction was further confirmed by a comparison with the pathological data. The AUC of necrosis from the model prediction and ALT measurement based on the NIEHS validation set were 0.94 and 0.88, respectively, revealing that the model prediction was comparable and even slightly better indicator of necrosis, which serves as a gold standard for the diagnosis of liver injury in rats.

The identified gene set used by the model provided insightful information and shed some light on DILI mechanisms, such as mitochondrial toxicity and cholestasis as shown in Figures 2-4. Mitochondrial dysfunction is a major mechanism whereby drugs can induce liver injury due to its critical role in respirationation to supply energy, glucose and fat metabolism, and generation of reactive oxygen species. Many liver-related side effects were caused by mitochondrial dysfunction, such as microvesicular steatosis, lethal lactic acidosis, hepatic necrosis and/or apoptosis, cytolytic hepatitis, and liver failure.

Fatty acids are one of the main sources of cellular energy. Inhibition of fatty acid oxidation is expected to cause a distinct effect on ATP concentrations in liver slice studies [36], since hepatocytes are highly reliant on ATP for ureagenesis, gluconeogenesis, and many other metabolic processes [37]. Not surprisingly, many genes involved in the fatty acid metabolism were affected, like ACSL1, HADH, SLC27A5, and CES3.

Glutathione is one of the most important molecules in the detoxification of reactive metabolites or reactive oxygen species, providing protection as a free radical scavenger [38]. Glutathione is used in the biosynthesis of glutathione conjugates of foreign compounds, such as reactive metabolites [39]. Endogenous glutathione is exhausted if its biosynthesis cannot be maintained, which can lead to a decrease of the covalent binding capacity of the toxic metabolites and result in hepatic damage. Three genes (GNMT, NNT, and CDO1) in our list are involved in the biosynthesis of cysteine and the conversion of cysteine into glutathione. Cysteine levels are maintained within a narrow range in order to regulate the biosynthesis of glutathione in rat livers [40], and it is known that cysteine catabolism is tightly regulated via regulation of cysteine dioxygenase (CDO) levels in the liver. S-adenosylmethionine is the primary source of the sulfur atom in glutathione, and GNMT and NNT are two important genes involved in the trans-sulfuration pathway from S-adenosylmethionine to glutathione [41]. Besides those genes, others within the antioxidant system, such as catalase, was also changed by the hepatotoxins.

We observed that many genes involved in the drug metabolism processes were changed, such as phase I cytochrome P450 genes (CYP2A2, CYP2C, CYP2C23, and CYP27A1). The regulation of these P450 genes may be related to the findings that the hepatotoxins form reactive intermediates capable of irreversibly inhibiting some of the P450 enzymes [42].

Cholestasis is another important mechanism for DILI. Drug-induced cholestasis not only impairs the bile duct, but also mediates the excretion of toxic metabolites and the uptake of bile acids and bilirubin. The impairment of bile secretion results in cholestasis (including intrahepatic and extrahepatic cholestasis) and causes accumulation of bile acids in the blood and within hepatocytes [43]. In recent years, it has also been found that bile acids induce apoptosis and necrosis by damaging the mitochondria when the concentration of bile acids exceed the binding capacity of the binding protein located in the cytosol of the hepatocyte [44]. Several genes (SLCO01B2, SLC10A1, SLC27A5, BAAT, and AQP9) in the selected gene list are involved in bile acids synthesis and transport. BAAT and SLC27A5 are involved in primary bile acids biosynthesis. AQP9 and SLC10a1 are involved in hepatic bile acids and bile salt transport, while SLCO1B2 is related to organic anion transport.

Cumulative evidence suggests that calcium homeostasis plays an important role in lethal cell injury [45]. Calcium ions have been proposed as mediators of the functional consequences of plasma membrane injury, but an influx of extracellular calcium ions across a damaged permeability barrier and steep concentration gradient may convert potentially reversible injury into irreversible injury [45, 46]. Three genes (RGN, F10, and PON3) in our list were related to calcium homeostasis. RGN plays an important role in the calcium homeostasis [47] by activating the Ca^{2+} pump enzymes in the plasma membrane, microsomes, and mitochondria. RGN also had a suppressive effect on Ca^{2+} signaling from the cytoplasm to the nucleus in proliferative cells [48]. Moreover, reduced liver RGN mRNA levels and elevated serum RGN concentrations have been observed in liver injury from galactosamine administration in rats [49], which is consistent with the findings in current studies.

In summary, the proposed phenotypic anchoring model has been developed using a set of regulated genes involved in diverse biological processes related to liver injury induced by chemicals. In addition, the model has shown the potential to enhance the specificity of the ALT measurement for detecting DILI and improve correlation with histological examination. The methodologies applied in this chapter have the potential to be used to improve the identification of DILI liability with toxicogenomics data in pre-clinical studies.

Acknowledgments

We thank Don Ding for his careful English editing. Jürgen Borlak receives funding from the German Federal Ministry for Education and Research as part of the Virtual Liver Network initiative (Grant number 031 6154). Furthermore, Jürgen Borlak is a recipient of an ORISE Stipend of the FDA which is gratefully acknowledged.

References

[1] Afshari CA, Nuwaysir EF, Barrett JC. Application of complementary DNA microarray technology to carcinogen identification, toxicology, and drug safety evaluation. *Cancer Res.* 1999;59:4759-4760.

[2] Chen M, Zhang M, Borlak J, Tong W. A decade of toxicogenomic research and its contribution to toxicological science. *Toxicol. Sci.* 2012;130:217-228.

[3] Hamadeh HK, Knight BL, Haugen AC, Sieber S, Amin RP, Bushel PR et al. Methapyrilene toxicity: anchorage of pathologic observations to gene expression alterations. *Toxicol. Pathol.* 2002;30:470-482.

[4] Shi W, Bessarabova M, Dosymbekov D, Dezso Z, Nikolskaya T, Dudoladova M et al. Functional analysis of multiple genomic signatures demonstrates that classification algorithms choose phenotype-related genes. *Pharmacogenomics J.* 2010;10:310-323.

[5] Paules R. Phenotypic anchoring: linking cause and effect. *Environ. Health Perspect* 2003;111:A338-339.

[6] Chen M, Borlak J, Tong W. High lipophilicity and high daily dose of oral medications are associated with significant risk for drug-induced liver injury. *Hepatology* 2013;58:388-396.

[7] Lasser KE, Allen PD, Woolhandler SJ, Himmelstein DU, Wolfe SM, Bor DH. Timing of new black box warnings and withdrawals for prescription medications. *JAMA* 2002;287:2215-2220.

[8] Wysowski DK, Swartz L. Adverse drug event surveillance and drug withdrawals in the United States, 1969-2002: the importance of reporting suspected reactions. *Archives of internal medicine* 2005;165:1363.

[9] Chen M, Vijay V, Shi Q, Liu Z, Fang H, Tong W. FDA-approved drug labeling for the study of drug-induced liver injury. *Drug Discov. Today* 2011;16:697-703.

[10] Ostapowicz G, Fontana RJ, Schiodt FV, Larson A, Davern TJ, Han SH, McCashland TM, et al. Results of a prospective study of acute liver failure at 17 tertiary care centers in the United States. *Ann. Intern. Med.* 2002;137:947-954.

[11] Chen M, Zhang J, Wang Y, Liu Z, Kelly R, Zhou G, Fang H, et al. Liver Toxicity Knowledge Base (LTKB) – A Systems Approach to a Complex Endpoint. *Clinical Pharmacology & Therapeutics* 2013;95:409-412.

[12] Navarro VJ, Senior JR. Drug-related hepatotoxicity. *N. Engl. J. Med.* 2006;354:731-739.

[13] Kaplowitz N. Idiosyncratic drug hepatotoxicity. *Nat. Rev. Drug Discov.* 2005;4:489-499.

[14] Ozer J, Ratner M, Shaw M, Bailey W, Schomaker S. The current state of serum biomarkers of hepatotoxicity. *Toxicology* 2008;245:194-205.

[15] Senior JR. Monitoring for hepatotoxicity: what is the predictive value of liver "function" tests? *Clin. Pharmacol. Ther.* 2009;85:331-334.

[16] Nathwani RA, Pais S, Reynolds TB, Kaplowitz N. Serum alanine aminotransferase in skeletal muscle diseases. *Hepatology* 2005;41:380-382.

[17] Lobenhofer EK, Auman JT, Blackshear PE, Boorman GA, Bushel PR, Cunningham ML et al. Gene expression response in target organ and whole blood varies as a function of target organ injury phenotype. *Genome Biol.* 2008;9:R100.

[18] Boorman GA, Blackshear PE, Parker JS, Lobenhofer EK, Malarkey DE, Vallant MK et al. Hepatic gene expression changes throughout the day in the Fischer rat: implications for toxicogenomic experiments. *Toxicological Sciences* 2005;86:185-193.

[19] Huang J, Shi W, Zhang J, Chou JW, Paules RS, Gerrish K et al. Genomic indicators in the blood predict drug-induced liver injury. *Pharmacogenomics J.* 2010;10:267-277.

[20] Fang H, Harris SC, Su Z, Chen M, Qian F, Shi L et al. ArrayTrack: an FDA and public genomic tool. *Methods Mol. Biol.* 2009;563:379.

[21] Fan X, Lobenhofer EK, Chen M, Shi W, Huang J, Luo J et al. Consistency of predictive signature genes and classifiers generated using different microarray platforms. *Pharmacogenomics J.* 2010;10:247-257.

[22] Chen HY, Yu SL, Chen CH, Chang GC, Chen CY, Yuan A et al. A five-gene signature and clinical outcome in non-small-cell lung cancer. *N. Engl. J. Med.* 2007;356:11-20.

[23] Hastie T, Tibshirani R, Eisen MB, Alizadeh A, Levy R, Staudt L et al. 'Gene shaving' as a method for identifying distinct sets of genes with similar expression patterns. *Genome Biol.* 2000;1(2): 1-0003.

[24] Haaland DM, Thomas EV. Partial least-squares methods for spectral analyses. 1. Relation to other quantitative calibration methods and the extraction of qualitative information. *Analytical Chemistry* 1988;60:1193-1202.

[25] Mevik Br-H, Wehrens R. The pls package: principal component and partial least squares regression in R. *J. Stat. Softw.* 2007;18:1-24.

[26] Boulesteix AL. PLS dimension reduction for classification with microarray data. *Stat. Appl. Genet. Mol. Biol.* 2004;3(1):1075.

[27] Chen M, Shi L, Kelly R, Perkins R, Fang H, Tong W. Selecting a single model or combining multiple models for microarray-based classifier development? - A comparative analysis based on large and diverse datasets generated from the MAQC-II project. *BMC bioinformatics* 2011;12:S3.

[28] Simon R. Roadmap for developing and validating therapeutically relevant genomic classifiers. *J. Clin. Oncol.* 2005;23:7332-7341.

[29] Greiner M, Pfeiffer D, Smith RD. Principles and practical application of the receiver-operating characteristic analysis for diagnostic tests. *Preventive veterinary medicine* 2000;45:23-41.

[30] Sing T, Sander O, Beerenwinkel N, Lengauer T. ROCR: visualizing classifier performance in R. *Bioinformatics* 2005;21:3940-3941.

[31] Gerecke DR, Chen M, Isukapalli SS, Gordon MK, Chang Y-C, Tong W et al. Differential gene expression profiling of mouse skin after sulfur mustard exposure: extended time response and inhibitor effect. *Toxicol. Appl. Pharmacol.* 2009;234:156-165.

[32] Ding Y, Chen M, Liu Z, Ding D, Ye Y, Zhang M, Kelly R, et al. atBioNet - an integrated network analysis tool for genomics and biomarker discovery. *BMC Genomics* 2012;13:325.

[33] Sun H, Fang H, Chen T, Perkins R, Tong W. GOFFA: Gene Ontology For Functional Analysis - A FDA Gene Ontology tool for analysis of genomic and proteomic data. *BMC Bioinformatics* 2006;7:S23.

[34] Cullen JM, Miller RT. The role of pathology in the identification of drug-induced hepatic toxicity. *Expert Opinion on Drug Metabolism & Toxicology* 2006;2:241-247.

[35] Zhang M, Chen M, Tong W. Is toxicogenomics a more reliable and sensitive biomarker than conventional indicators from rats to predict drug-induced liver injury in humans? *Chem. Res. Toxicol.* 2012;25:122-129.

[36] Mortishire-Smith RJ, Skiles GL, Lawrence JW, Spence S, Nicholls AW, Johnson BA et al. Use of metabonomics to identify impaired fatty acid metabolism as the mechanism of a drug-induced toxicity. *Chem. Res. Toxicol.* 2004;17:165-173.

[37] Jaeschke H, Gores GJ, Cederbaum AI, Hinson JA, Pessayre D, Lemasters JJ. Mechanisms of hepatotoxicity. *Toxicol. Sci.* 2002;65:166-176.

[38] Mutlib AE, Shockcor J, Espina R, Graciani N, Du A, Gan LS. Disposition of glutathione conjugates in rats by a novel glutamic acid pathway: characterization of unique peptide conjugates by liquid chromatography/mass spectrometry and liquid chromatography/NMR. *J. Pharmacol. Exp. Ther.* 2000;294:735-745.

[39] Meister A. Glutathione deficiency produced by inhibition of its synthesis, and its reversal; applications in research and therapy. *Pharmacol. Ther.* 1991;51:155-194.

[40] Stipanuk MH, Dominy JE, Jr., Lee JI, Coloso RM. Mammalian cysteine metabolism: new insights into regulation of cysteine metabolism. *J. Nutr.* 2006;136:1652S-1659S.

[41] Schnackenberg LK, Chen M, Sun J, Holland RD, Dragan Y, Tong W et al. Evaluations of the trans-sulfuration pathway in multiple liver toxicity studies. *Toxicol. Appl. Pharmacol.* 2009;235:25-32.

[42] Zhou S, Koh HL, Gao Y, Gong ZY, Lee EJ. Herbal bioactivation: the good, the bad and the ugly. *Life Sci.* 2004;74:935-968.

[43] Palmer RH. Bile acids, liver injury, and liver disease. *Arch. Intern. Med.* 1972;130:606-617.

[44] Palmeira CM, Rolo AP. Mitochondrially-mediated toxicity of bile acids. *Toxicology* 2004;203:1-15.
[45] Farber JL. The role of calcium in lethal cell injury. *Chem. Res. Toxicol.* 1990;3:503-508.
[46] Schanne FA, Kane AB, Young EE, Farber JL. Calcium dependence of toxic cell death: a final common pathway. *Science* 1979;206:700-702.
[47] Yamaguchi M. Role of regucalcin in maintaining cell homeostasis and function. *Int. J. Mol. Med.* 2005;15:371-389.
[48] Yamaguchi M. Role of calcium-binding protein regucalcin in regenerating rat liver. *J. Gastroenterol. Hepatol.* 1998;13 Suppl:S106-112.
[49] Isogai M, Oishi K, Yamaguchi M. Serum release of hepatic calcium-binding protein regucalcin by liver injury with galactosamine administration in rats. *Mol. Cell Biochem.* 1994;136:85-90.

In: Microarrays: Principles, Applications and Technologies
Editor: James V. Rogers
ISBN: 978-1-62948-669-7

Chapter 12

Application of Microarray Technology in Toxicology

Abir T. El-Alfy*
Department of Pharmaceutical Sciences, College of Pharmacy,
Chicago State University Chicago, IL, US

Abstract

The advent of microarray technology has revolutionized research in the pharmacogenomic and toxicogenomic fields. The strength of this technique lies within the ability to simultaneously analyze the expression of a large number of genes hence providing an invaluable research tool. Recent years have witnessed tremendous applications of microarray technology to the field of toxicology. These applications include: 1) screening chemicals for specific target toxicity, 2) determining pathways involved in the mechanism of toxicity, 3) understanding the physiology to help identify targets and pathophysiology of various disease states, and 4) identification of biomarkers of toxicity. The current book chapter provides a comprehensive overview and summary of the recent findings and advances regarding the utilization of microarray technology in the field of toxicology.

Introduction

Since the discovery of DNA structure by Watson and Crick [1], the scientific quest to fully translate the genetic code to better understand protein, cell, organ, and organismal function has been incessant. Early on it became clear that technological advances are necessitated in a myriad of areas to enable researchers to: 1) track and follow changes in gene transcription, protein translation, as well compensatory changes in protein levels, 2) be able to measure these changes in the whole genome within a feasible time frame and with the use of

* Email: ael-alfy@csu.edu.

small sample material, and 3) to fully integrate the vast data knowledge gathered, especially in a collaborative manner. The development of Southern followed by Northern blotting techniques [2, 3] highlights the first breakthrough that enabled researchers to track changes at the gene transcript level. The use of both Southern and Northern blots was short-lived because of the fast advancement to microarray chips, first reported in the mid 1990s [4-6].

The term "microarray" describes a collection of microspots arranged in an orderly (array) manner and immobilized on a solid support, a glass microscopic slide, a silicon chip, or a nylon membrane. The microspots, or probes, can be DNA, cDNA, or oligonucleotides, and are designed to hybridize with specific nucleic acid sequences. Such sequences are termed "targets" and are usually labeled with a fluorophore to allow for quantitation of the level of binding of the target and the probe. The signal measured thus reflects the expression of the gene containing the target sequence [7]. As such, DNA microarray technology provided a rapid, high-throughput method that allowed researchers to monitor alterations in genome-wide patterns of gene expression. The application of microarray technology in toxicology in the late 1990s led to the materialization and advancement of the field of toxicogenomics [8]. The potential for simultaneous analysis of expression of thousands of genes offered a great prospect in the field of toxicology particularly with a myriad of applications determining the effect of xenobiotics on biological systems as well as providing insights into the molecular pathology of disease states and its implication in drug targeting. Such advancement paved the way for the evolution of the field of toxicology from the study of dose-related effects of individual chemicals to the integration and compilation of massive experimental data to generate databases that will enhance the understanding of toxicant-induced effects. The emergence of toxicogenomics led to several national and international programs to coordinate and standardize the technological applications of DNA microarrays as well as maintain a database for the knowledge provided by various research studies in toxicogenomics. The National Center of Toxicogenomics (NCT) was established by the National Institute of Environmental Health Sciences (NIEHS) to develop a reference scheme for genome wide gene expression data, to construct a knowledge base for the effects of xenobiotics on biological systems, and to apply such knowledge for toxicological risk assessment in humans [9]. The NCT also focused its efforts on developing the Chemical Effects in Biological Systems database [10].

Databases for DNA Microarray Experiments

With a plethora of data collected from microarray experiments, researchers worldwide quickly recognized the necessity to: 1) standardize methods used to record and report microarray gene expression data and 2) to establish a set of databases to effectively utilize and compress the data into a manageable form. Capitalizing on computational algorithms and resources available for the genome sequencing project, several analytical tools were developed for the analysis of microarray expression data. Analytical tools widely used in toxicogenomics have focused on clustering algorithm, statistical comparisons of classes, mechanistic analysis, and biomarker discovery, validation, and quantification [11]. The Minimum Information about a Microarray Experiment (MIME) guidelines, established by the European Bioinformatics Institute (EBI), depicts the minimum information needed for microarray data to be analyzed, interpreted, and verified [12]. Such information included the supporting raw data, types of samples assayed, how they were assayed, the nucleotide sequences used on the array, the data file obtained, and how the results were processed. The availability of such information allowed independent verification of microarray data and made microarray data available "*in a way that makes these data usable for others*" [12]. Following the MIAME guidelines, similar ones were created for an array of high throughput technologies, including the MISFISHE (Minimum Information Specification For In Situ Hybridization and Immunohistochemistry Experiments) [13], MIAPE (Minimum Information About a Proteomics Experiments) [14], and MIAME-tox (MIAME for toxicology) [15]. Several databases were established to help standardize and manage microarray data. The Microarray Gene Expression Data (MGED) Society, founded in 1999, established submission of MIAME-compliant data to public repositories in order to allow open access to genomic data sets as well as help archiving such data [16]. A MIAME checklist was created and adopted by most scientific journals (www.mged.org/miame). Public repositories were established worldwide such as Gene Expression Omnibus (GEO) [17], Array Express [18], the Center for Information Biology Gene Expression Database (CIBEX) [19], followed by other secondary databases as GENEVESTIGATOR [20], Array Express Gene Expression Atlas [21], and ONCOMINE [22].

The application of microarray technology in the field of toxicology led to the emergence of toxicogenomics, and with it the founding of several public databases of microarray data. These databases aim at genomic applications in identification of hazards, monitoring chemical and stressor-induced alterations in physiological parameters, systems toxicology, and integration of system response data to organismal functioning. Two international consortia, EBI and the National Institutes of Health (NIH), spearheaded several initiatives to institute public databases that incorporate biological and toxicological endpoints. The Chemical Effects in Biological Systems (CEBS) and the Tox-MIAMExpress databases are examples of these initiatives.

Application of Microarray Technology in Toxicology

Since the inception of the field of toxicogenomics, the application of microarray technology to toxicology has rapidly progressed from the proof-of-principle phase to applications. The NIEHS established the NCT [9], which conducted proof-of principle experiments. It was essential to validate the hypothesis that transcript changes induced by exposure to a chemical or environmental insult can be utilized to group compounds based on similar effects and to elucidate unknown mechanisms of toxicity. Hammadeh et al. [23] studied patterns of altered gene expression induced by chemical exposure using high density microarray analysis. Using compounds belonging to different classes, results revealed that different agents belonging to a common class of compounds yielded similar patterns of gene expression profiles, while these patterns were very different from compounds belonging to a different class.

The data collected from this study was the first to support the hypothesis that microarray profiling could be used to classify compounds based on similar effects, and that such profile could provide insights into the mechanism of toxicity. The Health and Environmental Sciences Institute (HESI) has been at the forefront of the experimental validation of the utility of microarray technology in the field of toxicology since its inception in 1999.

In an unprecedented collaborative research effort between academic, governmental, and corporate scientists, the research team conducted a series of experiments that examined the transcriptional change profiles produced by exposure to prototypical toxicants [24]. The results collected confirmed that transcriptional profiles can be used to differentiate between compounds belonging to different classes and their associated toxicity. Additionally, dose-dependent changes were documented for the observed transcriptional changes. Finally, the experiments also investigated sensitivity, reproducibility, and limitations of microarray technology as well as the interpretation of such data when integrated with biological and pathological data. The studies also explored the potential for using collected data for mechanism-based risk assessment of toxicants. The experiments also unraveled the challenges and sources of variability facing this new technology. This chapter focuses on key applications of microarray technology in the field of toxicology. Table 1 lists the applications discussed.

Table 1. Key Examples of microarray technology applications in toxicology

Grouping compounds based on transcriptional changes profile
Prediction toxicology
Elucidation of pathways of toxicity mechanisms
Understanding the physiology to help identify targets and pathophysiology of various disease states
Identify, evaluate, and validate biomarkers of exposure and signature biomarkers of toxicity

Table 2. Summary of studies using microarray technology to help cluster compounds and predict class toxicity

Reference	Finding
29	Identification of genes that discriminate cytotoxic anti-inflammatory agents and DNA damaging ones
30	Microarray studies in animals can provide unique gene expression signatures that can be used to classify toxicants
23	Microarray studies can provide signature patterns for toxicant classes
31	Application of microarray data to predictive toxicology
33	Call for chemical indexing and availability of public toxicogenomics capacity for predictive toxicology
34	Application of prediction analysis of microarrays (PAM) to carcinogenic hepatotoxic compounds
35	Correlation of species-specific differences in gene expression response to chemicals and their toxicity
36	Development of a neural crest migration functional assay to screen compounds for developmental neurotoxicity

Table 3. Summary of studies using microarray technology to elucidate toxicity mechanisms pathways

Reference	Findings
39	Identification of genes involved in the mechanism of hepatotoxicity
9	Applying NCT toxicogenomics data for mechanistic toxicology
40	Application of genomics for the determination of NOTEL
41	Application of microarray analysis to elucidate the mechanism of benzene-induced leukomogenicity and hematotoxicity
42	Application of microarray analysis to elucidate toxicity of acrylamide in juvenile rat model
43	Application of microarray analysis to the hepatotoxicity of cadmium
44	Application of microarrays to the hepatoxoxic injury induced by ritodrine
45	Application of microarray technology to determine gene signature profile in archival tissues isolated from aflatoxin-exposed rats
10	Establishment of CEBS database
47	Development of Iconix and GeneLogic databases
48	Compiling toxicogenomics data into EDGE database

Grouping Compounds Based on Transcriptional Changes Profile and Predictive Toxicology

Following validation of the hypothesis that transcriptional profile changes induced by compounds or stressor exposure could be used to classify compounds, a series of microarray experiments were conducted. The amount of data generated from such experiments necessitated the concerted efforts of mathematicians, computer scientists, statisticians, and engineers alongside toxicologists to develop analytical tools for microarray gene expression data. Clustering algorithm and tree-based visualization are the most widely used analysis tools for microarray data. In a typical clustering figure, gene expression profiles are grouped based on similarity on the x- and y-axes. Individual animal gene expression profile is represented on individual columns, while each row represents a single gene in each profile. Comparison between control- and compound-treated tissues is represented by different colors. A red color typically represents a gene with increased expression in the compound-treated samples as compared to the control, while a green color indicates decreased expression. Such hierarchical clustering method was developed by Eisen et al. [25]. However, early on, common clustering analysis tools proved to be inadequate for toxicogenomic data analysis. The complexity of the data stemmed from the use of dose- and time-response series, in addition to multiple compounds used in the studies. These challenges were addressed by several research efforts. Biclustering analysis method developed by Cheng and Church [26] enabled researchers to divide the two-dimensional gene expression data into subsets of genes with common expression patterns across different sample subsets. Chou et al. [27] utilized gene expression structure to identify patterns and co-expressed genes, thus establishing the Extracting microarray gene expression Patterns and Identifying coexpressed Genes (EPIG) analytical approach. The cc-biclustering approach further enhanced the analysis, whereby biclustering could be additionally based on correlation with phenotypic response of the samples [28]. The next few paragraphs will describe a few examples of studies that employed microarray data for compound clustering and providing early predictors of toxicity.

The first application of toxicogenomics in class clustering was a study published by Burczynski et al. [29]. The studies conducted evaluated the utility of gene expression analysis to differentiate toxicants belonging to different classes with distinct mechanisms of action. Using HepG2 human cells, approximately 100 compounds were used. The cytotoxic agent cisplatin was used as a positive control. All compounds were initially evaluated for cytotoxicity using the HepG2 cells reductase activity assay following 72 h of exposure. Based on the cytotoxicity assays, the ED30 dose was selected to examine the genomic effects of the compounds at 24 h. The 24 h time point was chosen for minimal cytotoxic effect. Initial analysis of the data revealed no significant correlation between all the genes in the array used and database compounds with similar toxic effects. To ensure that lack of replication in the study did not obscure any significant correlations, the investigators obtained multiple replicates (13 individual treatments) for the DNA damaging agent, cisplatin, and two non-steroidal anti-inflammatory drugs, diflumisal, and flufenamic acid. Additionally, the non-genotoxic compound, transplatin, was used in microarray experiments to avoid non-specific cellular responses to chemical stimuli. Such experiments revealed a set of genes that consistently responded to cisplatin. These genes included p53-responsive genes (p21, GADD45, and PCNA), Fas, c-jun, and rad6, confirming the role of apoptosis in cisplatin-induced cytotoxicity. Furthermore, the study revealed that cisplatin repressed the expression of several genes of the chaperone family, primarily regulating Ca^{2+}-dependent cell death. Additional analysis of the data was performed after introducing biological and technical replications. The analysis disclosed a set of discriminators that distinguished between the database of DNA-damaging and cytotoxic anti-inflammatory agents. As such, these studies initiated further validation of transcriptional profiling for in vivo toxicity testing and predictive toxicity in the field of drug discovery.

The study by Bartosiewicz et al. [30] was among the first in vivo studies that attempted to categorize chemicals based on tissue, time, and dose-specific gene expression profile. The study used ten compounds that belong to five chemical classes: benzo(a)pyrene and 3-methylcholanthrene (polycyclic aromatic hydrocarbons), bromobenzene and carbon tetrachloride (CCl_4) (oxidative stressors), dimethylnitrosamine and ethylnitrosourea (DNA alkylators), $CdCl_2$ and $HgCl_2$ (heavy metals), and clofibrate and diethylhexylphthalate (DEHP) (peroxisome proliferators). The study focused on these compounds because of their environmental impact as well as extensive mechanistic understanding of their toxicity. The studies administered the chemicals to mice at doses determined in previous studies, followed by conducting microarray studies using liver and kidney tissues collected at various time points. The microarray chip used included 260 genes involved in phase I and II metabolism, DNA repair, inflammation, repair, and housekeeping processes. Using a 2-fold cut-off for significant change in gene expression, data analysis showed that chemicals belonging to each class produced a distinct pattern of gene expression changes correlating with their respective mechanism of toxicity. The oxidative stressors, benzene and CCl_4, significantly altered the expression of liver genes involved primarily involved in DNA damage, heat shock, oxidative stress, and immediate early and acute phase responses. Similar genes were affected in kidney tissues in addition to genes involved in phase II metabolism. The data also showed that in addition to the shared genes affected by both benzene and CCl_4, each compound altered the expression of another set of unique genes. The data confirmed that gene expression profiles could not only be used to categorize compounds, but also to fingerprint particular compounds. Genes affected by both heavy metal compounds in liver and kidney tissues were mainly concerned with phase I and II metabolism, metal binding, DNA damage response, acute phase, and heat shock. The kidney response to both metals was more pronounced than the liver response. A similar pattern of gene alterations were reported for DNA alkylators, polycyclic aromatic hydrocarbons, and peroxisome proliferators. Each class of compounds altered the expression of a set of genes unique to the class, a few genes were even distinctive to each compound within the class, and chemical-induced gene expression profile varies by time, dose, and tissue. The study also showed that proper replication using multiple animals, as well assessing each gene multiple times, led to consistent data that might be effectively used for chemical screening. One challenge this study pointed out is that typically high doses of exposure are needed to induce significant alteration in gene expression sufficient to be used as a chemical signature.

Hamadeh et al. [23] conducted a similar study using structurally unrelated compounds belonging to the same chemical classes for the purpose of obtaining a microarray expression signature profile for these chemicals. The study aimed at investigating whether compounds belonging to the same class would produce similar, yet distinct gene expression profiles. Male Sprague Dawley rats were administered one of three peroxisome proliferators, gemfibrozil (5-2[2,5-dimethylphenoxy]2-2dimethylpentanoic acid), Wyeth 14,643 (4-chloro-6[2,3-xylidino)-2-pyrimidinylthio]acetic acid, or clofibrate (ethyl-p-chlorophenoxyisobutyrate) for either 24 h, 1 week, or 2 weeks. Another group of animals was exposed to the enzyme inducer, phenobarbital. Liver tissues were processed for histopathological analysis and RNA extracted was used for microarray studies. The results of the study showed that peroxisome proliferators induced significant gene expression changes in the liver after 24 h and 2 weeks of exposure. However, the compounds did not cause significant histopathological alterations except after 2 weeks of exposure. Mapping of these genes to molecular pathways revealed that pathways altered by peroxisome proliferators included triglyceride hydrolysis, β-oxidation, and fatty acid synthesis/transport/metabolism. The data not only supported previous findings but also revealed that transcriptional changes preceded the hepatocellular hypertrophy induced by these compounds. The findings of this study also validated previous reports of phenobarbital-induced upregulation of CYP450, GST, epoxide hydrolase, and diaphorase. Additional novel genes effects were reported for phenobarbital, including downregulation of the carnitine palmitoyl transferase 1 (CPT1) gene. The data provided an explanation for the clinically observed reduction of carnitine constituents in serum of patients treated with phenobarbital. In summary, these studies produced gene expression profiles for phenobarbital and peroxisome proliferators. The application of recognition algorithms and computational analyses in these studies proved that these profiles are different, hence providing robust evidence that such profiles can be used as signature patterns for toxicant classes. The studies also supported the idea of generating databases for gene expression profiles for drugs as well as environmental toxicants. The utilization of these expression profiles to provide an "effector pathway map" for each compound was valuable in discerning cellular and molecular pathways characteristic for each chemical class. For further validation of the utility of gene expression profiles for chemical responses, Hamadeh et al. [31] expanded upon the hypothesis and applied the technology to predictive toxicology evaluating the gene expression profiles from hepatic samples dissected from animals treated with a variety of chemicals. The gene profiling experiment was conducted blindly with 23 coded samples in order to test the hypothesis in an unbiased manner. Data were analyzed using the genetic algorithm/K-nearest neighbor (GA/KNN) method, which applies the GA as a searching tool and the KNN for pattern recognition. The linear discriminant analysis (LDA) was used to recognize gene pairs that could be used to distinguish between the chemical classes by using discriminative genes reported in earlier studies [32]. Out of 23 samples tested, the method successfully predicted the class of 22, supporting the possibility of using gene profile expression data for the classification of chemicals.

The field of predictive toxicology has benefited from the technological advances in genomic technology. The technology offered a shift in the field from correlating individual chemicals to one toxicological end point to the unraveling of genome-wide changes induced by chemical exposure. For genomics technology to be successfully applied to predictive toxicology a large gene expression database is needed. The database should provide adequate chemical indexing that would support predictive inferences. The past decade has witnessed significant advances in quantitative high-throughput screening, toxicology data models and standards, chemically-indexed bioassay data, and the establishment of public toxicogenomic resources. Williams-Devane et al. [33] surveyed the state of available resources and concluded that they are inadequate for prediction of chemical safety. The deficiency lies within the lack of chemical annotation and chemical reporting standards, coming up with the conclusion that chemical indexing of public genomics databases is a crucial and invaluable step for the proper assimilation of toxicological, genomic, and chemical data into predictive toxicology. A predictive analysis approach was then utilized in a study by Uehara et al. [34] in which thioacetamide and methapyrilene were used as positive non-genotoxic hepatotoxic carcinogens and compared to six negative compounds: acetaminophen, aspirin, phenylbutazone, amiodarone, alphanaphthylisothiocyanate, and rifampicin. Using prediction analysis of microarray (PAM) and a classifier of 112 probe sets, the prediction success rate was 95%. The study also examined the validity of the classifier for 30 chemicals where the majority of noncarcinogenic samples were predicted correctly, and the PAM score revealed that chemical treatment caused a time-dependent increase in gene expression. Another study [35] assessed the role of species-specific differences in the application of microarray technology in predictive toxicology. Using rat in vitro and in vivo, as well as human in vitro systems, the study showed that species differences between human and rat hepatocyte gene expression responses to coumarin corresponded to species differences in toxicity.

Another example of the application of microarray technology for toxicity prediction is the recent study conducted by Zimmer et al. [36] that focused on the development of a functional assay to screen toxicants for potential developmental toxicity. Accordingly, the need for a better functional assay for chemical screening was the driving force behind the study. Exposure to environmental toxicants has been recognized as a major risk factor for the emergence of developmental disorders [37]. Evaluation of chemicals for developmental toxicity is currently based on animal testing as delineated by the OECD Test Guideline 426 [38]. Such tests are both costly, time and personnel demanding, as well as have exhibited species differences. Since neural crest cells are associated with approximately one third of human congenital birth defects, Zimmer et al. [36] developed a screening assay using neural crest cells. Following generation of the cells, a functional assay based on the migration of neural crest cells (MINC) was established, whereby chemicals are evaluated for their potential to disturb or inhibit such migration as a measure of their potential for neurodevelopmental toxicity. The functional assay was validated by examining lead acetate and methyl mercury chloride as positive neurodevelopmental toxicants. Using 13 independent replicate assays, both compounds significantly inhibited the migration of neural crest cells in the assay. The study tested a myriad of other chemicals ranging from triazole fungicides and valproic acid, both proven in vivo as developmental toxicants, to aspirin, acetaminophen, and mannitol that do not cause such toxicity. The MINC assay showed high sensitivity in detecting in vivo neural crest toxicants. Gene expression analysis revealed several signaling pathways involved in the process of neural crest cell migration. The study provided an important step in the evaluation of toxicants for potential neurodevelopmental hazards.

Elucidation of Toxicity Mechanisms and Pathways

Recognizing the potential contribution of the advances in genomics technology to the field of toxicology, a mission of the NCT is to effectively apply gene expression technologies to the assessment of toxicologic hazards to human health. In addition to its goal of establishing worldwide knowledge base of chemical effects in biological systems, the center embarked on further profiling chemical effects, not just at the gene level, but at the protein expression (proteomics) and metabolic (metabonomics) levels. A major goal of the NCT is to utilize such data to define gene interactions and pathways that determine the mechanistic action(s) of chemical toxicants. However, such a task proved to be far more complicated; unraveling mechanistic pathways of toxicant effects necessitated the proper understanding of the intricate biological systems. Intense research efforts contributed to the application of microarray technology to mechanistic understanding of toxicology.

Bulera and coworkers [39] used high density DNA microarray to examine gene expression profile of six established hepatotoxicants (carbon tetrachloride, thioacetamide, microcystin-LR, lipopolysaccharide, cyproterone acetate, and phenobarbital), which exert their hepatotoxicity by different mechanisms. The study examined the transcriptional effects of these compounds on livers of adult male rats. The gene expression profile changes was correlated with hepatotoxicity confirmed by both histopathological and biochemical assays. The study showed that hepatoxicants induce gene expression changes involved in tumor liver promotion, and such expression profile was similar for compounds causing comparable hepatic injury. The researchers also applied the technique to identify a blinded hepatotoxicant sample in which data were analyzed using cluster analysis and dendogram representation. However, the study did not effectively utilize the data for mapping pathways involved in the toxicity.

One of the earliest studies that applied genomics to mechanistic toxicology and extended the application to further determining the No Observed Transcriptional Effect Level (NOTEL) was conducted by Lobenhofer et al. [40]. The NOTEL is the dose of a compound or stressor that does not exert a change in gene transcription, taken as a measure of mechanistic activity. Stimulation of the hormone responsive breast cancer cell line (MCF-7) with estrogen at various concentrations was used to determine the NOTEL. The study showed that estrogen elicits transcriptional changes only at physiologically relevant (or higher) concentrations, confirming the potential use of NOTEL in regulatory toxicology.

Several studies have applied genomics into understanding toxicity mechanisms. Among these studies, the application of gene expression analysis was used to unravel the mechanism underlying benzene-induced toxicity. Benzene-induced leukomogenicity and hematotoxicity mechanism is a complex and puzzling toxicity with unclear mechanism. Examining benzene-induced toxicity in bone marrow cellularity using p53 knockout mice, microarray studies revealed several signaling pathways implicated in benzene toxicity. These pathways centered around three processes: 1) cell cycle regulation, 2) DNA repair, and 3) oxidative stress response [41]. The cell cycle regulation genes p53, p21, Rb-related genes, and elongation factor 1-delta were all up-regulated, while cyclin D1 and Bim B genes were down-regulated. Such changes explain the cell cycle suppression observed in bone marrow tissues in response to benzene exposure. The second pathway affected was the DNA repair process. Genes encoding the ADP-ribosylation factor like protein 1 and Rad51 were upregulated. Among the oxidative stress-related genes altered expression by benzene exposure was the upregulation of mitogen-activated protein kinase 2, GST-1, UDP-glucuronosyl transferase, C3h-dioxin-inducible cytosolic aldehyde dehydrogenase-3, lactate dehydrogenase, and cytochorme c oxidase Vb. These genes are involved in stress and inflammation response as well as in the scavenging of reactive oxygen species. This study elucidated the mechanism by which benzene induces bone marrow cell proliferation suppression.

Acrylamide is a type-2 alkene monomer with established human neurotoxic effects. While the primary source of human exposure to acrylamide is occupational, other exposure sources include food, drinking water, and smoking. To better understand the mechanism underlying acrylamide neurotoxicity, neurobehavioral assays coupled with transcriptional profiling analysis were conducted to assess both behavioral and gene expression effects induced by acrylamide neurotoxicity in rats when administered during early postnatal life [42]. Acrylamide administration in rat pups induced significant characteristic neurotoxic symptoms including increased heel splay, decrease in grip strength, and decrease in locomotor activity. Transcriptome analysis with the Affymetrix Rat Genome 230 2.0 array indicated that acrylamide treatment caused a significant alteration in the expression of a few genes that are involved in muscle contraction, pain, and dopaminergic neuronal pathways. First, expression of the Mylpf gene involved in muscle contraction was downregulated in the spinal cord in response to acrylamide. Second, in sciatic nerves, acrylamide repressed the expression of the opioid receptor gene Oprk1 that is known to play a role in neuropathic pain regulation. Finally, in the cerebellum, acrylamide treatment caused a decrease in the expression of the nuclear receptor gene Nr4a2 that is required for development of dopaminergic neurons. Thus, the work examining the effect of acrylamide at the whole-genome level on a developmental mammalian model has identified a few genes previously not implicated in acrylamide neurotoxicity that might be further developed into biomarkers for assessing the risk of acrylamide exposure.

Elucidation of the mechanism of xenobiotic-induced hepatotoxicty has long been a major interest of scientific investigation. The advent of genomics techniques has provided an opportunity to expand the research query of molecular mechanisms of chemically-induced hepatotoxicity. Application of DNA microarray to the hepatotoxic effect of cadmium [43] demonstrated that cadmium induces the expression of 2 genes only, the zinc finger and BTB domain containing 4 (Zbrb4) and replication protein A2 (Rpa2), while 15 genes were significantly down-regulated. Serum amyloid A 2 (Saa2) and eyes absent 1 homolog (Eya1) genes showed the lowest level of expression in response to cadmium. However, the study did not elaborate on potential pathways that might be involved in cadmium toxicity. Future studies are still needed to elucidate a better mechanistic understanding of hepatotoxicity. A recent study showed that the acute phase reactant serum amyloid A (SAA) significantly increased following injection with ritodrine, a tocolytic agent used in preterm labor, and commonly associated with liver damage [44]. Other known hepatotoxic agents as acetaminophen, metformin, and valproic acid did not alter the expression of SAA. Thus, the study suggested that SAA elevation might be used as an early biomarker of hepatic injury caused by ritodrine. Using rat archival tissues, Merrick et al. [45] conducted a unique retrospective study that evaluated the use of qPCR for uncovering of genomic changes revealed by microarray data conducted on tissues recovered from rats exposed to the liver carcinogen aflatoxin B1. The study tested the accuracy of gene signatures using archived tissue samples as compared to fresh frozen tissues. Samples included liver, kidney, and lung tissues. Hepatic microarray profiling was first conducted in rats exposed to aflatoxin B1 subacutely and subchronically; the obtained transcript profile was then used to derive computational models to be used for hepatocarcinogen classification. Clustering probes resulted in a subset of genes overexpressed in different malignancies. The genes cluster signature chosen for comparison represented cell cycle progression, response to DNA damage, as well as detoxification processes. The study provided evidence that RNA isolated form archival tissue samples can be used for retrospective evaluation of gene signatures, and can thus be employed to determine molecular events underlying toxicity.

Much attention has focused on delineating molecular mechanisms of hepatic toxicity and such interest stems from the fact that a primary reason for drugs failing late stages of clinical trials is liver injury. Thus, it is crucial to develop tools to recognize liver toxicity issues in an early stage. The National Center for Toxicological Research (NCTR) within the FDA developed a liver toxicity knowledge database (LTKB) [46]. The database compiles data including histopathology, metabolism, drug targets, therapeutic uses, adverse effects, gene signatures, and liver specific biomarkers. Over the past decade, several databases have been developed to help integrate the massive data collected to enable for cross species extrapolation and the combination of gene annotation with mechanistic information. Iconix and GeneLogic provided databases for molecular profiles of chemical perturbations [47] and established models after analyzing databases for thousands of gene expression profiles induced by reference compounds. The NCT studied the expression profiles of various toxicants and made the data publically available in the CEBS database [10]. A major feature of CEBS is that the data generated from toxicogenomics experiments are integrated with other toxicological data. DrugMatrix was another toxicology database acquired by the National Toxicology Program at the NIEHS. The database along with the toxicogenomic analysis software ToxFX helped establish the application of toxicogenomics into risk characterization. The database contains toxicology profiles accumulated for approximately 640 compounds. The EDGE database provides a centralized source for toxicogenomics data collected according to a standardized protocol [48]. Other commercially available knowledge databases available include the Ingenuity Pathway Analysis (IPA) and Netherlands Organization for Applied Scientific Research's T-profile. Both databases bridge the gap between toxicogenomics data and their assessment in view of toxicological pathways, biological processes, and signaling cascades. Despite these efforts, our understanding of biological processes needs to be further advanced in order to better utilize toxicogenomics in mechanistic toxicology.

Understanding the Physiology to Help Identify Targets and Pathophysiology of Various

Disease States

Soon after the start of applying genomics technology to the field of toxicology, it was realized that such application will not be feasible without clearly understanding the mechanisms and pathways involved in toxic injuries and various disease states. Few studies have emerged that examined gene expression profiles of disease conditions [49-54]. In a review article, Farris et al. [55] summarized recent genomics findings to the field of alcohol research using quantitative trait locus (QTL) mapping was the method of choice to study alcoholism.

Table 4. Summary of studies using microarray technology for alcoholism and ethanol behavioral response

Reference	Findings
54	Expression analysis of mice strains inbred for alcohol sensitivity revealed 81 cDNA with differential expression. Genes affected are primarily involved in glycolysis. Follow up RT-PCR showed a splice variant that might account for alcohol sensitivity.
53	The study examined gene expression profiles for inbred strains of mice known to have a high alcohol preference as compared to other strains with low preference. The study identified a set of candidate genes on chromosome 9 that might be related to alcohol preference. Genes recognized include transcription regulation and mitogen activated protein kinase pathways.
51	Employing a genetics behavior approach, the researchers identified a set of candidate genes associated with complex behavioral phenotype differences between inbred stains in response to alcohol. Phenotypes investigated were anxiety, seizure susceptibility, and ethanol-induced distance traveled across a grid floor. The study recognized a subset of brain region specific genes mostly associate with the glutamatergic signaling pathway. Genetic differences were both brain region specific as well as phenotype-dependent.
49	The study compared gene expression profiles of wild type and mutant PKCgamma mice. It was been previously established the wild type PKC gamma mice develop tolerance to the sedative hypnotic effect of ethanol following chronic treatment while the mutant ones do not. Using these models, gene expression patterns and follow up conformation studies revealed that the Twik-1 gene, encoding a K-channel, was associated with tolerance and was dependent on the PKCgamma.
50	The study applied microarrays to develop expression profiles on various brain regions of different in-bred mouse strains that exhibit differences in acute ethanol response phenotypes. Multivariate analysis of various brain regions revealed basal and ethanol-induced differences. Acute ethanol-induced gene expression patterns were mostly associated with neuroplasticity pathways. Region specific gene expression profiles were further examined for a subset of signaling pathways including the Bdnf signaling in the nucleus accumbens, retinoic acid signaling in the ventral tegmental area, myelination and glucocorticoid signaling in the prefrontal cortex. Myelin gene dysregulation was of interest due to its involvement in schizophrenia as well as cocaine addiction.
52	Using an alcohol-preferring rat strain, the study examined temporal alterations of gene expression profiles in two regions of the extended amygdala following 8 weeks of binge alcohol drinking. Genes identified as significantly different included genes involved in regulation of synaptic plasticity and nerve impulse transmission, neurite development, protein transport, homo- and hetero-dimerization, and long term potentiation.

The QTL method relies on mapping the phenotype to identify alleles that significantly cosegregate with the disease state examined. Using inbred mice derived from parental strains that differ significantly in the phenotype examined, the studies typically characterize a panel of F2 progeny for the phenotype. The progeny is genotyped, mapped, and inbred to obtain recombinant inbred (RI) strains in which one carries the unique combination of the progenitor genomes. A panel of over 80 RI inbred strains is currently available for alcohol research [56]. The phenotypes derived from these lines are deposited at the GeneNetwork web resource. Table 4 provides a summary of some of the findings of microarray studies on ethanol.

Dysregulation of intracellular Ca^{2+} signaling has been recognized as a common pathway involved in various stress conditions. Defects in Ca^{2+} homeostasis leads to a disruption of endoplasmic reticulum (ER) function and hence the accumulation of misfolded and unfolded proteins in the ER lumen, leading to events that activate apoptosis pathways. Because of its implication in disease states, Vardarajan et al. [57] recently examined gene expression profile alterations induced by ER membrane reorganization induced by a myriad of chemicals, followed by connectivity mapping to predict potential ER membrane dysregulation by a series of chemicals. The studies elaborated on the role of ER reorganization as a contributor to perturbations of cellular Ca^{2+} homeostasis. The application to predictive toxicology will help screen chemicals for their effects on intracellular Ca^{2+}-induced cellular stress.

Identification, Evaluation, and Validation of Biomarkers of Exposure and Signature Biomarkers of Toxicity

The availability of toxicogenomics data raised the question whether such bioinformatics tools can be used to recognize indicators of toxicity and early markers of a toxic response, hence may be applied into biomarker safety assessment. The HESI established a committee in 1999 to explore such possibility. Initial studies focused to transcriptional profiling of three established nephrotoxicants (cispaltin, puromycin, and gentamicin). Results of the studies conducted by the Nephrotoxicity Work Group were published in a mini-monograph [58]. The three nephrotoxicants chosen exert their toxicity by different mechanisms: cisplatin is actively metabolized by the S3 proximal tubular epithelial cells into cytotoxic intermediates that cause severe tubular damage. Puromycin, on the other hand, is an aminonucleoside antibiotic that causes necrosis of glomerular podocytes. Gentamicin inhibits cell lysosomal function leading to tubular degeneration caused by phospholipidosis. Using these three toxicants, the researchers identified gene expression profiles that correlate with different types of kidney toxicity. Since renal toxicity limits the clinical uses of these agents, proper understanding of the mechanism of their toxicity, as well as early detection of such toxicity is of utmost importance, especially since the currently available biomarkers (urinary enzymes) exhibit considerable inter- and intra-individual variability.

The series of studies examined the relationship between renal gene expression changes induced by the nephrotoxicants and traditional toxicological endpoints. Among the potential molecular biomarkers identified in these studies are the genes encoding two enzymes involved the creatinine formation pathway. Amin et al. [59] showed that L-arginine-glycine amidinotransferase and guanidinoacetate methyltransferase were repressed as early as the first day of treatment with the nephrotoxicant cisplatin, and can thus serve as early markers of toxicity. If established, these enzymes could replace the serum creatinine levels of monitoring kidney status, particularly that creatinine levels are not sensitive to early stages of renal toxicity. In addition, the studies identified several other protein biomarkers, particularly kidney injury molecule-1 (KIM-1), TNF receptor, clusterin, and osteopontin. The studies confirmed that expression of these genes were upregulated in the kidney; however, follow up studies are still needed to validate the potential use of these as biomarkers of nephrotoxicity. KIM-1 is a membrane spanning protein that gets released at the extracellular surface following kidney injury. It is measured in human urine as an indicator of renal ischemic injury [60]. The TNF receptor superfamily member 12 was also significantly induced by renal injury. TNF receptors are involved in apoptotic signaling, and their upregulation are likely contributing to the nephrotoxicant-induced injury. The study also revealed that the neutrophil-associated lipocalin (LCN2) was also induced by nephrotoxicity. LCN2 functions as a carrier protein implicated in the regulation of cellular homeostasis and inducing the formation of renal epithelial cells. The final gene identified encodes the clusterin protein, which is a glycoprotein that circulates in the blood stream and its induction has been documented in various models of renal injury. Despite these studies, none of these gene expression profiles have been implemented clinically for monitoring nephrotoxicity. Osteopontin is a multifunctional protein involved in both physiological as well as pathological conditions. In addition to mineralization, osteopontin expression increases during inflammation, and the protein is involved in the recruitment of leukocytes, function, and cell survival [61]. A recent study by Wadey et al. [62] confirmed that the expression changes and urinary levels of KIM-1, osteopontin, and clusterin correlate with cisplatin-induced renal tissue injury. Despite such studies, the clinical application of these markers necessitates further elucidation and validation of these markers as well as sensitivity prior to their use.

In the field of toxicogenomics, identification of biomarkers of liver toxicity has been a major research interest. In 2004, the FDA released a whitepaper that reported a significant decrease in the number of new molecular entities since 2000 [63]. A major reason behind such decline despite the increased cost of R&D in pharmaceutical companies is the unexpected adverse effect elicited in early clinical trials. Thus, preclinical toxicological evaluation became a crucial component of the “critical path” of drug development. The liver emerged as a research target because clinical manifestation of hepatotoxicity has been a major attrition to drug development and it is established that liver exposure to drugs following treatment is higher than other organs. Animal models for hepatotoxicity using classical toxicants like acetaminophen and carbon tetrachloride, have identified a set of genes that associated with the hepatic injury. These genes include primary response cell death effects as well as secondary inflammatory responses. Waring et al. [64] used 15 different known hepatotoxins, including carbon tetrachloride, methotrexate, monocrotaline, etoposide, indomethacin, dimethylformamide, diquat, allyl alcohol, amiodarone, Aroclor 1254, arsenic, carbamazepine, diethylnitrosamine, methapyrilene, and 3-methylcholanthrene, all exerting hepatocellular injury by different mechanisms. The study was among the first to demonstrate the utility of microarray techniques for development of biomarkers of hepatotoxicity. Using a gene specific oligonucleotide microarray, Zidek et al. [65] identified 64 potential marker genes that can be used for class prediction of hepatotoxicants. A set of five genes was identified that were commonly affected by all used hepatotoxicants following 24 h of exposure. These genes include hepatic glycogen phosphorylase (carbohydrate metabolism), thioredoxin reductase (oxidative stress), flavin monooxygenase 1 and monoamine oxidase B (xenobiotic mectabolism/phase 1), and ATP binding cassette (xenobiotic metabolism/phase 2). The study also demonstrated that gene expression changes can be used to discriminate between hepatotoxic and non-hepatotoxic compounds as early as 24 h after exposure. The genes affected include phase 1 metabolism, lipid and carbohydrate metabolism, and cell proliferation genes. Similar to renal efforts, further studies are still needed to validate the clinical relevance of these markers. Table 5 provides a list of the potential biomarkers recognized so far for renal and hepatic injury.

Table 5. Summary of potential biomarkers of renal and hepatic injury identified by microarray studies

Potential Biomarkers of Renal Toxicity	Potential Biomarkers of Hepatic Toxicity
L-Arginine-glycine amidinotransferase	Glycogen phosphorylase
Guanidinoacetate methyltransferase	Thioredoxin reductase
Kidney injury molecule 1	Flavin monooxygenase 1
TNF receptor superfamily member 12	Monoamine oxidase B
Lipocalin 2	ATP-binding cassette
Clusterin	
Osteopontin	

Challenges of Toxicogenomics and its Application

There is no doubt that toxicogenomics offers great promise for risk assessment, improvement of clinical intervention, as well as drug safety evaluation. However, despite technological advancement, it seems that progression in the field of toxicogenomics is rather slow. Several challenges need to be addressed for the field to advance further. Concerns have been raised regarding the quality control and reproducibility of microarray datasets, and the inconsistency of data reported across laboratories and array platforms. Additionally, the variability in animal baseline data caused by gender, strain, fasting, state, or vehicle used for dosing complicates the analysis of datasets across studies conducted by different laboratories. As such, the MIAME guidelines are not enough to standardize array studies, and additional standards are definitely required. Problems associated with clinical application of microarray data also stem from the use of experimental animals in preclinical toxicity studies. Species difference issues have been reported by several research groups, particularly the surfacing of species specific toxicity response. Such species differences, coupled with minimal understanding of the complex biology of various physiological processes, as well as toxicity responses in different species hinder full utilization of omics data. Thus, efforts should be focused on the integration and interpretation of the complex data sets obtained across research groups to fully characterize molecular perturbations induced by disease states as well as chemical exposure. Finally, the development of high-throughput screening omics assays should of the focus of intensive research. Current microarray assays are low throughput, expensive, and sometimes use samples derived from cells or tissues exposed to chemical doses that are much higher than those encountered in human exposure scenarios. Translational research that bridges the gap between experimental animal and human data is crucial for the full implementation of microarray data for drug and chemical safety assessment.

In conclusion, the development of computational and technological tools has led to the emergence of the field of toxicogenomics. However, the full potential of the field and its application to clinical and environmental toxicology cannot be achieved without proper integration and interpretation of the data collected to the unraveling of underlying molecular pathways that mediate responses to chemical exposure as well as disease states.

References

[1] Watson JD, Crick FH. Molecular structure of nucleic acids; a structure for deoxyribose nucleic acid. *Nature* 1953;171:737-738.

[2] Southern EM. Detection of specific sequences among DNA fragments separated by gel electrophoresis. *J Mol Biol* 1975;98:503-517.

[3] Alwine, Kemp DJ, Stark GR. Method for detection of specific RNAs in agarose gels by transfer to diazobenzyloxymethyl-paper and hybridization with DNA probes. *Proc Natl Acad Sci USA* 1977;74:5350-5354.

[4] Chan V, Graves D, McKenzie SE. The biophysics of DNA hybridization with immobilized oligonucleotide probes. *Biophys J* 1995;69:2243-2255.

[5] Chee M, Yang, Hubbell, Berno A, Huang XC, Stern D et al. Accessing genetic information with high-density DNA arrays. *Science* 1996;274:610-614.

[6] Chetverin AB, Kramer FR. Oligonucleotide arrays: new concepts and possibilities. *Biotechnology* 1994;12:1093-1099.

[7] Jaluria P, Konstantopoulos K, Betenbaugh M, Shiloach J. A perspective on microarrays: current applications, pitfalls, and potential uses. *Microb Cell Fact* 2007;6:4.

[8] Afshari CA, Nuwaysir EF, Barrett JC. Application of complimentary DNA microarray technology to carcinogen identification, toxicology, and drug safety evaluation. *Cancer Res* 1999;59:4759-4760.

[9] Tennant RW. The National Center for Toxicogenomics: using new technologies to inform mechanistic toxicology. *Environ Health Perspect* 2002;110:A8-A10.

[10] Waters M, Boorman G, Cunningham M, Irwin R, Merrick A, Olden K et al. Systems toxicology and the chemical effects in biological systems (CEBS) knowledge database. *EHP Toxicogenomics* 2003;111:15-28.

[11] Afshari CA, Hamadeh HK, Busher PR. The evolution of bioinformatics in toxicology: Advancing toxicogenomics. *Toxicol Sci* 2011;10:S225-S237.

[12] Brazma A, Hingamp P, Quachkenbush J, Sherlock G, Spellman P, Stoeckert C et al. Minimum information about a microarray experiment (MIAME)-toward standards for microarray data. *Nature Genet* 2001;29:365-371.

[13] Deutsch EW, Ball CA, Bova GS, Brazma A, Bumgarner RE, Campbell D et al. Development of the Minimum Information Specification for In Situ Hybridization and Immunohistochemistry Experiments (MISFISHIE). *OMICS* 2006;10:205-208.

[14] Taylor CF, Paton NW, Garwood KL, Kirby PD, Stead DA, Yin Z et al. A systematic approach to modeling, capturing, and disseminating proteomics experimental data. *Nature Biotechnol* 2003;21:247-254.

[15] Tong W, Cao X, Harris S, Sun H, Fang H, Fuscoe J et al. ArrayTrack--supporting toxicogenomic research at the U.S. Food and Drug Administration National Center for Toxicological Research. *Environ Health Perspect* 2003;111:1819-1826.

[16] Ball CA, Sherlock G, Parkinson H, Rocca-Sera P, Brooksbank C, Causton HC et al. Standards for microarray data. *Science* 2002;298:539.

[17] Edgar R, Domrachev M, Lash AE. Gene Expression Omnibus: NCBI gene expression and hybridization array data repository. *Nucl Acids Res* 2002;30:207-210.

[18] Brazma A, Parkinson H, Sarkans U, Shojatalab M, Vilo J, Abeygunawardena N et al. ArrayExpress--a public repository for microarray gene expression data at the EBI. *Nucl Acids Res* 2003;31:68-71.

[19] Ikeo K, Ishi-i J, Tamura T, Gojobori T, Tateno Y. CIBEX: center for information biology gene expression database. *C R Biol* 2003;326:1079-1082.

[20] Zimmermann P, Hirsch-Hoffmann M, Hennig L, Gruissem W. GENEVESTIGATOR. Arabidopsis microarray database and analysis toolbox. *Plant Physiol* 2004;136:2621-2632.

[21] Parkinson H, Kapushesky M, Kolesnikov N, Rustici G, Shojatalab M, Abeygunawardena N et al. ArrayExpress update--from an archive of functional genomics experiments to the atlas of gene expression. *Nucl Acids Res* 2009;37:D868-872.

[22] Rhodes DR, Kalyana-Sundaram S, Mahavisno V, Varambally R, Yu J, Briggs BB et al. Oncomine 3.0: Genes, Pathways, and Networks in a Collection of 18,000 Cancer Gene Expression Profiles. *Neoplasia* 2007;9:166–180.

[23] Hamadeh HK, Bushel PR, Jayadev S, Martin K, DiSorbo O, Sieber S et al. Gene expression analysis reveals chemical-specific profiles. *Toxicol Sci* 2002;67:219-231.

[24] Pennie W, Pettit SD, Lord PG. Toxicogenomics in risk assessment: an overview of an HESI collaborative research program. *Environ Health Perspect* 2004;112:417-419.

[25] Eisen MB, Spellman PT, Brown PO, Botstein D. Cluster analysis and display of genome-wide expression patterns. *Proc Natl Acad Sci USA* 1998;95:14863-14868.

[26] Cheng Y, Church GM. Biclustering of expression data. *Proc Int Conf Intell Syst Mol Biol* 2000;8:93-100.

[27] Chou JW, Zhou T, Kaufmann WK, Paules RS, Bushel PR. Extracting gene expression patterns and identifying co-expressed genes from microarray data reveals biologically responsive processes. *BMC Bioinformatics* 2007;8:427.

[28] Chou JW, Bushel PR. Discernment of possible mechanism of hepatotoxicity via biological processes over-represented by co-expressed genes. *BMC Genomics* 2009;10:272.

[29] Burczynski ME, McMillian M, Ciervo J, Li L, Parker JB, Dunn RT 2nd et al. Toxicogenomics based discrimination of toxic mechanism in HepG2 human hepatoma cells. *Toxicol Sci* 2000;58:399-415.

[30] Bartosiewicz MJ, Jenkins D, Penn S, Emery J, Buckpitt A. Unique gene expression patterns in liver and kidney associated with exposure to chemical toxicants. *J Pharmacol Exp Ther* 2001;297:895-905.

[31] Hamadeh HK, Bushel PR, Jayadev S, DiSorbo O, Bennett L, Li L et al. Prediction of compound signature using high density gene expression profiling. *Toxicol Sci* 2002;67:232-240.

[32] Bushel PR, Hamadeh H, Bennett L, Sieber S, Martin K, Nuwaysir EF et al. MAPS: a microarray project system for gene expression experiment information and data validation. *Bioinformatics* 2001;17:564-565.

[33] Williams-Devaen CR, Wolf MA, Richard AM. Towards a public toxicogenomics capability for supporting predictive toxicology: survey of current resources and chemical indexing of experiments in GEO and ArrayExpress. *Toxicol Sci* 2009;109:358-371.

[34] Uehara T, Hirode M, Ono A, Kiyosawa N, Omura K, Shimizu T et al. A toxicogenomics approach for early assessment of potential non-genotoxic hepatocarcinogenicity of chemicals in rats. *Toxicology* 2008;250:15-26.

[35] Uehara T, Kiyosawa N, Shimizu T, Omura K, Hirode M, Imazawa T, Mizukawa Y, Ono A, Miyagishima T, Nagao T, Urushidani T. Species-specific differences in coumarin-induced hepatotoxicity as an example of toxicogenomics-based approach to assessing risk of toxicity to humans. *Hum Exp Toxicol* 2008; 27:23-35.

[36] Zimmer B, Lee G, Balmer NV, Meganathan K, Sachinidis A, Lorenz S et al. Evaluation of developmental toxicants and signaling pathways in a functional test based on the migration of human neural crest cells. *Environ Health Perspect* 2012;120:1116-1122.

[37] Hass U. The need for developmental neurotoxicity studies in risk assessment for developmental toxicity. *Reprod Toxicol* 2006;22:148-156.

[38] OECD (Organization for Economic Co-operation and Development). OECD guideline for testing of chemicals. Developmental Neurotoxicity Study. Test No426. 2007. Available: http://www.oecd-ilibrary.org/environment/test-no-426-developmental-neurotoxicity-study_9789264067394-en (accesssed June 30, 2013).

[39] Bulera SJ, Eddy SM, Ferguson E, Jatkoe TA, Reindel JF, Bleavins MR, De La Iglesia FA. RNA expression in the early characterization of hepatotoxicants in Wistar rats by high-density DNA microarrays. *Hepatology* 2001;33:1239-1258.

[40] Lobenhofer EK, Cui X, Bennett L, Cable PL, Merrick BA, Churchill GA, Afshari CA. Exploration of low-dose estrogen effects: identification of No Observed Transcriptional Effect Level (NOTEL). *Toxicol Pathol* 2004;32:482-492.

[41] Hirabayashi Y, Yoon B, Li G, Kanno J, Inoue T. Mechanism of benezene-induced hematotoxicity and leukomogenicity: current review with implication of microarray analyses. *Toxicol Pathol* 2004;32:12-16.

[42] Seale SM, Feng Q, Agarwal AK, El-Alfy AT. Neurobehavioral and transcriptional effects of acrylamide in juvenile rats. *Pharmacol Biochem Behav* 2012;101:77-84.

[43] Tokumoto M, Ohtsu T, Imai S, Honda A, Nagase H, Satoh M. DNA microarray analysis of hepatic gene expression in mice exposed to cadmium for 30 days. *J Toxicol Sci* 2013;38:155-157.

[44] Tsuchiya H, Sato J, Tsuda H, Fujiwara Y, Yamada T, Fujimura A et al. Serum amyloid A upsurge precedes standard biomarkers of hepatotoxicity in ritodrine-injected mice. *Toxicology* 2013;305:79-88.

[45] Merrick BA, Auerbach SS, Stockton PS, Foley JF, Malarkey DE, Sills RC et al. Testing an aflatoxin B1 gene signature in rat archival tissues. *Chem Res Toxicol* 2012;25:1132-1144.
[46] LTKB; http://www.fda.gov/ScienceResearch/BioinformaticsTools/LiverToxicityKnowledgeBase/ucm2024036.htm (accessed June 30, 2012).
[47] Ganter B, Tugendreich S, Pearson CI, Ayanoglu E, Baumhueter S, Bostian KA et al. Development of a large-scale chemogenomics database to improve drug candidate selection and to understand mechanisms of chemical toxicity and action. *J Biotechnol* 2005;119:219-244.
[48] Hayes KR, Vollrath AL, Zastrow GM, McMillan BJ, Craven M, Jovanovich S et al. EDGE: a centralized resource for the comparison, analysis, and distribution of toxicogenomic information. *Mol Pharmacol* 2005;67:1360-1368.
[49] Bowers BJ, Radcliffe RA, Smith AM, Miyamoto-Ditmon J, Wehner JM. Microarray analysis identifies cerebellar genes sensitive to chronic ethanol treatment in PKC gamma mice. *Alcohol* 2006;40:19-33.
[50] Kerns RT, Miles MF. Microarray analysis of ethanol-induced changes in gene expression. *Methods Mol Biol* 2008;447:395-410.
[51] Letwin NE, Kafkafi N, Benjamini Y, Mayo C, Frank BC, Luu T et al. Combined application of behavior genetics and microarray analysis to identify regional expression themes and gene-behavior associations. *J Neurosci* 2006;26:5277-5287.
[52] McBride WJ, Kimpel MW, Schultz JA, McClintick JN, Edenberg HJ, Bell RL. Changes in gene expression in regions of the extended amygdala of alcohol-preferring rats after binge-like alcohol drinking. *Alcohol* 2010;44:171-183.
[53] Mulligan MK, Ponomarev I, Hitzemann RJ, Belknap JK, Tabakoff B, Harris RA et al. Toward understanding the genetics of alcohol drinking through transcriptome meta-analysis. *Proc Natl Acad Sci USA* 2006;103:6368-6373.
[54] Xu Y, Ehringer M, Yang F, Sikela JM. Comparison of global brain gene expression profiles between inbred long-sleep and inbred short-sleep mice by high-density gene array hybridization. *Alcohol Clin Exp Res* 2001;25:810-818.
[55] Farris SP, Wolen AR, Miles MF. Using expression genetics to study the neurobiology of ethanol and alcoholism. *Int Rev Neurobiol* 2010;91:95-128.
[56] Peirce JL, Lu L, Gu J, Silver LM, Williams RW. A new set of BXD recombinant inbred lines from advanced intercross populations in mice. *BMC Genet* 2004;5:7.
[57] Varadarajan S, Tanaka K, Smalley JL, Bampton ET, Pellecchia M, Dinsdale D et al. Endoplasmic reticulum membrane reorganization is regulated by ionic homeostasis. *PLoS One* 2013;8:e56603.
[58] Kramer JA, Petti SD, Amin RP, Bertram TA, Car B, Cunningham M et al. Overview of the application of transcription profiling using selected nephrotoxicants for toxicology assessment. *Environ Health Perspect* 2004;112:460-419.
[59] Amin RP, Vickers AE, Sistare F, Thompson KL, Roman RJ, Lawton M et al. Identification of putative gene-based markers of renal toxicity. *Environ Health Perspect* 2004; 112:465-479.
[60] Bailly V, Zhang Z, Meier W, Cate R, Sanicola M, Bonventre JV. Shedding of kidney injury molecule-1, a putative adhesion protein involved in renal regeneration. *J Biol Chem* 2002;277:39739-39748.

[61] Mazzali M, Kipari T, Ophascharoensuk J, Wesson JA, Johnson R, Hughes J. Osteopontin - a molecule of all seasons. *Q J Med* 2002;95:3-13.

[62] Wadey RM, Pinches MG, Jones HB, Riccardi D, Price SA. Tissue expression and correlation of a panel of urinary biomarkers following cisplatin-induced kidney injury. *Toxicol Pathol* 2013. doi: 10.1177/0192623313492044.

[63] FDA; USA Food and Drug Administration. Innovation or stagnation? Challenge and opportunity on the critical path to new medicinal products, 2004. http://www.fda.gov/ScienceResearch/SpecialTopics/CriticalPathInitiative/CriticalPathOpportunitiesReports/ucm077262.htm.

[64] Waring JF, Jolly RA, Ciurlionis R, Lum PY, Praestgaard JT, Morfitt DC, Buratto B, Roberts C, Schadt E, Ulrich RG. Clustering of hepatotoxins based on mechanism of toxicity using gene expression profiles. *Toxicol Appl Pharmacol* 2001;175:28-42.

[65] Zidek N, Hellmann J, Kramer PJ, Hewitt PG. Acute hepatotoxicity: a predicitive model based on focused illumina microarrays. *Toxicol Sci* 2007;99:289-302.

In: Microarrays: Principles, Applications and Technologies
Editor: James V. Rogers
ISBN: 978-1-62948-669-7

Chapter 13

Microarray Analysis of Gene Transcription Profiles in Chronic Rhinosinusitis

Marcin Fraczek*[1,*]*, Beata Rostkowska-Nadolska*[1] *and Malgorzata Kapral*[2]

[1]Department of Otolaryngology, Wroclaw Medical University, Wroclaw, Poland
[2]Department of Biochemistry, Medical University of Silesia, Sosnowiec, Poland

Abstract

DNA microarray analysis through simultaneous measurement of expression of thousands of genes in the same tissue sample has enormous discovery potential. Comparison of gene expression profiles in disease with healthy tissues may highlight the involvement of both expected and unsuspected pathways leading to the formation of the pathology. Thus, genome wide expression array analysis has been successfully used to characterize a variety of different abnormal processes. Chronic rhinosinusitis (CRS) is a general term for multiple subsets of inflammatory conditions affecting paranasal cavities. Distinct clinical presentations and laboratory findings together with incomplete understanding of CRS pathophysiology make that disease an ideal candidate for genome wide expression array analysis. The potential development of the technique in CRS would be the definition of disease specific gene profiles enabling better understanding disease etiology. The purpose of this chapter is to describe DNA microarray contributions toward achieving a better understanding of CRS. The outcomes of DNA microarray analysis of sinus mucosa in patients with chronic rhinosinusitis with nasal polyps (CRSwNP) compared to healthy epithelium are described. We identified 556 genes, which were differentially expressed in CRS tissues compared to the control. Among them, 217 showed significantly higher expression, whereas 339 lower expression in CRSwNP than in the controls. In CRSwNP, significantly higher transcriptional activity of MMP10, NOS2A, ALOX15 and IL-8 genes was observed. In the control group, significantly higher expression of DMBT1, ALOX12 and LTF genes was detected. Those

* For correspondence (cedorau@gmail.edu.

findings were confirmed via quantitative reverse transcription polymerase chain reaction (qRT-PCR). It might be concluded that the described technology can be successfully used to identify genes implicated in the pathogenesis of CRS. The results are discussed with the outcomes of other research in that field. The analysis of gene expression by microarray is a promising research method in the field of CRS. Some reports demonstrated that DNA microarray analysis is a feasible tool for studying different disease subclassifications in CRS. Moreover, molecular alterations obtained using that technique may indicate changes during the clinical course of CRS. Defining gene expression profiles with DNA microarray may help elucidate new prognostic factors and support development of new therapeutic alternatives.

Introduction

Chronic rhinosinusitis (CRS) is a chronic inflammatory condition affecting the nose and sinuses manifesting with rhinorhea, nasal congestion, anosmia, facial pain, and headache for more than 12 weeks duration and refractory to medical management. The process may exacerbate, and in contrast to the acute form of rhinosinusitis, symptoms do not usually disappear completely. CRS is now considered one of the major health problems in developed countries. This is mainly due to the severe medical costs of treatment, proven negative effect on lower airway, cardiovascular system, and general health. CRS, besides heart diseases and diabetes, reduces a patient's quality of life. The incidence of rhinosinusitis as other lifestyle illnesses is constantly increasing. It is estimated that in western countries all the forms of CRS occurs in up to 10-15% of the population, making it the second most common chronic disease.

The term rhinosinusitis represents multiple different diseases with or without various stages of severity along their spectra. CRS in 80% of cases has a form of edematous-eosinophilic, hyperplastic rhinosinusitis (chronic rhinosinusitis with nasal polyps; CRSwNP). This group includes the subjects with the worst course of CRS with concomitant intolerance of NSAIDs and bronchial asthma. Typical histomorphological characteristics of nasal polyps (NP) include edema, inflammatory cell infiltration with numerous activated eosinophils, and various degree of tissue remodeling in epithelium, glands, connective tissue, and vessels [1]. Less common occurring in clinical practice form of chronic rhinosinusitis with nasal polyps, the neutrophilic fibroinflammatory type constitutes approximately 7% cases. Medical experience shows that NP demonstrate both a different clinical course and therapy reaction [2].

CRS is a multifactorial disease of unknown etiology, which should take into account the interaction of genetic background, environmental, immunological, and anatomical abnormalities. These factors trigger a local inflammatory process that lead in most of the patients to NP formation. The dominant infiltrative inflammatory cells in nasal polyps are eosinophils, which are the driving force of inflammatory process and they affect the function of structural cells of the mucosa through the whole spectrum of mediators and cytokines.

In eosinophilic CRSwNP, cytokine profiles of the ongoing inflammatory process is referred to as lymphocytic mixed (Th1/Th2), with substantial participation of cytokines characteristic of Th2 cells (IL-4, IL-5, IL-9, and IL-13). Th2 interleukins and the chemokines, RANTES and eotaxin, show the strongest chemotactic properties for eosinophils and are responsible for eosinophil influx into NP. Elevated levels of RANTES and eotaxin have been commonly demonstrated in patients with CRS. Other cytokines and chemokines (IL-3, IL-8, TNFα, and GM-CSF) levels, as well as expression of adhesion molecules ICAM-1, VCAM-1, and selectins have been also increased in NP. However, in the fragments of sinus mucosa obtained from patients with chronic rhinosinusitis without dominant eosinophilia, higher concentrations of IL-1, IL-6, IL-8, TNF, IL-3, GM-CSF, and ICAM-1 were observed.

Eosinophils, along with infiltrating lymphocytes, neutrophils, macrophages, and mast cells have an impact on the structural cells of mucosa (fibroblasts and epithelial cells). Besides Th2 lymphocytes and eosinophils, fibroblasts play an important role in tissue remodeling the upper airways. The secretion of IL-4 and IL-13 are also important elements of the process of tissue reconstruction.

CRS usually develops due to a number of factors that predispose an individual to the induction of an inflammatory response not necessarily associated with infection. Such factors considered are abnormalities within the mechanisms of innate immunity as well as genetic background. Evidence of this relationship includes the coexistence of CRS with cystic fibrosis and Kartagener syndrome as confirmed by hereditary background playing a role in 60% of CRS cases. Lastly, a patient's predisposition and environmental exposures underpin the development of chronic inflammation.

Structure and function of cells are determined by the expression of many genes. Aberrant gene expression is critical for the initiation and progression of diseases. Currently, there are available many methods that enable the identification of changes in mRNA and protein levels. Traditionally, studies were limited to examining only a few genes. However, with the appearance of DNA microarrays, large-scale gene expression analysis became possible. DNA microarrays consist of slides containing DNA probes of known sequences enabling simultaneous analysis of thousands of genes. In the oligonucleotide microarray technique, gene expression (transcription) is analyzed based on hybridization of single-stranded RNA fragments with complementary DNA probes [3]. Comparison of gene expression profiles in disease with healthy tissues often highlights the involvement of both expected and unsuspected pathologic pathways.

The aim of this chapter is to present the outcomes of studies conducted that determine mRNA expression in NP compared with normal nasal mucosa, as well as characterizing differences in tissue-specific gene expression.

Materials and Methods

Subjects

Fifty-three patients (32 males and 21 females; mean age 51.23 ± 15.21) with NP were treated surgically at our department and included in the study group according to diagnostic criteria for chronic rhinosinusitis established by the Task Force on Rhinosinusitis (AAO-HNS) [4]. The extent of disease was assessed by computed tomography (CT) and endoscopically. Endoscopic staging of NPs were performed according to Lildholdt et al. [5]. Findings on CT scans were graded according to the Lund-Mackay score [6].

Patients had been free of any medication for at least 2 weeks before surgery and had bilateral polyps in the nasal cavities. Atopy was defined by positive personal history of allergic respiratory symptoms and positive skin prick tests (SPT; wheal > 3 mm) to the standard panel of inhalant allergens, including *Dermatophagoides pteronyssinus*, *D. farinae*, mixed grass pollen, mixed tree pollen, cat and dog epithelium, and *Alternaria alternata* provided by Allergopharma/Nexter. Atopy and bronchial asthma was diagnosed in 14 (26%) and 12 (22%) patients with NP, respectively.

The control group included 28 healthy individuals (19 males, 9 females; mean age 36.4 ± 16.2). All control tissue samples were taken from unchanged nasal mucosa of uncinate process during nasal trauma surgery or dacrocystorhinostomy. The absence of NP was assessed through clinical history, endoscopic examination, and imaging. Negative prick tests and absence of allergy in personal history were required. Other diseases were also excluded.

Tissue samples were flash frozen in liquid nitrogen immediately after harvesting and stored at -70°C before RNA extraction. A portion of each sample was routinely fixed in 10% neutral-buffered formalin and embedded in paraffin wax for subsequent immunohistochemical examination to visualize inflammatory cells and to exclude other pathologies. The presence of edema or inflammation was ruled out in healthy mucosa. The diagnosis of eosinophilic NP was determined if the percentage of eosinophils was greater than 8.0% of all leukocytes or the presence of clusters of eosinophils was seen. When the dominant cells in the tissue were lymphocytes and plasmocytes, the diagnosis of neutrophilic NP was established. On the basis of that criterion 36 (68%) cases were classified as eosinophilic NP and 17 (32%) as neutrophilic NP.

Seventeen patients (11 males and 6 females; mean age 50.8 ± 16.3) were selected to microarray analysis. Based on histological examination, 11 (65%) had eosinophilic NP and 6 (35%) had neutrophilic NP. The control group for microarrays consisted of 8 healthy individuals (6 males and 2 females; mean age 32.4 ± 13.2) (Table 1).

Table 1. Characteristics of patients with rhinosinusitis and nasal polyps (NP) included in the microarray study. Data are expressed as mean ± SD

Age	Eosinophilic NP	Neutrophilic NP
	52.7 ± 13.1	47.3 ± 22.2
Sex (men/women)	8/3	3/3
Patients with bronchial asthma	5	0
Patients with atopy	6	0
Number of endonasal procedures	2.2 ± 1.8	1.8 ± 1.3
Nasal endoscopic score	4.8 ± 1.1	3.8 ± 0.9
Lund-Mackay score	18.1 ± 2.9	16.2 ± 2.0

RNA Extraction

Total RNA was extracted from frozen tissue specimens (about 40 mg) with the use of TRIzol® reagent (Invitrogen, Carlsbad, CA) according to the manufacturer's protocol. Tissue samples were homogenized in 1 mL TRIZOL® reagent using a Polytron tissue homogenizer (Kinematica AG, Lucerne, Switzerland). All RNA extracts were treated with DNase I and cleaned with Mini-spin columns using the RNeasy Mini Total RNA Purification kit (Qiagen, Valencia, CA).

The RNA extracts were qualitatively checked by electrophoresis in 1.0% agarose gel stained with ethidium bromide. RNA concentration was determined spectrophotometrically using a GeneQuant pro (Biochrom, Cambridge, UK). RNA purity was judged by the ratio of absorbance at 260 and 280 nm (A_{260}/A_{280}) (ratios between 1.9 and 2.1 were acceptable).

Microarray Procedure

Gene expression profiles were determined using commercially available oligonucleotide microarrays HG-U133A (Affymetrix, Santa Clara, CA) according to the manufacturer's recommendations [7]. Each gene chip contains 22,238 probe sets that correspond to more than 18,400 transcripts and 14,500 well-characterized human genes. Briefly, 8 μg of total RNA from each sample was used for the cDNA synthesis using SuperScript Choice System (Gibco BRL Life Technologies, Paisley, UK). In the next step, cDNA templates were used to generate biotin-labeled cRNA using BioArray HighYield RNA Transcript Labeling Kit (Enzo Life Sciences, Farmingdale, NY). cRNA was purified on RNeasy Mini Kit columns (Qiagen). Next, 16 μg biotin-labeled cRNA was fragmented by using the Fragmentation Buffer (Qiagen) at 94°C before hybridization to gene chips for 16 h at 45°C. The hybridized cRNA probe was stained with streptavidin-phycoerythrin conjugate and scanned using GeneArray Scanner G2500A.

Microarray Data Analysis

The resulting images *were* analyzed *using* Microarray Suite 5.0 software (Affymetrix). Results were normalized with RMAExpress version 0.5 Release (http://rmaexpress.bmbolstad.com). Statistical analysis was performed with the use of Statistica 6.0 software (StatSoft, Poland). Cluster analysis (Ward's method) was applied to normalized data. To analyze microarray data, the t-test for two unpaired groups was used. The criterion for estimating the differentially expressed genes was fold change (FC) of ≥ 2 or greater. Fold change is calculated from the signal log ratio (SLR) (values following log2 used as the ratio between 2 groups). Genes are upregulated if $SLR \geq 0$ and $FC = 2^{SLR}$ and downregulated by $SLR < 0$ and $FC = 2^{-(SLR)}$, *respectively*. Genes were considered differentially expressed if $p<0.05$ and there was at least a 2-fold change in the mean expression level between NP and controls. All genes of unknown function, hypothetical genes, and expressed sequence tags were omitted.

Real-time qRT-PCR

The microarray results were confirmed by real-time qRT-PCR for 8 randomly selected genes: MMP10, MMP3, IL8, DMBT1, LTF, NOS2A, ALOX12, and ALOX15. Oligonucleotide primers specific for examined genes were designed using Primer Express™ v. 2.0 software (PE Applied Biosystems, CA) (Table 2).

Table 2. Characteristics of primers used for amplification

Gene	GenBank	Sequence of primers	Amplicon Length (bp)	Tm (ºC)
MMP10	NM_002425	Fwd:5'TCTTGCATTCCTTGTGCTGTTG3' Rev:5'ATTGCTGGGCAAGATCCTTGT3'	101	81
MMP3	NM_002422	Fwd:5'CCGAGAAATGCAGAAGTTCCTTGGATTGG3' Rev:5'TCGGGATGCCAGGAAAGGTTCTGAAGTG3'	130	82
IL8	NM_000584	Fwd:5'CCCCAAATTTATCAAAGAACTGAGA3' Rev:5'CCTCTGCACCCAGTTTTCCTT3'	127	78
LTF	NM_002343	Fwd:5'AAGTGTGAATGGCAAGGAGGAT3' Rev:5'AGCCAAGGTACAGCCCAGAA3'	191	81
NOS2A	NM_000625	Fwd:5'CAAGCCTACCCCTCCAGATG3' Rev:5'CATCTCCCGTCAGTTGGTAGGT3'	161	79
ALOX12	NM_000697	Fwd:5'GGAGATGGCCCTCAAACGT3' Rev:5'GAGGAACTGGTAGCTGAACAACTCA3'	151	82
ALOX15	NM_001140	Fwd:5'GATGGCTGAGGTCATTGTTGTG3' Rev: 5'GCCCGGACGTTAATTTCCA3'	111	80
DMBT1	NM_007329	Fwd:5'CCAATGACTTCACGTCTTTGACTT3' Rev:5'CGAAGAGACGGCGAGGAGTA3'	91	79

Transcriptional activity of studied genes were evaluated based on mRNA copy number per 1 µg of total RNA by the use of real time qRT-PCR with a SYBR Green I chemistry (SYBR Green Quantitect RT-PCR Kit, Qiagen, Valencia, CA). Analysis was carried out using an Opticon™ DNA Engine Continuous Fluorescence detector (MJ Research, Watertown, MA). Reaction mixtures consisted of 25 µL 2x QuantiTect SYBR Green RT-PCR Master Mix, 0.5 µL QuantiTect RT Mix, 0.5 µM primers forward and reverse and 0.1 µg RNA. The thermal cycling conditions for one-step RT-PCR were as follows: 50°C for 30 min for reverse transcription and 95°C for 15 min, 50 cycles at 94°C for 15 s, 60°C for 30 s and 72°C for 30 s. Standards, from 400 to 8000 copies cDNA β-actin (TaqMan® DNA Template Reagents Kit, PE Applied Biosystems) were used to generate a standard curve. Finally, specificity of RT-PCR reaction was confirmed by determining characteristic temperature of melting for each amplicon. Additionally, the RT-PCR products were separated on 6% polyacrylamide gels and visualized with silver salts.

Statistical Analysis

Statistical analysis was performed with the use of Statistica 6.0 software. The differences in expression of examined genes detected by real-time RT-PCR between NP and normal mucosa were analyzed using the Mann-Whitney U test. The results of mRNA copy number per µg of total RNA were expressed as medians. Significance was assumed at the $p<0.05$ level.

Results

Microarray Analysis

Oligonucleotide microarray enabled the assessment of gene expression profiles in NP tissues and healthy nasal mucosa. Comparison of samples taken from 17 patients with NP as well as 8 normal nasal mucosa showed distinct global gene expression patterns using cluster analysis. The examined samples were clustered into two distinct groups. The NP tissues clustered together, indicating different gene expression profiles compared to the control (Figure 1).

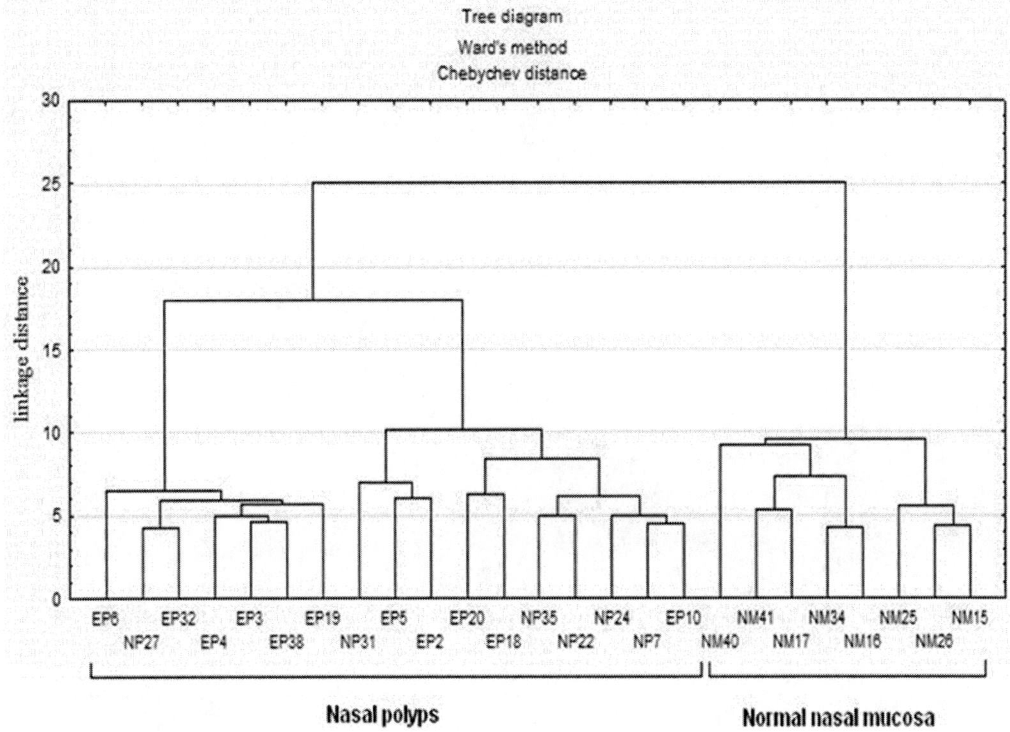

Figure 1. Dendrogram of 17 nasal polyps and 8 normal mucosa samples clustered using Ward's method with 22,238 probe sets represented on the HG-U133A GeneChip.

Simultaneously, dendrogram analysis revealed a more heterogenic gene expression profile among NP compared to the more homogeneous control group. In the NP group, two subgroups emerged. The first subgroup included NP with predominance of eosinophils and single case of neutrophilic NP (NP27). All eosinophilic NP patients showed similar clinical presentations in which five of the 7 patients had atopy. Coexistence of bronchial asthma was present in 4 of 7 cases in which the course of disease in these groups were more serious as evidenced by nasal endoscopy and Lund-Mackay scores of 5.0 ± 1.0 and 18.4 ± 2.6, respectively.

The second subgroup was characterized by differences in terms of cellular composition of the inflammatory lesion (both eosinophilic and neutrophilic NP). Among those, lesion clusters with predominance of neutrophilic polyps was revealed and these patients were characterized by chronic rhinosinusitis with purulent discharge and nonresponsive to pharmacological treatment. Nasal endoscopy and Lund-MacKay scores for this group were 3.6 ± 1.0 and 16.0 ± 2.2, respectively. Mean number of surgical procedures was 2.5 ± 1.5. The rest of the patients in that subgroup presented eosinophilic infiltration with the exception of one neutrophilic NP (NP31). Most of cases (3 of 4) were nonallergic eosinophilic chronic rhinosinusitis with NP having nasal endoscopy and Lund-MacKay scores of 4.5 ± 0.8 and 17.8 ± 2.9, respectively. Mean number of surgical procedures was 1.4 ± 1.0. In most of those cases the course of disease was mild. There was a not statistical difference between the subgroups considering age, sex rate, nasal endoscopy score, Lund-MacKay score, and number of surgeries performed.

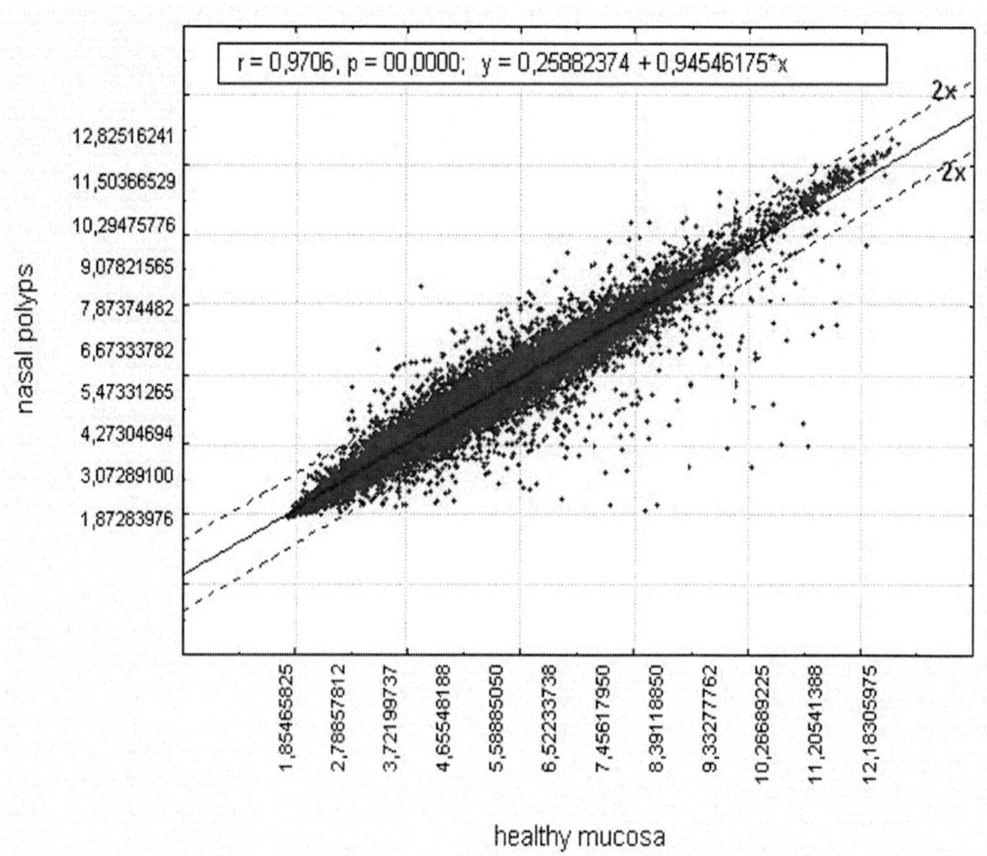

Figure 2. Scatter plots of gene expression in NP tissues and normal tissues (x-axis represents the normalized signal value of normal nasal mucosa, and y-axis represents the normalized signal value of NP).

In the next step of analysis, genes showing differential expression in NP in comparison with normal nasal tissues were identified (Figure 2). We identified 556 genes that were differentially expressed between the two groups. Among them, 217 showed significantly higher expression, whereas 339 showed lower expression in the NPs than in controls. Of the 217 genes, expression level in NP ranged from 2- to 3-fold (161 genes; 74.2%), 3-4-fold (28;

12.9%), 4-5-fold (16; 7.4%), 5-6-fold (8; 3.7%), or at least 6-fold (4; 1.8%). The largest increase was revealed for the gene encoding MMP10 (20-fold higher expression). In healthy mucosa, there were 198 genes (58.4%) of at least 2-fold but less than 3-fold higher expression than in NP. Moreover, 70 genes (20.6%) had 3-5-fold higher expression, 38 (11.2%) 5-10-fold, 25 (7.4%) from 10-fold to 50-fold, 6 (1.8%) from 50-100-fold, and 2 genes (0.6%) had over 100-fold higher expression.

Validation of Microarray Data by Real-Time RT-PCR

To confirm data obtained by microarray analysis, qRT-PCR was performed for assessing the expression of 8 genes. For verification, 4 genes showing higher expression in NP (IL8, MMP10, NOS2A, and ALOX15) and 4 genes with lower expression in NP (MMP3, ALOX12, LTF, and DMBT1) were chosen. In order to assess mRNA level of selected genes, larger cohorts of samples (53 nasal polyps and 28 healthy mucosa) were studied. Specificity of RT-PCR for the target genes was experimentally confirmed by PAA electrophoresis and the amplicons melting temperatures. For each RT-PCR product, a single peak was obtained by melting curve analysis at the expected temperatures. Gel electrophoresis revealed the presence of single products of the predicted size (data not shown).

The microarray and qRT-PCR results were compatible for seven of the eight evaluated genes. In NP, significantly higher transcriptional activity of MMP10, NOS2A, ALOX15, and IL8 genes than in normal mucosa were observed. Contrary, DMBT1, ALOX12, and LTF genes were highly expressed in the control group. The expression of MMP3 gene was similar to the microarray results. Comparative analysis revealed the similar level of MMP3 mRNA in NP and healthy mucosa (Table 3).

Table 3. Quantitative comparison of examined genes × 10^3 mRNA copy number per 1 μg of total RNA between nasal polyps and normal nasal mucosa. Date are presented as median (range). [a]Mann-Whitney U test

Gene	Nasal polyps	Normal nasal mucosa	p-value[a]
IL8	5.0 (0.023 – 8699.5)	0.8 (0.013 – 107.5)	p = 0.0032
MMP10	27.3 (1.0 – 2858.3)	2.4 (0.3 – 55.5)	p <0.001
NOS2A	25.8 (0.61 – 991.4)	1.83 (0.13 – 55.9)	p <0.001
ALOX15	120.5 (1.07 – 4463.3)	4.07 (0.026 – 404.8)	p <0.001
MMP3	5.46 (0.03 – 68.0)	6.72 (0.35 – 269.8)	p = 0.0528
ALOX12	1.42 (0.01 – 30.3)	2.77 (0.2 – 87.2)	p = 0.0116
LTF	9.85 (0.05 – 333.9)	17.75 (0.85 – 6726.6)	p = 0.0493
DMBT1	0.22 (0.01 – 442.0)	2.4 (0.01 – 3591.4)	p = 0.0413

Conclusion

Medical experience indicates that nasal polyps show various clinical courses and different treatment results. Possible changes in gene expression between NP and normal tissues may point to differences in transcriptional activity of genes depending on histopathological examination, frequency of recurrence, clinical course, and response to treatment. Gene microarrays are now a standard experimental platform for investigating the differences in gene expression in diseased tissues compared to the control and to identify differential gene expression. In the past years, microarray techniques for gene expression profiling have been applied to a wide range of biological problems.

In this chapter, DNA microarrays were used to study expression of 14,500 genes in 17 NP and 8 healthy nasal mucosa samples. Gene expression patterns were significantly different in normal mucosa in comparison with NP. Consequently, all examined tissues could be classified either as NP or healthy mucosa. Serial samples taken from patients with NP tended to cluster together (in the control group the same tendency was observed), indicating that aberrant gene expression is critical for NP to arise. Previously, distinct gene expression fingerprints in NP compared with normal tissues was revealed also by Stankovic et al. [8]. Recognition of subgroups of samples seems to be the first step to understanding the molecular changes underlying the disease development. To find similarities between the 17 NP cases hierarchical clustering analysis was performed. Grouping based on gene expression fingerprints revealed two separate subgroups. The length of the "arms" is proportional to the similarities between the profiles of transcript expression of the studied gene. Shorter arms indicate a closer resemblance. Hierarchical clustering of the nasal polyps corresponded with histological classification of tissues. Our results show similarities between clustering and clinical course of disease. Thus, future gene expression analysis could serve as additional diagnostic tool to the standard histological examination of NP.

Next, we defined marker genes that were differentially expressed in NP in comparison with healthy mucosa. Among 556 such genes, 217 showed significantly higher expression in NP than in control tissues, whereas 339 exhibited lower expression. These genes may be potentially associated with NP development. Few studies have applied microarray analysis to the problem of rhinosinusitis. Liu et al. [9] used oligonucleotide microarrays to examine gene expression profiles in NP (n=10) in comparison to sinus mucosa. Among 10,500 genes, 192 were identified as up-regulated and 156 as down-regulated in NP. Fritz and colleagues [10] examined nasal polyps, comparing them to adjacent tissue within the nasal cavity and sinuses, and found 34 genes with two-fold overexpression or decreased expression. Payne et al. [11] and Wu et al. [12] found 120 and 283 differentially expressed genes in NP compared with controls, respectively. According to our results most of genes revealed 2-fold, 3-fold differences in expression.

Oligonucleotide microarray technology as initial screening stage could result in identification of novel disease-related genes. Genes detected by gene chip assays as potentially involved in pathogenesis of NP should be confirmed by real-time PCR and/or immunohistochemistry. Among the genes differentially transcribed in NP and normal mucosa, eight (IL-8, ALOX15, NOS2A, MMP10, DMBT1, LTF, ALOX12, and MMP3) were selected for further analysis with real-time qRT-PCR. Interleukin 8 (IL-8) is one of the major mediators of the inflammatory response and a strong chemotactic factor for neutrophils, leukocytes, T lymphocytes, and eosinophils. However, challenge of primed nasal mucosa with IL-8 resulted in a significant influx of neutrophils, but not eosinophils [13]. Neovascularization and migration of leukocytes may be stimulated by IL-8 and are important mechanisms in the development of NP. Thus, locally produced IL-8 contributes to the progression of chronic inflammation in the sinus mucosa. In the airway epithelium, IL-8 is one activator of MMP9 and MMP8 expression and secretion during airway remodeling [14]. IL-8 and MMP8 may form a pivotal inductive cytokine-proteinase cascade in the pathogenesis of NP. On the basis of those data, IL-8 is considered as nonspecific marker of sinus inflammation. Our results show significantly higher expression of IL-8 in NP compared to normal mucosa.

Lipoxygenases (ALOXs) are enzymes that catalyse dioxygenation of polyunsaturated fatty acids (mammalian arachidonate lipoxygenases). The biological role of most ALOXs, including 15-ALOX and 12-ALOX, are not yet fully understood [15, 16]. Alterations in the arachidonic acid metabolism pathway have been implicated in the pathogenesis of NP [17]. Currently, both DNA microarray and QRT-PCR analysis revealed higher expression of 15-ALOX and lower expression of 12-ALOX in NP compared to normal mucosa. Previously, significantly increased expression of 15-ALOX was also shown in patients with bronchial asthma and chronic rhinosinusitis [18, 19]. Owens et al. [17] demonstrated increased in expression of both 15-ALOX and 12-ALOX in patients with NP.

Nitrogen oxide (NO) is an important neurotransmitter and mediator of inflammatory processes. It increases transudation of blood from the vessels and glandular secretion, leading to edema typical for NP [20]. NO is produced by nitric oxide synthase (NOS) that uses the amino acid arginine as the substrate. A potential source of NOS could be from eosinophils that are present in large amounts in NP [21]. Herein, a 5-fold higher NOS-2A expression in NP was observed, which is consistent with reported increased NOS activity [22]. Our findings support previous suggestions that NOS-2 can be co-responsible for developing inflammatory process in the lateral nasal wall [23]. Moreover, Benson et al. [24] observed a decrease in expression of inflammatory genes IL-8 and NOS2A after glucocorticoid treatment.

Metalloproteinases are the family of endopeptidases playing a role in degradation of extracellular matrix and its remodeling. MMP1, MMP2, MMP7, MMP8, and MMP9 have been detected in NP. In this chapter, different MMP1, MMP3, MMP9, and MMP10 mRNA expression levels in NP and normal mucosa was observed. For subsequent analysis, we chose MMP3 and MMP10 (belonging to stromelysins), which degrade the basement membrane components, type IV collagen, nidogen, and fibronectin [25]. Real-time QRT-PCR technique confirmed significantly higher transcriptional activity of MMP10, but also showed insignificant lower MMP3 expression in NP compared with normal mucosa. Wu et al. [12] observed significantly up-regulated MMP1 gene in NP, which is similar to our results presented in this chapter.

Lactotransferrin (LTF) is an iron binding glycoprotein expressed in most biological fluids and also a major component of the innate immune system. Its protective effect ranges from direct antimicrobial activities against a large panel of microorganisms to anti-inflammatory and anticancer activities [26]. DMBT1 (Deleted in Malignant Brain Tumours 1) is expressed mainly by mucosal epithelia and glands, in particular in the respiratory tract, and plays a role in local innate immune defense. In adults, research interest in DMBT1 results from its possible role in tumor suppression [27]. We detected significantly lower expression of these two genes in NP compared with normal mucosa. Those data are in agreement with that of Stankovic et al. [8], where underexpression of LTF and DMBT1 in polyps was found.

Our results of real-time RT-PCR are in accordance with those of DNA microarray except for MMP3, indicating the data obtained by gene chip analysis are valid, and these genes could potentiality serve as NP related genes.

The results presented in this chapter as well as other reports examined sinonasal polyps in general in comparison to the unchanged control tissues (Table 4). Those results should serve as a starting point to identify potential pathophysiologic similarities and differences by comparing the two-gene expression profiles. The identification of novel disease-related genes may help to understand the biology of NP and indicate new targeted therapies. In a study by Plager et al. [28], the most prominent class of up-regulated genes identified in CRSwNP appeared to be the chemokines. Most of the up-regulated genes were located at chromosome 17q11.2, which is in accordance with previous observation that gene cluster contributes to eosinophilic epithelial inflammation; increased transcription of CCL18 (PARC), CCL22 (MDC), and CXCL13 was observed in CRS [28]. These genes are induced in inflammatory dendritic cells that include monocytes, macrophages, and dendritic cells. CCL18 and CCL22 are associated with chemoattraction of Th2 cells, which further amplifies eosinophilic inflammation. The CCL18 gene is also located near the eotaxin-1 gene cluster at chromosome 17q11.2. Thus, the 17q11.2 genes likely contribute to the eosinophilic inflammatory response.

Table 4. List of genes identified as upregulated by cDNA microarray technique in chronic rhinosinusitis

Reference	Genes Upregulated in CRS
42	MMP10, NOS2A, ALOX15, IL-8
28	CCL18, CLC, CST1, CCL13, IL1RL1, CST2, POSTN, EMR1, EMR3, F13A1
33	IL-6, IL-12A, IL-13, TNFα
31	INDO, IL1R2, CLECSF6, CCL11, CD163, TNFSF10, AIF1, NCF2, ALOX5AP, FPR1, TYROBP, CSC, IFI30, MICB, LCP2NCK1, LST1, TLR2, PTAFR, CKLF, EDG6
9	STATH, PIP, LTF, DMBT1
11	IL-6, IL-8, CXCL1, CCL2, HIF-1α, Tenascin

It could be highly interesting to analyze the impact of an applied therapy on the gene expression profile in CRS. Benson and colleagues [29] investigated the gene expression changes associated with glucocorticoid treatment using microarray technique. They compared polyp tissue to the same tissue in the same patients following six weeks of topical therapy. They found alterations in 134 known genes, many of which were associated with inflammation. Uteroglobulin (CC10) which poses anti-inflammatory and immunomodulatory properties demonstrated the greatest increase in response to corticosteroids. Similarly, Bolger et al. [30] found that also systemic glucocorticoids decreased expression of several chemokine and leukotriene receptor genes in nasal polyp tissues.

The worst prognosis for the patient with CRS is associated with coexisting asthma and intolerance of non-steroidal anti-inflammatory drugs. Those patients are characterized by an aggressive mucosal inflammation in both lower airway and the upper airways. The molecular pathogenesis of aspirin-intolerant asthma (AIA), also known as Samter's triad, is still unclear. Sekigawa et al. [31] analyzed genetic susceptibilities underlying CRS with aspirin-intolerant asthma. The authors used microarrays to compare global gene expression of NP from AIA patients with nasal polyps from eosinophilic sinusitis patients without aspirin sensitivity. Similar expression profiles were expected in both groups due to the similar histological and biochemical characteristics. Unexpectedly, two discrete clusters appeared corresponding to the AIA and classic eosinophilic sinusitis, indicating that those conditions appear to exhibit distinct expression profiles. INDO and IL1R2, which had significant associations with AIA, were suggested to represent susceptibility genes for aspirin-intolerant asthma. In contrast, Stankovic et al. [8] analyzed aspirin-sensitive polyps and aspirin-tolerant polyps and did not find substantial differences between two polyp phenotypes. They detected increased expression of periostin and met protoncogene, and decreased expression of PIP compared to the control. Periostin is a protein that is highly expressed in the airway epithelium of asthma patients believed to play a role in TGF-β activation, collagen deposition, fibrosis, and remodeling.

Currently, it remains unclear to what extent allergic reaction influences the development of chronic rhinosinusitis. Plager et al. [28] examined the differences between asthmatic CRSwNP with and without allergic rhinitis. Transcript levels of the five genes in CRSwNP were relatively distinct when compared to the allergic rhinitis samples. However, this observation is not in accordance with data presented by Fritz et al. [10] where patients with allergic rhinitis with/without nasal polyps showed great similarity with nasal polyposis. Thus, it remains unclear if transcript profiles between allergic rhinitis and CRSwNP are similar, or whether nasal polyposis in these two diseases is mediated by two separate mechanisms.

To date, microarray techniques have not yet been widely applied in the pediatric population. Wu et al. [32] evaluated gene expression by microarray analyses of inflammatory mediators in the sinus mucosa of children with or without chronic rhinosinusitis. They demonstrated significant expression changes in several genes involved in the inflammatory and immune responses. The chemokines CXCL13 and CXCL5 (both potent chemoattractants of B lymphocytes), serum amyloid A, serpinB4, and defensin B1 were highly upregulated. However, expression levels of Th2 associated cytokine (IL-5) and those involved in innate immunity (IL-6 and IL-8) were similar in both cohorts. The microarray gene profiles generated by this study indicate that the sinus mucosa in children with CRS is immunocompetent and that both arms of the immune system are involved in pathogenesis of CRS. The authors suggested that due to different cytokine gene expression pattern, pediatric CRS is not the same entity as the adult form. That data are consistent with previously described differences between children and adult CRS regarding exacerbation of eosinophilic infiltration, mucus gland hyperplasia, and clinical pictures.

Several inflammatory cytokines typically involved in CRS were not found to have a significant change in gene expression in this chapter. Most compelling was the lack of evidence for expression of genes encoding Th2-associated cytokines (e.g., IL-4 and IL-13). That observation remains in accordance with outcomes of other research where none of these cytokines exhibited increased expression in microarray analyses. Payne et al. [11] reported significant down regulation of IL-4, IL-13, and IFN-γ with up-regulation of IL-6, IL-8, SCF (Stem Cell Factor), CXCL1, and hypoxia inducible factor 1α (HIF-1α) in polyps versus control tissues. Both IL-6 and CXCL1 chemokines play a role of chemoattractants for neutrophils, suggesting their involvement in the etiology of CRS. However, in a study published by Anand et al. [33], the inflammatory genes for IL-13, IL-6, IL-12A, and TNF-α, were upregulated in the ethmoid mucosa of patients with CRS without NP when compared to the control group. However, the authors did not confirm the data by RT-PCR. Figueiredo et al. [34] compared polyp tissue and surrounding non-polypoid tissue from non-atopic patients with control tissues and observed increased IL-5 expression in the polyps and increased TGF-β expression in the adjacent inflamed mucosa.

The inflammatory process underlying nasal polyposis is induced and perpetuated by the enhanced activity of several agents including transcription factors. It has been demonstrated that nuclear factor-kappa B (NF-κB) is implicated in the regulation of multiple pro-inflammatory genes. Although NF-κB has been considered the "holy grail" in the targeting of new anti-inflammatory drugs, initial attempts did not provide satisfactory outcomes. It could be partially explained by the fact that NF-κB influences the expression of a broad spectrum of genes, some of which may have an opposite action during inflammation. The aim of our recent study was to use microarrays to identify NF-κB-dependent genes that are activated in NP compared to the control mucosa [35]. Among 582 human NF-κB-dependent genes, 25 showed a significantly higher expression in NPs compared to the control. The largest increase was for the gene encoding TFF3 (5-fold higher expression), followed by NOS2A (5-fold), SERPINA1 (4-fold), UCP2 (4-fold), OXTR (4-fold), IL-8 (3-fold), CXCL1 (3.5-fold), CCL11/eotaxin1 (2.7-foldx), Tenascin C (2.7-fold), and IL-32 (2-fold).

Similarly, we observed enhanced expression of many chemokines including eotaxin 1 (CCL11), CXCL1, and CCL5. Eotaxins are the most potent chemoattractant for eosinophils, and are thought to play a key role in tissue eosinophilia. In such a mechanism eotaxins increase local inflammatory status and are strongly involved in the development of NPs [36]. Similar to IL-8, CXCL1 is of particular interest as a chemoattractant and activator of neutrophils. This growth-related oncogene is crucial for the recruitment of other inflammatory cells including monocytes, dendritic cells, and T cells. Both transcripts and protein levels for CXCL1 were noted to be significantly elevated in NP and also released by infiltrating eosinophils [11]. Besides IL-8, we revealed another significantly up-regulated interleukin in NP tissues. IL-32 is a recently described pro-inflammatory cytokine produced mainly by epithelial cells, T lymphocytes, natural killer cells, and monocytes. IL-32 induces the expression of other important pro-inflammatory cytokines (IL-1β, IL-6, TNF-α, and IL-8) and may further amplify inflammatory reaction. To date, IL-32 has been reported to be involved in several chronic inflammatory diseases like asthma, atopic dermatitis, and rheumatoid arthritis [37, 38].

Similarly to Payne et al. [11], we observed enhanced expression of tenascin, which is a pro-inflammatory extracellular matrix glycoprotein expressed during embryogenesis and not normally present in most adult tissues. Tenascin-C reappears in association with wound healing, cancer invasion, tissue remodeling, and chronic inflammations. It is a prognostic marker for asthma patients [39]. Expression of tenascin-C is induced upon activation of cell surface TLRs, but also subsequent activation of TLR4 by tenascin-C stimulates the synthesis of pro-inflammatory cytokines and MMPs [40].

In this chapter, microarrays enabled us to identify the differential expression of NF-κB-dependent genes in NP and control mucosa. Gene expression patterns we obtained were significantly different in normal mucosa in comparison with NP. Consequently, all examined tissues could be classified either as NPs or healthy mucosa (Figure 3). Simultaneously, analysis of NF-κB-dependent gene transcription failed to differentiate eosinophilic from non-eosinophilic NP, suggesting a similar expression pattern of those genes in different types of NP.

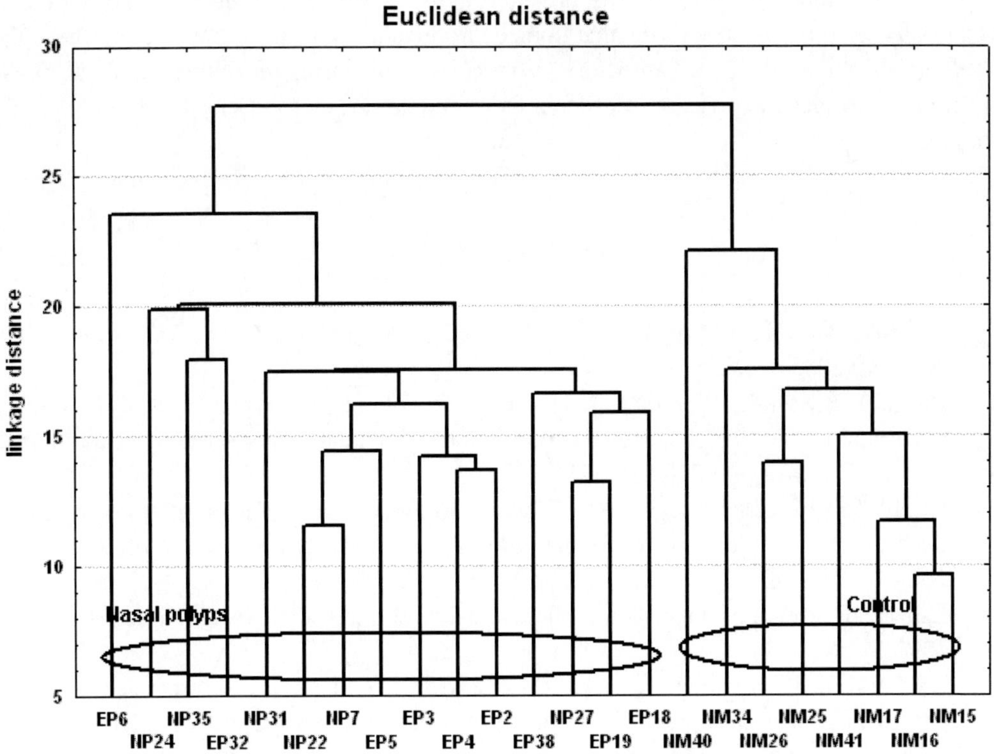

Figure 3. Dendrogram of 15 nasal polyps (eosinophilic polyps (EP), non-eosinophilic polyps (NP) and 8 normal mucosa (NM)) samples using cluster analysis based on the Euclidean distance with 582 human NF-κB-dependent genes represented on the HG-U133A GeneChip [35].

The outcomes of the studies mentioned above showed many differences, despite apparent similarities in clinical phenotypes of patients included into those studies. Therefore, the comparison of results is also difficult due to differences in methodology – especially in terms of microarrays used and sequences of probes. We agree with the Benson et al. [41] opinion that microarray analysis involves a considerable risk of spuriously identified genes. Additionally, a large number of genes are analyzed for a small patient group. Moreover, there is a lack of standards for data collection, analysis and validation. Further, choice of different subjects and tissue types may be the cause of inconsistent results demonstrated in previous reports [9]. This also applies to some difficulty to obtain optimum control sinus mucosal tissues for comparison. Turbinate mucosa frequently used as control might not be the most suitable as NP originate from the sinus mucosa. The heterogeneity among CRS patients can be supported by the relatively poor correlation of microarray data with RT-PCR results. Lastly, differences between studies may be explained by the required use of intranasal glucocorticoids before tissue collection. All of these problems seem to be common to microarray analysis and can lead to both spurious data as well as not identifying all of the important biological differences.

In conclusion, gene expression analysis in inflamed NP tissues may become a supplementary research method. Microarrays can be successfully used to identify previously unexplored inflammatory mechanisms involved in the pathophysiology of chronic rhinosinusitis. Despite the enormous initial promise, the introduction of microarray technology has failed to result in any breakthroughs for understanding of etiology of CRS. Our results in this chapter show that molecular mechanisms may have an influence on development and type of inflammatory process as well as on the clinical course of NP. Determining the gene expression profiles in NP is a concept that may lead to the preparation of a molecular map indicative of the type of polyps.

References

[1] Pawankar R. Nasal polyps: an update: editorial review. *Curr Opin Allergy Clin Immunol* 2003;3:1-6.

[2] Burgel PR, Escudier E, Coste A, Dao-Pick T, Ueki IF, Takeyama K et al. Relation of epidermal growth factor receptor expression to goblet cell hyperplasia in nasal polyps. *J Allergy Clin Immunol* 2000;106:705-712.

[3] Goldsmith ZG, Dhanasekaran N. The microrevolution: applications and impacts of microarray technology on molecular biology and medicine (review). *Int J Mol Med* 2004;13:483-495.

[4] Lanza DC, Kennedy DW. Adult rhinosinusitis defined. *Otolaryngol Head Neck Surg* 1997;117:S1-7.

[5] Lildholdt T, Rundcrantz H, Bende M, Larsen K. Glucocorticoid treatment for nasal polyps: the use of topical budesonide powder, intramuscular betamethasone, and surgical treatment. *Arch Otolaryngol Head Neck Surg* 1997;123:595-600.

[6] Lund VJ, Mackay IS. Staging in rhinosinusitis. *Rhinology* 1993;31:183-184.

[7] Affymetrix GeneChip®. *Expression analysis technical manual.* Santa Clara, CA: Affymetrix, 2001

[8] Stankovic KM, Goldsztein H, Reh DD, Platt MP, Metson R. Gene expression profiling of nasal polyps associated with chronic sinusitis and aspirin-sensitive asthma, *Laryngoscope* 2008;118:881-889.

[9] Liu Z, Kim J, Sypek JP, Wang IM, Horton H, Oppenheim FG et al. Gene expression profiles in human nasal polyp tissues studied by means of DNA microarray. *J Allergy Clin Immunol* 2004;114:783-790.

[10] Fritz SB, Terrell JE, Conner ER, Kuowska-Latallo JF, Baker JR. Nasal mucosal gene expression in patients with allergic rhinitis with and without nasal polyps. *J Allergy Clin Immunol* 2003;112:1057-1063.

[11] Payne SC, Han JK, Huyett P, Negri J, Kropf EZ, Borish L et al. Microarray analysis of distinct gene transcription profiles in non-eosinophilic chronic sinusitis with nasal polyps. *Am J Rhinol* 2008;22:568-581.

[12] Wu J, Bing L, Jin H, Jingping F. Gene expression profiles of nasal polyps associated with allergic rhinitis. *Am J Otolaryngol* 2009;30:24-32.

[13] Bocheńska-Marciniak M, Kupczyk M, Górski P, Kuna P. The effect of recombinant IL-8 on eosinophils' and neutrophils' migration in vivo and in vitro. *Allergy* 2003;58(8):795-801.

[14] van den Steen P, Wuyts A, Husson S, Proost P, van Damme J, Opdenakker G. Gelatinase B/MMP-9 and neutrophil collagenase/MMP-8 process the chemokines

human GCP-2/CXCL6, ENA-78/CXCL5 and mouse GCP-2/LIX and modulate their physiological activities. *Eur J Biochem* 2003;270:3739-3749.

[15] Kuhn H, Walther M, Kuban RJ. Mammalian arachidonate 15-lipoxygenases structure, function, and biological implications. *Prostaglandins Other Lipid Mediat* 2002;68-69:263-290.

[16] Yoshimoto T, Takahashi Y. Arachidonate 12-lipoxygenases. *Prostaglandins Other Lipid Mediat* 2002;68-69:245-262.

[17] Owens JM, Shroyer KR, Kingdom TT. Expression of cyclooxygenase and lipoxygenase enzymes in nasal polyps of aspirin-sensitive and aspirin-tolerant patients. *Arch Otolaryngol Head Neck Surg* 2006;132:579-587.

[18] Laprise C, Sladek R, Ponton A, Bernier MC, Hudson TJ, Laviolette M. Functional classes of bronchial mucosa genes that are differentially expressed in asthma. *BMC Genomics* 2004;5:21-31.

[19] Pérez-Novo CA, Watelet JB, Claeys C, Van Cauwenberge P, Bachert C. Prostaglandin, leukotriene, and lipoxin balance in chronic rhinosinusitis with and without nasal polyposis. *J Allergy Clin Immunol* 2005;115:1189-1196.

[20] Furukawa K, Harrison DG, Saleh D, Shennib H, Chagnon FP, Giaid A. Expression of nitric oxide synthase in the human nasal mucosa. *Am J Respir Crit Care Med* 1996;153:847-850.

[21] Saleh D, Ernst P, Lim S, Barnes PJ, Giaid A. Increased formation of the potent oxidant peroxynitrite in the airways of asthmatic patients is associated with induction of nitric oxide synthase: effect of inhaled glucocorticoid. *FASEB J* 1998;12:929-937.

[22] Escobar P, Marquez FJ, Carrasco C, Alonso-Orgaz S, Fortes J, Cenjor C et al. Endothelial nitric oxide synthase/soluble guanylate cyclase system in human nasal polyps. *Eur Arch Otorhinolaryngol* 2005;262:512-516.

[23] Watkins DN, Lewis RH, Basclain KA, Fisher PH, Peroni DJ, Garlepp MJ et al. Expression and localization of the inducible isoform of nitric oxide synthase in nasal polyp epithelium. *Clin Exp Allergy* 1998;28:211-219.

[24] Benson M, Carlsson L, Adner M, Jernås M, Rudemo M, Sjögren A et al. Gene profiling reveals increased expression of uteroglobin and other anti-inflammatory genes in glucocorticoid-treated nasal polyps. *J Allergy Clin Immunol* 2004;113:1137-1143.

[25] Westermarck J, Kahari VM. Regulation of matrix metalloproteinase expression in tumor invasion. *FASEB J* 1999;13:781-792.

[26] Legrand D, Pierce A, Elass E, Carpentier M, Mariller C, Mazurier J. Lactoferrin structure and functions. *Adv Exp Med Biol* 2008;606:163-194.

[27] Mollenhauer J, Herbertz S, Holmskov U, Tolnay M, Krebs I, Merlo A et al. DMBT1 encodes a protein involved in the immune defense and in epithelial differentiation and is highly unstable in cancer. *Cancer Res* 2000;60:1704-1710.

[28] Plager DA, Kahl JC, Asmann YW, Nilson AE, Pallanch JF, Friedman O et al. Gene transcription changes in asthmatic chronic rhinosinusitis with nasal polyps and comparison to those in atopic dermatitis. *PLoS ONE* 2010;5:e11450.

[29] Benson M, Carlsson L, Adner M, Carén H, Carlsson B, Carlsson LM et al. Gene profiling reveals increased expression of uteroglobin and other anti-inflammatory genes in glucocorticoid-treated nasal polyps. *J Allergy Clin Immunol* 2004;113:1137-1143.

[30] Bolger WE, Joshi AS, Spear S, Nelson M, Govindaraj K. Gene expression analysis in sinonasal polyposis before and after oral corticosteroids: a preliminary investigation. *Otolaryngol Head Neck Surg* 2007;137:27-33.

[31] Sekigawa T, Tajima A, Hasegawa T, Hasegawa Y, Inoue H, Sano Y et al. Gene-expression profiles in human nasal polyp tissues and identification of genetic susceptibility in aspirin-intolerant asthma. *Clin Exp Allergy* 2009;39:972-981.

[32] Wu X, Ghimbovschi S, Aujla PK, Rose MC, Peña MT. Expression profiling of inflammatory mediators in pediatric sinus mucosa. *Arch Otolaryngol Head Neck Surg* 2009;135:65-72.

[33] Anand VK, Kacker A, Orjuela AF, Huang C, Manarey C, Xiang J. Inflammatory pathway gene expression in chronic rhinosinusitis. *Am J Rhinol* 2006;20:471-476.

[34] Figueiredo CR, Santos RP, Silva ID, Weckx LL. Microarray cDNA to identify inflammatory genes in nasal polyposis. *Am J Rhinol* 2007;21:231-235.

[35] Frączek M, Rostkowska-Nadolska B, Kapral M, Szota J, Kręcicki T, Mazurek U. Microarray analysis of NF-κB-dependent genes in chronic rhinosinusitis with nasal polyps. *Adv Clin Exp Med* 2013;22:209-217.

[36] Yao T, Kojima Y, Koyanagi A, Yokoi H, Saito T, Kawano K et al. Eotaxin-1, -2, and -3 immunoreactivity and protein concentration in the nasal polyps of eosinophilic chronic rhinosinusitis patients. *Laryngoscope* 2009;119:1053-1059.

[37] Meyer N, Zimmermann M, Bürgler S, Bassin C, Woehrl S, Moritz K et al. IL-32 is expressed by human primary keratinocytes and modulates keratinocyte apoptosis in atopic dermatitis. *J Allergy Clin Immunol* 2010;125:858-865.

[38] Joosten LA, Netea MG, Kim SH, Yoon DY, Oppers-Walgreen B, Radstake TR et al. IL-32, a proinflammatory cytokine in rheumatoid arthritis. *Proc Natl Acad Sci USA* 2006;103:3298-3303.

[39] Karjalainen EM, Lindqvist A, Laitinen LA, Kava T, Altraja A, Halme M et al. Airway inflammation and basement membrane tenascin in newly diagnosed atopic and nonatopic asthma. *Respir Med* 2003;97:1045-1051.

[40] Goh FG, Piccinini AM, Krausgruber T, Udalova IA, Midwood KS. Transcriptional regulation of the endogenous danger signal tenascin-C: a novel autocrine loop in inflammation. *J Immunol* 2010;184:2655-2662.

[41] Benson M, Olsson M, Rudemow M, Wennergren G, Cardellz LO. Pros and cons of microarray technology in allergy research. *Clin Exp Allergy* 2004;34:1001-1006.

[42] Rostkowska-Nadolska B, Kapral M, Fraczek M, Kowalczyk M, Gawron W, Mazurek U. A microarray study of gene expression profiles in nasal polyps. *Auris Nasus Larynx* 2011;38:58-64.

In: Microarrays: Principles, Applications and Technologies ISBN: 978-1-62948-669-7
Editor: James V. Rogers

Chapter 14

Transcriptomic Analysis of Oligomeric Amyloid-β Treated Primary Cortical Neurons Reveals Dysregulation of Genes Involved in the Biosynthesis of Steroids and Cholesterol

Bilal Malik[1,*], *Cathy Fernandes*[2], *Richard Killick*[1], *Alessia Usardi*[1], *Ritchie Williamson*[3], *Stuart Kellie*[4,5], *Brian H. Anderton*[1] *and C. Hugh Reynolds*[1]

[1]King's College London, Department of Neuroscience, Institute of Psychiatry, De Crespigny Park, Denmark Hill, London, UK

[2]MRC Social, Genetic and Developmental Psychology Centre, King's College London, Institute of Psychiatry, De Crespigny Park, Denmark Hill, London, UK

[3]Bradford School of Pharmacy, School of Life Sciences, University of Bradford

[4]The University of Queensland, School of Chemistry and Molecular Biosciences; Australian Infectious Diseases Research Centre and institute for Molecular Bioscience, St Lucia, Brisbane, Australia

[5]Yamanouchi Research Institute, Armstrong Road, Littlemore Park, Oxford, UK

Abstract

Alzheimer's disease (AD) is a neurodegenerative disorder and is the most common cause of dementia. It is characterized by synaptic and neuronal loss, and the hallmark of AD is the presence of plaques in the brains of patients, which are primarily composed of the amyloid-β peptide (Aβ). However, the mechanisms by which Aβ gives rise to pathology have not been fully defined as yet. Evidence suggests that soluble oligomers of

* Present corresponding address: Sobell Department of Motor Neuroscience and Movement Disorders, UCL Institute of Neurology, Queen Square, London WC1 N 3BG. Tel.: +44 (0)20 3448 4318; fax: +44 (0)20 7813 3107. Email: b.malik@ucl.ac.uk.

the peptide are the most potent and toxic form of Aβ. To uncover the potential mechanism by which Aβ may lead to toxicity in neurons, a transcriptomic profile of primary mouse cortical neurons in response to oligomeric Aβ (1-42) treatment was performed. The results show the differential regulation of genes involved in the biosynthesis of steroids and cholesterol, as well as other genes whose products are associated with lipid metabolism. Cholesterol may be a risk factor for development of the disease and several cholesterol-related genes linked to AD have been reported. The findings in this chapter highlight the importance of interactions between Aβ and cholesterol, and have potential implications for pathology in AD brain.

Introduction

Alzheimer's disease (AD) is a neurodegenerative disorder and the most common cause of dementia leading to progressive cognitive decline, marked selective neuronal degeneration and synaptic loss, particularly in the hippocampus, amygdala and the association cortices of the frontal, temporal and parietal lobes [1, 2]. The characteristic hallmarks of the disease are defined pathologically by neuritic amyloid plaques, which develop extracellularly and contain amyloid–β (Aβ) peptides [3] and also the formation of intraneuronal neurofibrillary tangles comprised of the abnormally hyperphosphorylated cytoskeletal tau protein [4-6].

The amyloid cascade hypothesis stresses the role of Aβ in initiating the development of AD pathology [7], leading to the formation of neurofibrillary tangles and to widespread changes of a more general nature including cellular and regional atrophy [8], mitochondrial dysfunction [9-12] and loss of synapses [13-15]. Aβ found deposited in plaques is principally in two forms, $A\beta_{1\text{-}40}$ and $A\beta_{1\text{-}42}$, with the longer peptide having the greater propensity to fibrillize, aggregate, and induce toxicity. Furthermore, support has gathered for soluble oligomers of $A\beta_{1\text{-}42}$, termed amyloid-β derived diffusible ligands (ADDLs), as the primary toxic species and it has been reported that Aβ oligomers contribute to neuronal cell death and are involved in early stage pathogenesis [16, 17]. Importantly, the levels of ADDLs have been found to be elevated in brains of AD patients [18]. Furthermore, the evidence suggests a role for smaller soluble oligomers in the development of AD pathology, which have been shown to inhibit long-term potentiation (LTP), a mechanism that underlies learning and memory [19, 20]. Several studies have also provided evidence that low-n oligomers of Aβ are proximate neurotoxins and are indeed much more potent than monomers [20, 21].

Although the mechanisms initiating and driving AD are as yet unknown, several pathways have been proposed including axonal transport, mitochondrial abnormalities, hyperphosphorylation of cytoskeletal proteins, oxidative stress, inflammation, excitotoxicity, as well as altered cholesterol metabolism [22, 23]. Furthermore, changes in the expression levels of genes have been found to occur in brain of late stage AD patients [24-26], as well as brain tissue from transgenic mouse models [27-29]. The findings implicated several cellular components that may be important for pathological changes in AD, however it is important to note that in these studies the brains from AD patients may have been harboring long-term aberrant physiology leading to alterations in gene expression. The mechanisms by which Aβ gives rise to pathology are therefore not obvious, and an alternative approach is to treat cultured cells with preparations of Aβ to investigate the mechanisms of toxicity. Differential expression changes of mRNA transcripts have been shown to occur in neurons after the addition of fibrillar Aβ [30, 31]. Transcription factors such as NF-κB [32, 33], c-jun [34, 35], c-fos, and egr1 have all been reported to be differentially regulated by fibrillar Aβ [36], indicating the regulation of gene expression. However, up to now there have been relatively few studies examining the role of Aβ oligomers on gene expression in neurons. Therefore, we examined the mechanism underlying the toxicity of Aβ oligomers by undertaking a gene expression transcriptomic profiling study on ADDL-treated cultured neurons using Affymetrix oligonucleotide microarrays [37].

Our findings demonstrate a co-ordinated down-regulation in neurons of several genes involved in steroid and cholesterol biosynthesis and furthermore the regulation of other genes which are associated with the lipid metabolism pathway. In AD, several genes related to the cholesterol pathway have been linked to the disease, with the ApoE ε4 allele in particular being a strong risk factor for the development of pathology [38-40]. We have therefore shown that oligomeric Aβ may affect cholesterol and other lipids, and previous studies have also suggested the potential importance of interactions between Aβ and cholesterol [41-44]. Significantly, our results emphasise the role of ADDLs and the cholesterol and steroid biosynthesis pathways in the development of the pathology of AD. Therefore, these findings merit further examination of the relationship of cholesterol and lipids with oligomeric Aβ.

Methods

Primary Mouse Cortical Neuronal Culture

Primary mouse cortical neuronal cultures were prepared using E16 embryos from time mated C57BL/6JOlaHsd mice (Harlan, UK), using a papain dissociation kit (Worthington Biochemical Corporation, UK) according to the manufacturer's instructions, and cultured as described previously [45; 46].

Preparation of ADDLs

Amyloid beta ($A\beta_{1\text{-}42}$) (California Peptide Research, California, USA) was used to prepare oligomeric amyloid-β derived diffusible ligands (ADDLs) according to the method described by Klein [16], except that phenol red free Neurobasal medium was used in place of phenol red free F12 medium. The preparations of ADDLs were characterised by western blot [47] and contained oligomers that resolved into primarily monomer, dimmer, trimer, and tetramer, and importantly is distinct from fibrillar preparations [16, 18, 48, 49]. The concentration was determined by the Bradford assay [50].

ADDLs Toxicity

The lactate dehydrogenase (LDH) colorimetric assay was used according to the manufacturer's instructions to determine ADDL toxicity (Promega, UK). Caspase activation was measured by western blot analysis using antibodies against cleaved (activated) caspase 3

(Cell Signaling, UK) and β-actin (Sigma, UK). A pan-caspase CaspaTag© fluorescent stain (Intergen, UK) to assess caspase activation was also used according to the manufacturer's instructions and as described previously [51]. Apoptosis was further monitored by immunofluorescent labeling of activated caspase 3 (Cell Signaling, UK). Statistical analysis was performed using SPSS v15 (SPSS Inc, USA) with the unpaired t-test for two independent samples (two-tailed) or ANOVA with corresponding post-hoc tests used to determine significance of data ($p\text{-value} < 0.05$).

RNA Extraction and Quantitative RT-PCR (qPCR)

Total RNA was extracted using the guanidine isothiocyanate based method [52] with Trizol solution (Invitrogen) according to the manufacturer's instructions and was reverse transcribed to cDNA using Superscript III reverse transcriptase (Invitrogen). Real-time quantitative RT-PCR (qPCR) was performed with QuantiTect SYBR Green PCR master mix (Qiagen) in a total volume of 20 μL as described previously [53]. A dissociation curve was generated to ensure absence of primer dimers. Reactions were performed in triplicate and the values were normalized using the geometric mean of the housekeeping genes, β-actin, Gapdh and hypoxanthine guanine phosphoribosyl transferase-1 (Hprt1) [54]. The comparative threshold cycle (ddCt) method [55] was used for the relative quantification of gene expression and an unpaired t-test was used to determine statistical significance.

Affymetrix Microarrays and Data Analysis

Total RNA was reverse transcribed to cDNA using Superscript II (Affymetrix kit) and the T7 oligo(dT) promoter primer. The double stranded cDNA containing T7 promoter sequences served as a template for in vitro transcription (IVT), using a T7 RNA polymerase with a biotinylated ribonucleotide mix to generate amplified and labelled cRNA. The cRNA was fragmented and then hybridized onto Murine U74Av2 chips, which contain 12,488 gene transcripts. Microarrays were stained and washed with the EukGE-WS2v4 fluids script program using the Affymetrix Gene Chip fluidics station 400 (Affymetrix, UK). Each array was scanned with the Agilent GeneArray scanner and image acquisition was performed using the Microarray Analysis Suite 5.0 software (MAS 5.0). Hybridization was performed using three biological replicates of control (untreated) and Aβ treated samples from three independent primary mouse cortical neuronal cultures. Prior to full analysis, one of the six arrays used for the first RNA sample from treated cells did not meet all the standard checks of internal consistency with the other data sets. It was decided therefore to omit this array from the analysis.

The raw data were transferred to an R v2.6.0 environment (http://www.r-project.org/) [56], and analysed using the BioConductor packages (http://www.bioconductor.org/) [57]. The Robust Multichip Average (RMA) algorithm was implemented to normalize the microarray data sets and the Simpleaffy package [58] was used to assess the quality of the arrays. The empirical Bayes moderated t-statistics test from the Limma package [59] was used to identify the differentially expressed probe sets, and then the q-values were calculated by using the false discovery rate (FDR) method of Benjamini and Hochberg [60] and of

Storey [61] to correct for multiple hypothesis testing. Gene ontology pathway and statistical analysis of over-representation of genes in gene ontology categories was performed using DAVID [62]. GenMAPP was also used to assist in the visualization of gene expression data within specific biological pathways [63].

Results

Neurotoxicity of ADDLs

In the first instance, to establish the toxicity of the ADDL preparation, cultured primary mouse cortical neurons were treated with a range of concentrations of ADDLs up to 20 μM and over a period of 5 days. This allowed the determination of the optimal conditions that produced a measurable effect but with low levels of toxicity, enabling the transcriptomic analysis of cultured neurons after ADDL treatment. Therefore, LDH release was measured in order to determine cell lysis and death (Figure 1A). Induction of apoptosis was established by assessing caspase 3 activation using western blot (Figure 1B), as well as quantification of the number of cells stained for activated caspase 3 and those exhibiting apoptotic morphology using the Hoechst nuclear stain (Figure 2A). Further confirmation of the initiation of toxicity by ADDLs was sought by using a pan-casapse CaspaTag stain, as well as the uptake of propidium iodide to stain cells that were undergoing a cell death response (Figure 2B).

Our results show that at a concentration of 5 μM ADDLs, both LDH release and caspase activation, including caspase 3, showed a small but significant increase after 24 h, compared to control cultures (Figures 1 and 2). Therefore, treatment of primary cortical neurons with 5 μM ADDLs for 24 h was adopted as the optimal condition, as ADDLs are clearly producing a response but did not lead to substantial cell loss, permitting the investigation of differential changes in gene expression.

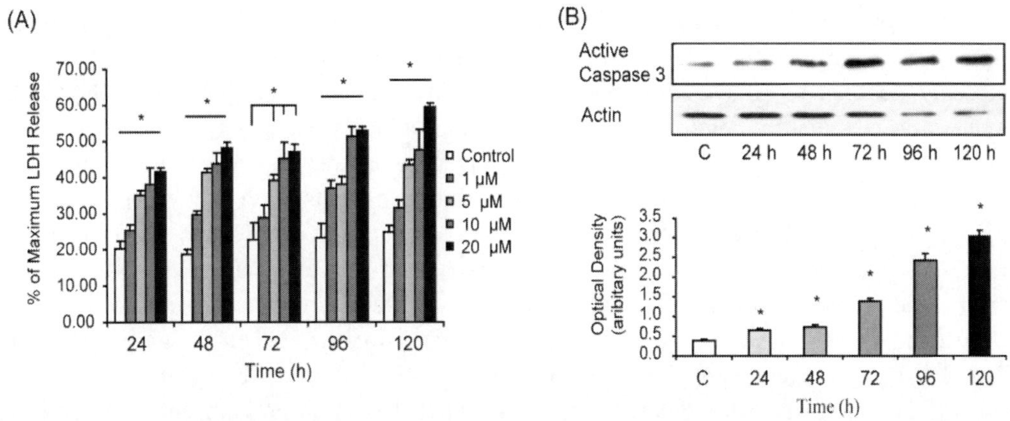

Figure 1. ADDLs are neurotoxic to primary cortical neurons. (A) Primary mouse neuronal cortical cultures were incubated with 5 μM ADDLs for 24 h and LDH release was measured to determine cell death. (B) Activation of caspase 3 was monitored using western blot.

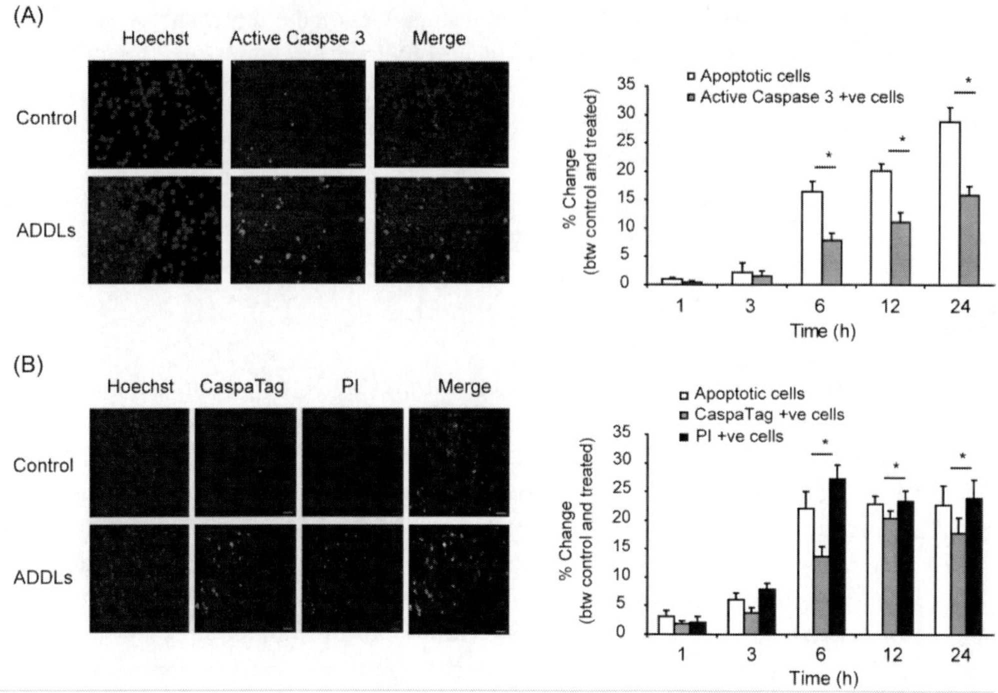

Figure 2. Activation of apoptosis by ADDLs. (A) Caspase 3 activation was quantified by counting caspase 3 positive cells and cells exhibiting apoptotic morphology after visualisation using the Hoechst nuclear stain. (B) Cell death and caspase activation was also measured using the pan-caspase CaspaTag stain and propidium iodide (PI).

Analysis of Gene Expression in Primary Cortical Neurons after ADDLs Treatment

The expression level of genes was determined by undertaking a transcriptomic profile of primary mouse cortical neuronal cultures in response to 5 μM ADDL treatment for 24 h. By exposing neurons to ADDLs for this period of time, we were confident that the ADDL preparation was mainly comprised of oligomers and which were stable under our culturing conditions, not resulting in dissociation or the conversion into a fibrillar conformation [47].

Several potentially interesting and significant transcripts were shown to be up or downregulated by ADDL treatment (Table 1 and Table 2). The data for the 50 most strongly downregulated and upregulated probe sets are shown in Table 1 (representing 44 different genes) and Table 2 (48 genes), respectively. It should be noted that some genes were represented by two or more probe sets; therefore, increasing the statistical significance of the changes. The gene encoding Nsdhl (NAD(P) dependent steroid dehydrogenase-like) was found to be the most downregulated gene after ADDL treatment and indeed had three probe sets in the 50 most strongly downregulated probe sets (Table 1). The p-values of change for the three probe sets ranged between 0.0137 and 0.0199, but the combined probability for all three was 4.97×10^{-6}. The Bonferroni correction for false discovery (by multiplying by the number of probe sets) gave p=0.062, we can therefore be confident the down-regulation of Nsdhl was statistically significant. Abca1 (ATP-binding cassette transporter A1) was the most strongly upregulated gene and had two probe sets in the list of the top 50 most upregulated probe sets, (Table 2). The combined p-value of change was 7.01×10^{-7}, or with the Bonferroni correction p=0.0088 and so expression of this gene was significantly upregulated. Tables 1 and 2 contain two probe sets each for several other genes, and so their expression changes might well be relevant. In Table 1 (downregulated), there are further genes that are represented by two probe sets: Fdps (farnesyl diphosphate synthase), Scd1 (stearoyl-coenzyme A desaturase), Ndrg1 (N-myc downstream regulated gene 1) and Gria1 (AMPA-1 alpha 1 glutamate receptor). Interestingly, down-regulation of the Gria1 gene expression has been reported in brain of patients with AD [64].

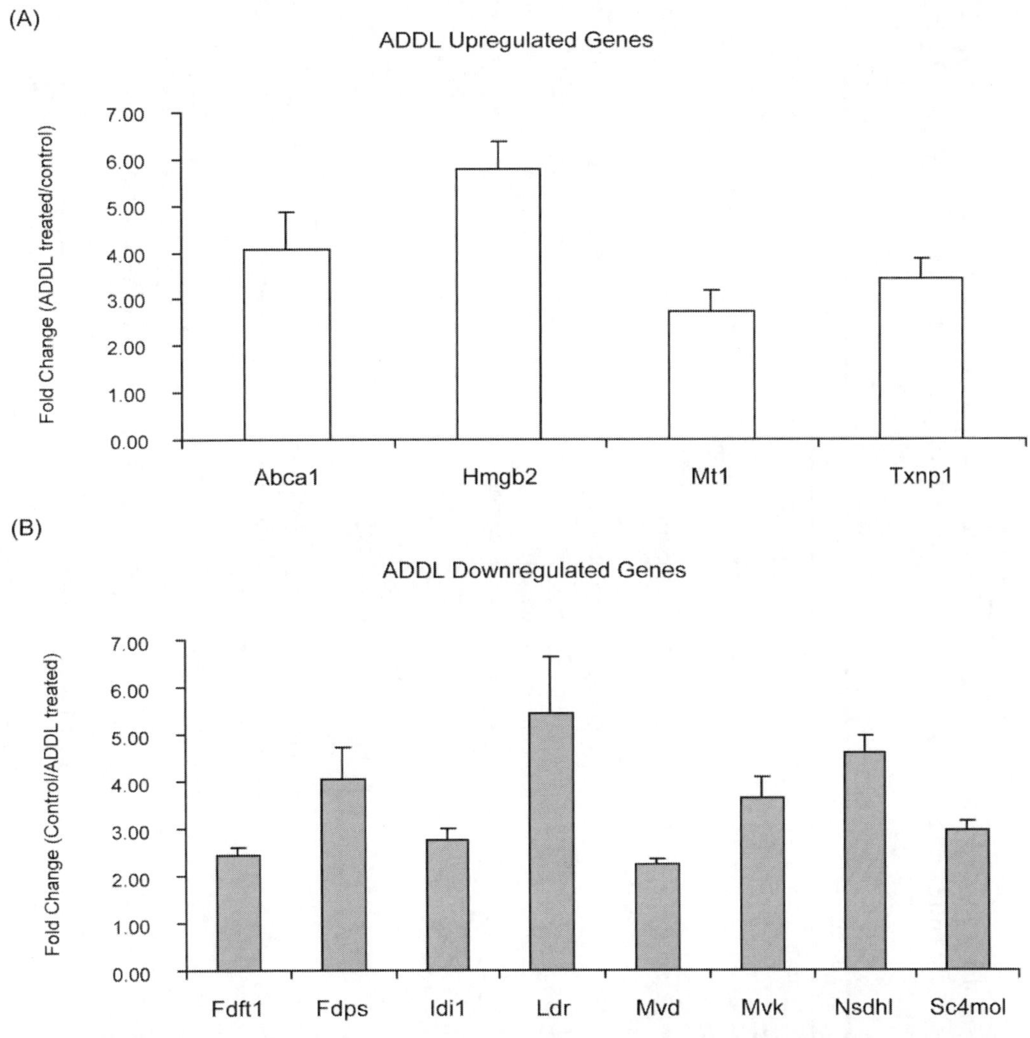

Figure 3. qPCR analysis of ADDL-induced gene expression changes. Real time qPCR was performed using primary cortical neurons treated with ADDLs. (A) The results show that qPCR corroborated the expression levels of genes found to be differentially regulated using microarray. (B) Genes involved in the steroid and cholesterol biosynthesis pathways were found to be downregulated.

Table 1. Fifty most highly down-regulated probe sets after treatment with 5 μM ADDLs for 24 h, with corresponding p-values (P) and fold-change (FC). Probe set results for genes classified as 00100 Biosynthesis of Steroids and for 00900 Terpenoid Biosynthesis are shown in bold

Probe	Symbol	Description	Pathway	*P*	FC
93868_at	Nsdhl	NAD(P) dependent steroid dehydrogenase-like	00100: Biosynthesis of steroids	0.0199	-4.771
98631_g_at	Nsdhl	NAD(P) dependent steroid dehydrogenase-like	00100: Biosynthesis of steroids	0.0137	-3.935
99098_at	Fdps	farnesyl diphosphate synthetase	00100: Biosynthesis of steroids, 00900: Terpenoid biosynthesis	0.0122	-3.785
160424_f_at	Fdps	farnesyl diphosphate synthetase	00100: Biosynthesis of steroids, 00900: Terpenoid biosynthesis	0.0097	-3.502
160388_at	Sc4mol	sterol-C4-methyl oxidase-like	00100: Biosynthesis of steroids	0.0464	-3.473
96269_at	Idi1	isopentenyl-diphosphate delta isomerase	00100: Biosynthesis of steroids, 00900: Terpenoid biosynthesis	0.0251	-3.284
93682_at	Ldb2	LIM domain binding 2		0.1355	-2.853
94325_at	Pbx1	pre B-cell leukemia transcription factor 1		0.0605	-2.820
98630_at	Nsdhl	NAD(P) dependent steroid dehydrogenase-like	00100: Biosynthesis of steroids	0.0182	-2.789
92403_at	St6gal-nac5	ST6 (alpha-N-acetyl-neuraminyl-2,3-beta-galactosyl-1,3)-N-acetylgalactosaminide alpha-2,6-sialyltransferase 5	00604: Glycosphingolipid biosynthesis - ganglioseries, 01031: Glycan structures - biosynthesis 2	0.0780	-2.781
160770_at	Mvd	mevalonate (diphospho) decarboxylase	00100: Biosynthesis of steroids	0.0110	-2.641
97518_at	Fdft1	farnesyl diphosphate farnesyl transferase 1	00100: Biosynthesis of steroids, 00900: Terpenoid biosynthesis	0.0763	-2.635
95036_at	Calb2	calbindin 2		0.0191	-2.608
160832_at	Ldlr	low density lipoprotein receptor		0.0065	-2.482
93568_i_at				0.0683	-2.359
103543_at	Sez6l	seizure related 6 homolog like		0.0048	-2.345

Probe	Symbol	Description	Pathway	*P*	FC
103282_at	Rasgrp2	RAS, guanyl releasing protein 2	04010: MAPK signaling pathway	0.0478	-2.325
94056_at	Scd1	stearoyl-Coenzyme A desaturase 1	03320: PPAR signaling pathway	0.0102	-2.252
99510_at	Prkcb1	protein kinase C, beta 1	(Many signaling pathways)	0.0199	-2.252
94057_g_at	Scd1	stearoyl-Coenzyme A desaturase 1	03320: PPAR signaling pathway	0.0063	-2.212
103224_at	Rab33a	RAB33A, member of RAS oncogene family		0.0136	-2.211
94322_at	Sqle	squalene epoxidase	00100: Biosynthesis of steroids, 00900: Terpenoid biosynthesis	0.0368	-2.196
93569_f_at				0.0960	-2.186
95079_at	Pdgfra	platelet derived growth factor receptor, alpha polypeptide	(Many signaling pathways)	0.3790	-2.152
92943_at	Gria1	glutamate receptor, ionotropic, AMPA1 (alpha 1)	04080: Neuroactive ligand-receptor interaction 04720: Long-term potentiation 04730: Long-term depression	0.1261	-2.091
160464_s_at	Ndrg1	N-myc downstream regulated gene 1		0.0713	-2.051
102144_f_at	Sfpq	splicing factor proline/glutamine rich (polypyrimidine tract binding protein associated)		0.1509	-2.047
100406_at	Ptpn5	protein tyrosine phosphatase, non-receptor type 5	04010: MAPK signaling pathway	0.0472	-2.041
160571_at	Idh1	isocitrate dehydrogenase 1 (NADP+), soluble	00020: Citrate cycle (TCA cycle), 00480: Glutathione metabolism, 00720: Reductive carboxylate cycle (CO2 fixation)	0.1186	-2.005
100019_at	Vcan	versican	04514: Cell adhesion molecules (CAMs)	0.1597	-2.001
96791_at	1500005 K14Rik	RIKEN cDNA 1500005K14 gene		0.0351	-2.000
160760_at	Ptprk	protein tyrosine phosphatase, receptor type, K		0.0611	-1.982
95436_at	Sst	somatostatin		0.1296	-1.977
93316_at	Osbpl1a	oxysterol binding protein-like 1A		0.1390	-1.940

Table 1. (Continued)

Probe	Symbol	Description	Pathway	*P*	FC
92944_at	Gria1	glutamate receptor, ionotropic, AMPA1 (alpha 1)	04080: Neuroactive ligand-receptor interaction, 04720: Long-term potentiation, 04730: Long-term depression	0.1241	-1.939
96596_at	Ndrg1	N-myc downstream regulated gene 1		0.0602	-1.933
103393_at	Pspc1	paraspeckle protein 1		0.1653	-1.912
98857_at	Neurod6	neurogenic differentiation 6		0.1186	-1.909
96135_at	3110003 A17Rik	RIKEN cDNA 3110003A17 gene		0.0434	-1.892
98967_at	Fabp7	fatty acid binding protein 7, brain	03320: PPAR signaling pathway	0.1375	-1.891
95135_at	Mid1ip1	Mid1 interacting protein 1 (gastrulation specific G12-like (zebrafish))		0.0174	-1.889
96346_at	Cdo1	cysteine dioxygenase 1, cytosolic	00272: Cysteine metabolism, 00430: Taurine and hypotaurine metabolism	0.0112	-1.878
100600_at	Cd24a	CD24a antigen	04640: Hematopoietic cell lineage	0.0778	-1.854
95632_f_at	Mvk	mevalonate kinase	00100: Biosynthesis of steroids	0.0054	-1.848
161059_at	Slc6a1	solute carrier family 6 (neurotransmitter transporter, GABA), member 1		0.0333	-1.832
160190_at	Syt4	synaptotagmin IV		0.2310	-1.827
94916_at	Cyp51	cytochrome P450, family 51	00100: Biosynthesis of steroids	0.0588	-1.826
95462_at	Bzw2	basic leucine zipper and W2 domains 2		0.0019	-1.822
99876_at	Sla	src-like adaptor		0.0619	-1.820
99425_at	Hmgcr	3-hydroxy-3-methylglutaryl-Coenzyme A reductase	00100: Biosynthesis of steroids	0.0643	-1.815

Table 2. Fifty most highly up-regulated probe sets after treatment with 5 μM ADDLs for 24 h, with corresponding p-values (P) and fold-change (FC)

Probe	Symbol	Description	Pathway	*P*	FC
160547_s_at	Txnip	thioredoxin interacting protein		0.0429	3.842
93573_at	Mt1	metallothionein 1		0.0233	2.359
95705_s_at	Actb	actin, beta, cytoplasmic	01430: Cell Communication, 04510: Focal adhesion, 04520: Adherens junction, 04530: Tight junction, 04670: Leukocyte transendothelial migration, 04810: Regulation of actin cytoskeleton	0.0035	2.180
94354_at	Abca1	ATP-binding cassette, sub-family A (ABC1), member 1	02010: ABC transporters - General	0.0007	2.103
99816_at	Hspa2	heat shock protein 2		0.0766	2.083
101587_at	Ephx1	epoxide hydrolase 1, microsomal	00980: Metabolism of xenobiotics by cytochrome P450	0.1051	2.022
100130_at	Jun	Jun oncogene	04010: MAPK signaling pathway, 04012: ErbB signaling pathway, 04310: Wnt signaling pathway, 04510: Focal adhesion, 04620: Toll-like receptor signaling pathway, 04660: T cell receptor signaling pathway, 04662: B cell receptor signaling pathway, 04912: GnRH signaling pathway, 05210: Colorectal cancer, 05211: Renal cell carcinoma	0.1316	1.958
93088_at	B2m	beta-2 microglobulin	04612: Antigen processing and presentation	0.0841	1.954
94967_at	Ermp1	endoplasmic reticulum metallo-peptidase 1		0.1713	1.954
97198_at	Abca1	ATP-binding cassette, sub-family A (ABC1), member 1	02010: ABC transporters - General	0.0010	1.953
100955_at	Ube2t	ubiquitin-conjugating enzyme E2T (putative)		0.1371	1.925
101561_at	Mt2	metallothionein 2		0.0063	1.921
104378_at	Pon2	paraoxonase 2	00361: gamma-Hexachlorocyclohexane degradation, 00363: Bisphenol A degradation	0.0284	1.890
101876_s_at	H2-T17	histocompatibility 2, T region locus 17		0.0282	1.874

Table 2. (Continued)

Probe	Symbol	Description	Pathway	*P*	FC
93250_r_at	Hmgb2	high mobility group box 2		0.0511	1.855
93595_at	Tpp1	tripeptidyl peptidase I		0.0272	1.855
104237_at	Pcbd2	pterin 4 alpha carbinolaminedehydratase/dimerization cofactor of hepatocyte nuclear factor 1 alpha (TCF1) 2		0.0531	1.854
100136_at	Lamp2	lysosomal-associated membrane protein 2		0.0473	1.853
94817_at	Serpinh1	serine (or cysteine) peptidase inhibitor, clade H, member 1		0.0214	1.845
95620_at	Dhrs7	dehydrogenase/reductase (SDR family) member 7	00361: gamma-Hexachlorocyclohexane degradation, 00363: Bisphenol A degradation, 00624: 1- and 2-Methylnaphthalene degradation, 00626: Naphthalene and anthracene degradation, 00632: Benzoate degradation via CoA ligation, 00642: Ethylbenzene degradation, 00903: Limonene and pinene degradation	0.1360	1.839
98144_f_at	Tspan31	tetraspanin 31		0.0824	1.834
93833_s_at	Hist1h2bc	histone cluster 1, H2bc		0.1749	1.832
93882_f_at	Tgoln2	trans-golgi network protein 2		0.0836	1.832
101578_f_at	Actb	actin, beta, cytoplasmic	01430: Cell Communication, 04510: Focal adhesion, 04520: Adherens junction, 04530: Tight junction, 04670: Leukocyte transendothelial migration, 04810: Regulation of actin cytoskeleton	0.0191	1.8307
160271_at	0610007 C21Rik	RIKEN cDNA 0610007C21 gene		0.0543	1.827
94840_at	Hexa	hexosaminidase A	00511: N-Glycan degradation, 00530: Aminosugars metabolism, 00531: Glycosaminoglycan degradation, 00603: Glycosphingolipid biosynthesis - globoseries, 00604: Glycosphingolipid biosynthesis - ganglioseries, 01032: Glycan structures - degradation	0.0962	1.826

Probe	Symbol	Description	Pathway	*P*	FC
96653_at	Rnaset2a	ribonuclease T2A		0.0086	1.813
161270_i_at	Wnk1	WNK lysine deficient protein kinase 1		0.1352	1.799
93731_at	Fkbp9	FK506 binding protein 9		0.0029	1.773
160493_at	Cd63	Cd63 antigen		0.0058	1.753
99009_at	Nnt	nicotinamide nucleotide transhydrogenase	00760: Nicotinate and nicotinamide metabolism	0.0587	1.725
160471_at	Slbp	stem-loop binding protein		0.1078	1.720
94813_at	Gas1	growth arrest specific 1	04340: Hedgehog signaling pathway	0.0200	1.702
101001_at	Gpr177	G protein-coupled receptor 177		0.0856	1.701
94275_at	Urod	Uroporphyrinogen decarboxylase	00860: Porphyrin and chlorophyll metabolism	0.0580	1.697
160482_at	Acaa1a	acetyl-Coenzyme A acyltransferase 1A	00071: Fatty acid metabolism, 00120: Bile acid biosynthesis, 00280: Valine, leucine and isoleucine degradation, 00362: Benzoate degradation via hydroxylation, 03320: PPAR signaling pathway	0.0744	1.6909
161076_at	BC023882	cDNA sequence BC023882		0.0541	1.684
100089_at	Ppic	peptidylprolyl isomerase C		0.0737	1.683
101966_s_at	Rnf13	ring finger protein 13		0.0372	1.678
99990_at	Rbbp6	retinoblastoma binding protein 6		0.0355	1.677
93511_at	Itm2a	integral membrane protein 2A		0.0452	1.671
92208_at	Caprin2	caprin family member 2		0.0777	1.668
104469_at	Pdpn	Podoplanin		0.1000	1.667
104386_f_at				0.0452	1.667
96316_at	LOC100043429	similar to SMT3B protein		0.0041	1.666
92256_at	Ctsb	cathepsin B	04612: Antigen processing and presentation	0.0728	1.664
103016_s_at	Cd68	CD68 antigen		0.0831	1.658
160200_at	3230401D17Rik	RIKEN cDNA 3230401D17 gene		0.0530	1.649
95350_at	Ttr	Transthyretin	Can bind to Aβ and retinoids	0.0648	1.649
102292_at	Gadd45a	growth arrest and DNA-damage-inducible 45 alpha	04010: MAPK signaling pathway, 04110: Cell cycle, 0415: p53 signaling pathway	0.1273	1.646

Table 3. Gene ontology analysis using DAVID

Category	Term	Count	%	*P*	Fold Enrichment	Bonferroni	Benjamini	FDR
SP_PIR_KEYWORDS	sterol biosynthesis	9	16.07%	4.55E-15	103.021	3.81E-12	3.81E-12	7.03E-12
GOTERM_BP_ALL	GO:0016126~sterol biosynthetic process	10	17.86%	6.23E-15	67.1816	2.94E-11	2.94E-11	1.18E-11
KEGG_PATHWAY	mmu00100:Biosynthesis of steroids	10	17.86%	1.58E-14	53.1826	3.07E-12	3.07E-12	1.97E-11
SP_PIR_KEYWORDS	cholesterol biosynthesis	8	14.29%	3.63E-14	129.7302	3.04E-11	1.52E-11	5.61E-11
GOTERM_BP_ALL	GO:0006695~cholesterol biosynthetic process	9	16.07%	6.44E-14	77.7387	3.04E-10	1.52E-10	1.22E-10
SP_PIR_KEYWORDS	steroid biosynthesis	9	16.07%	4.03E-13	64.8651	3.37E-10	1.12E-10	6.22E-10
GOTERM_BP_ALL	GO:0016125~sterol metabolic process	11	19.64%	4.47E-13	32.7097	2.11E-09	7.03E-10	8.45E-10
SP_PIR_KEYWORDS	lipid synthesis	10	17.86%	3.56E-12	36.0362	2.98E-09	7.46E-10	5.50E-09
GOTERM_BP_ALL	GO:0008299~isoprenoid biosynthetic process	7	12.50%	7.41E-12	115.4302	3.50E-08	8.75E-09	1.40E-08
GOTERM_BP_ALL	GO:0008202~steroid metabolic process	12	21.43%	8.81E-12	19.4347	4.16E-08	8.32E-09	1.67E-08
GOTERM_BP_ALL	GO:0006694~steroid biosynthetic process	10	17.86%	9.09E-12	32.3912	4.29E-08	7.16E-09	1.72E-08
GOTERM_BP_ALL	GO:0008203~cholesterol metabolic process	10	17.86%	9.09E-12	32.3912	4.29E-08	7.16E-09	1.72E-08
GOTERM_BP_ALL	GO:0006720~isoprenoid metabolic process	7	12.50%	6.00E-10	63.4866	2.84E-06	3.54E-07	1.14E-06
GOTERM_BP_ALL	GO:0008610~lipid biosynthetic process	11	19.64%	1.57E-08	11.6005	7.41E-05	8.23E-06	2.97E-05
GOTERM_BP_ALL	GO:0006066~alcohol metabolic process	11	19.64%	1.43E-07	9.1949	6.76E-04	6.77E-05	2.71E-04
GOTERM_BP_ALL	GO:0044255~cellular lipid metabolic process	13	23.21%	2.32E-07	6.5321	0.0011	9.97E-05	4.39E-04
GOTERM_BP_ALL	GO:0006629~lipid metabolic process	13	23.21%	1.10E-06	5.6549	0.0052	4.32E-04	0.0021
KEGG_PATHWAY	mmu00900:Terpenoid biosynthesis	4	7.14%	4.21E-06	97.856	8.20E-04	4.10E-04	0.0052
GOTERM_BP_ALL	GO:0009058~biosynthetic process	15	26.79%	1.10E-04	3.1025	0.4034	0.0390	0.2069
SP_PIR_KEYWORDS	Isoprene biosynthesis	3	5.36%	1.46E-04	145.9464	0.1155	0.0242	0.2258
COG_ONTOLOGY	Lipid metabolism	4	7.14%	1.75E-04	22.0522	0.0104	0.0104	0.1748
SP_PIR_KEYWORDS	sterol biosynthesis	9	16.07%	4.55E-15	103.021	3.81E-12	3.81E-12	7.03E-12
COG_ONTOLOGY	Lipid metabolism	4	7.14%	1.75E-04	22.0522	0.0104	0.0104	0.17478

Table 4. Genes with functions and fold-change (FC) related to steroid and lipid metabolism, but not classified as 00100 Biosynthesis of Steroids

Gene	Description	Notes	FC
Pbx1	pre B-cell leukemia transcription factor 1	Upstream regulator of gonadal (and adrenal) development	-2.820
St6gal-nac5	ST6 (alpha-N-acetyl-neuraminyl-2,3-beta-galactosyl-1,3)-N-acetylgalactosaminide alpha-2,6-sialyltransferase 5	00604: Glycosphingolipid biosynthesis	-2.781
Calb2	calbindin 2	Diminished in ApoE-deficient mice	-2.608
Ldlr	low density lipoprotein receptor	Uptake of cholesterol	-2.482
Scd1	stearoyl-Coenzyme A desaturase 1	03320: PPAR signalling pathway; unsaturated FA synthesis	-2.252, -2.212
Prkcb1	protein kinase C, beta 1	Knockouts are leaner (have less fat)	-2.252
Ndrg1	N-myc downstream regulated gene 1	Interacts with ApoA-I and A-II; possible role in HDL cholesterol transport	-2.051. -1.933
Sfpq	splicing factor proline/glutamine rich (polypyrimidine tract binding protein associated)	May regulate androgen receptor-mediated transcription; forms complex with Pspc1 (see below)	-2.047
Idh1	isocitrate dehydrogenase 1 (NADP+), soluble	May be involved in supply of NADPH for lipid and steroid biosynthesis	-2.005
Osbpl1a	oxysterol binding protein-like 1A	Sterol sensor; up-regulates SREBP-1c and enhances lipogenesis	-1.940
Pspc1	paraspeckle protein 1	Forms complex with Sfpq (see above); may affect androgen activity	-1.912
Fabp7	fatty acid binding protein 7, brain	03320: PPAR signalling pathway; lipid metabolism	-1.891
Slc6a1	solute carrier family 6 (neurotransmitter transporter, GABA), member 1	Knockouts have long-term decrease in testosterone production	-1.832

In order to extract further biologic meaning from the genes lists, we aimed to identify groups of genes that function in the same pathways using gene ontology analysis. Inspection of the 50 most strongly downregulated probe sets (Table 2) revealed that many are involved in steroid metabolism. Thirteen probe sets (representing 10 genes) out of a possible 35 (representing 22 genes) listed as in KEGG category of Biosynthesis of Steroids [65] were over-represented (Table 1, bold text). Further analysis by the DAVID bioinformatics software [62] (http://david.abcc.ncifcrf.gov/) confirmed that this KEGG pathway and several gene ontology categories involving steroids and lipids were over-represented with high statistical significance (Table 3). Various other genes related to steroid and lipid metabolism also appear in Table 1 (the list of most strongly downregulated probe sets) and are summarized in Table 4. Collectively, the results imply that along with its effect on steroid biosynthesis, a wider down-regulation of lipid metabolism was occurring after ADDL treatment. The down-regulation of Ldlr (low density lipoprotein receptor) for example is interesting, as a decrease in this protein would be predicted to lead to a decrease in cholesterol uptake.

In contrast, DAVID analysis of the list of most strongly upregulated probe sets (Table 2) did not reveal any significantly over-represented gene categories. However, the Abca1 gene product promotes cholesterol efflux from cells and functions to decrease levels of Aβ [66], underlining the potential importance of cholesterol in the development of AD pathology. The upregulation of Acaa1a (acetyl-CoA acyltransferase1A) also emphasise that alterations in lipid metabolism may be induced by ADDLs. Interestingly, two other steroid biosynthesis genes, Dhcr7 (7-dehydrocholesterol reductase) and Sc5d (sterol-C5-desaturase), were downregulated, although they did not appear in the 50 most downregulated probe sets. The levels of Acot7 were also decreased and is known to hydrolyze arachidonoyl-CoA to arachidonic acid, altering eicosanoid production [67].

ADDLs Down-regulate Genes Involved in the Steroid Metabolism and Cholesterol Biosynthesis Pathways

To confirm the results obtained in the microarray analysis, the expression levels of selected genes, Abca1, Hmgb2, Mt1 as well as Txnp, were determined using qPCR and were all shown to be upregulated after ADDL treatment of primary cortical neurons (Figure 3A). As the microarray analysis revealed that several downregulated probe sets were involved in steroid and lipid metabolism, the levels of genes classified within this pathway were also validated by qPCR. In line with the microarray data, qPCR analysis revealed that the expression of several genes (Fdft1, Fdps, Idi1, Ldr, Mvd, Mvk, Nsdhl, Sc4mol), was significantly decreased in primary cortical neurons after ADDL treatment (Figure 3B). As can be seen in Figure 4, these genes encode enzymes involved in cholesterol biosynthesis and remarkably the expression of every gene involved in this pathway was downregulated. All but one of the genes (Pmvk) were represented on the microarray. The overall results, therefore, strongly suggest a role for the cholesterol biosynthesis pathway in the neurotoxicity of ADDLs (Figure 4).

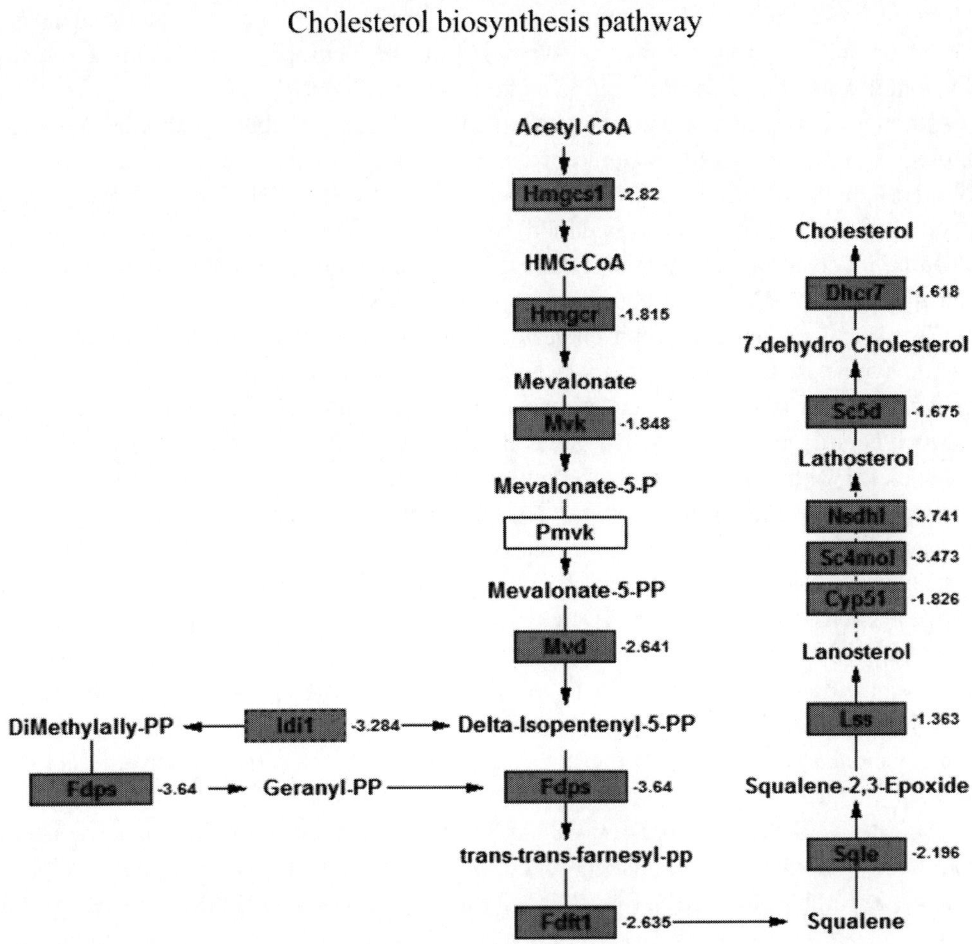

Figure 4. Downregulation of genes involved in the cholesterol biosynthesis pathway. GenMapp was used to visualise differentially expressed genes after treatment of primary mouse cortical neurons by ADDLs. Genes found in the cholesterol biosynthesis pathway downregulated by ADDL treatment are shown in grey, along with the average fold-changes. The array used in this study did not contain the probe set for Pmvk.

Conclusion

We have recently undertaken a transcriptomic screen of cultured primary cortical neurons treated with oligomeric Aβ (in the form of ADDLs) and our findings show marked effects in the downregulation of the expression levels of genes linked particularly with the steroid and cholesterol biosynthesis pathways [37]. Additionally, several genes implicated in lipid metabolism such as Ldlr, were also decreased. Notably, Abca1, a gene involved in cholesterol transport and previously implicated in AD pathology [68] was significantly upregulated. Collectively, the differential changes in gene expression indicate a co-ordinated decrease in the synthesis and uptake of cholesterol and lipid within neurons in response to ADDL treatment, as well as an increase in efflux of neuronal cholesterol mediated by an upregulation in Abca1 levels. The consequences of the downregulation of genes involved in the steroid and cholesterol biosynthesis pathways may not only lead to a decline in cholesterol levels in neurons but also a decrease in the synthesis of the neuroprotective neurosteroids, the reduction of which has been proposed to be involved in AD [69].

Although several mechanisms have been suggested to contribute to the development of pathology in AD, increasing epidemiological and molecular evidence suggests a key role of cholesterol in the initiation and progression of disease [23, 70]. Indeed, several studies support the idea that cholesterol oxidation products, known as oxysterols may interact with Aβ and be involved in AD pathogenesis [71, 72]. In addition, hypercholesterolemia is a risk factor for sporadic AD [73], although contradictory results have also been reported [74]. The brain is particularly enriched with cholesterol and contains 25% of total body cholesterol, which is synthesised primarily in glial cells, but also in neurons [44, 75]. The connections between cholesterol and AD, and between cholesterol and Aβ, are unclear and remain somewhat controversial [42, 74, 76]. Several studies have shown that high cholesterol levels are associated with amyloid deposition and increase the risk of developing AD later in life [71, 77, 78]. Low-density lipoprotein (LDL) plasma levels were found to be significantly increased in AD patients [79] and it was reasoned that higher brain cholesterol would increase the development of AD [67]. Trials on patients using cholesterol-lowering statins indicated some protection from developing AD [69, 75, 80-84] and may reduce the prevalence of the disease [85]. Statins act by lowering serum cholesterol through inhibition of HMG-CoA reductase, the first committed enzyme of isoprenoid synthesis [44, 85]. However, other studies have provided conflicting results and have suggested that any such effects are small, and may be due to a mechanism other than simple inhibition of cholesterol synthesis [38, 41, 86-88].

Cholesterol is thought to be involved in the development of AD in several ways. Cholesterol metabolism has been suggested to play a role in the formation of amyloid plaques and in tau hyperphosphorylation [89]. In addition, neurons containing NFTs have been shown to have higher levels of cholesterol [90] and cholesterol lowering drugs attenuate Aβ pathology in a mouse model of AD [91]. The ApoE gene and in particular its ε4 allele increase the risk of developing AD, and as it is the main transporter of cholesterol in the brain, it strongly implicates cholesterol transport in disease development [39, 40]. The precise mechanisms by which ApoE participates in pathogenesis remain largely undefined, although neuroinflammation, regulation of the cholinergic neurotransmitter system, its role in neuronal signaling, and maintenance of the integrity of the blood-brain barrier have all been reported [6]. However, the most prominent hypotheses of ApoE function are in mediating the metabolism, deposition and clearance of Aβ. Interestingly, other cholesterol related genes that may be associated with a risk of developing AD include Abca1, and CH25H, which converts cholesterol to 25-hydroxycholesterol, a transcriptional regulator of steroid and lipid metabolic enzymes [92]. The evidence for Abca1 is somewhat inconsistent [93-95]. However, it is important for cholesterol efflux and was found to be upregulated in our study [37]. Furthermore, Abca1 levels were found to be upregulated in AD brain and this dysregulation was associated with increasing severity of the disease [68]. Abca1 may regulate ApoE levels by stimulating efflux of intracellular cholesterol to extracellular lipid acceptors, including unlipidated ApoE [96]. Another gene, Osbpl1a (oxysterol binding protein-like1a), is likely to mediate transcriptional regulation by oxysterols and is downregulated by ADDLs (Tables 1 and 4). Finally, an enzyme of cholesterol biosynthesis, seladin-1 (encoded by the gene Dhcr1), has been linked to AD [38] but is also downregulated in brains of AD patients [97].

Additional candidate genes involved in lipid metabolism being investigated for any association with AD, include alpha-2-macroglobulin (A2M), LRP, insulin degrading enzyme (IDE), acyl-CoA cholesteryl acyl transferase (ACAT) and cytochrome P450, family 46 (CYP 46, converts cholesterol to 24S-hydroxycholesterol) [6].

The role of cholesterol in AD and its relationship and association with Aβ are significant [41, 42, 44]. Increased levels of Abca1 lower Aβ levels *in vitro* in cultured cells [98]. Furthermore, in mouse models of AD, deletion of the Abca1 gene decreased ApoE levels, which correlated with increased Aβ deposits [99, 100]. Evidence suggests that processing of the amyloid precursor protein (APP) by the γ and β-secretase enzymes (amyloidogenic pathway) and cleavage into Aβ occurs in cholesterol-rich membrane microdomains known as lipid rafts [42, 77, 101]. It has been reported that cholesterol promotes binding of and interaction of Aβ with the cell membrane, leading to enhanced generation and its subsequent conversion into soluble and toxic oligomers [102, 103]. We have demonstrated previously that ADDLs can be trafficked on neuronal plasma membranes leading to its enrichment in lipid rafts and its toxic effect is dependent on the protein kinase Fyn [47]. There is a suggestion that localization of APP and its secretases are affected by cholesterol levels as reduction of cholesterol in membranes can diminish Aβ production (by altering interaction with α-secretase), although modest reduction can displace β-secretase processing enzyme (BACE-1) from lipid rafts and promotes its co-localization with APP leading to enhanced processing via the amyloidogenic pathway [104]. Therefore, cholesterol can either inhibit or promote Aβ processing, depending on the molar ratio of lipid/cholesterol in the cell membrane [44].

Cholesterol and other lipids have the potential to increase the toxicity of already-deposited amyloid. Cholesterol depletion has additionally been observed to reduce the aggregation of Aβ *in vitro* and *in vivo* [105, 106]. Both cholesterol and 24S-hydroxycholesterol were reported to enhance the toxicity of fibrillar Aβ via oxidative stress [107] and also through calcium signalling [108]. However, other studies have also shown that increased cholesterol levels in the cell membrane inhibit the toxic effects of Aβ [109, 110]. There is evidence showing that lipids, including cholesterol, are able to dissociate the relatively non-toxic fibrillar form of $A\beta_{1\text{-}42}$ into oligomeric Aβ that is indistinguishable from ADDLs, and importantly this was able to interfere with learning in mice [111]. Additionally, increasing evidence links depletion of membrane cholesterol levels to the pathogenesis of AD. Increased cholesterol content in the plasma membrane was reported to protect against ADDL induced cytotoxicity by preventing membrane perturbation [112]. Enrichment of cholesterol within the membrane was shown to reduce the interaction of Aβ oligomers with the cell membrane of familial fibroblasts from AD patients, and as a result diminished intracellular reactive oxygen species (ROS) production and membrane lipoperoxidation, resulting in improved cell viability [113]. It was reasoned that Aβ oligomers were recruited to the plasma membrane in cholesterol depleted cells, increasing production of ROS causing membrane damage. Furthermore, the levels of cholesterol in the plasma membrane were shown to influence the neurotoxic pathway of Aβ oligomers, with depletion of levels leading to cell death in cortical neurons [114]. Crucially, enrichment of cholesterol within the plasma membrane offered protection in response to cell death induced specifically by oligomers only and not by Aβ fibrils.

Therefore, the implication is that Aβ oligomer-induced neurotoxicity is not only influenced by cholesterol levels within the plasma membrane but ADDLs may also act by downregulating the levels of genes fundamental in the pathways of cholesterol and steroid metabolism and biosynthesis. As Aβ may have a physiological function in regulating cholesterol and lipid metabolism [43, 115-117], the co-ordinated regulation of numerous genes would be consistent with a physiological process, either being part of such a process or of hijacking it to produce pathology [37].

This chapter provides evidence for the downregulation of genes involved in cholesterol and lipid metabolism by ADDLs, a pathogenic form of oligomeric Aβ. The findings may help to uncover and provide an insight into the mechanistic details underlying AD pathology. Further examination of the potential interaction between oligomeric Aβ and cholesterol and lipids is warranted and may help to establish future therapeutic targets.

Acknowledgments

This study was supported by the Medical Research Council and the Yamanouchi Research Institute. We thank Dr. Michael Minchin (of Astellas Pharma Europe Ltd (UK), formerly Yamanouchi UK Ltd) for advice and encouragement, and the Biostatistics Unit at the Institute of Psychiatry King's College London for advice.

References

[1] Selkoe DJ. Alzheimer's disease: genes, proteins, and therapy. *Physiol Rev* 2001 Apr;81(2):741-66.

[2] Gilbert BJ. The role of amyloid beta in the pathogenesis of Alzheimer's disease. *J Clin Pathol 2013* May;66(5):362-6.

[3] Jarrett JT, Berger EP, Lansbury PT, Jr. The carboxy terminus of the beta amyloid protein is critical for the seeding of amyloid formation: implications for the pathogenesis of Alzheimer's disease. *Biochemistry 1993* May 11;32(18):4693-7.

[4] Anderton BH, Betts J, Blackstock WP, Brion JP, Chapman S, Connell J, et al. Sites of phosphorylation in tau and factors affecting their regulation. *Biochem Soc Symp* 2001;(67):73-80.

[5] Reynolds CH, Garwood CJ, Wray S, Price C, Kellie S, Perera T, et al. Phosphorylation regulates tau interactions with Src homology 3 domains of phosphatidylinositol 3-kinase, phospholipase Cgamma1, Grb2, and Src family kinases. *J Biol Chem* 2008 Jun 27;283(26):18177-86.

[6] Chopra K, Misra S, Kuhad A. Neurobiological aspects of Alzheimer's disease. *Expert Opin Ther Targets* 2011 May;15(5):535-55.

[7] Hardy J, Selkoe DJ. The amyloid hypothesis of Alzheimer's disease: progress and problems on the road to therapeutics. *Science* 2002 Jul 19;297(5580):353-6.

[8] Swaab DF, Hofman MA, Lucassen PJ, Salehi A, Uylings HB. Neuronal atrophy, not cell death, is the main hallmark of Alzheimer's disease. *Neurobiol Aging* 1994 May;15(3):369-71.

[9] Hirai K, Aliev G, Nunomura A, Fujioka H, Russell RL, Atwood CS, et al. Mitochondrial abnormalities in Alzheimer's disease. *J Neurosci* 2001 May 1;21(9):3017-23.

[10] Bubber P, Haroutunian V, Fisch G, Blass JP, Gibson GE. Mitochondrial abnormalities in Alzheimer brain: mechanistic implications. *Ann Neurol* 2005 May;57(5):695-703.

[11] Eckert A, Hauptmann S, Scherping I, Rhein V, Muller-Spahn F, Gotz J, et al. Soluble beta-amyloid leads to mitochondrial defects in amyloid precursor protein and tau transgenic mice. *Neurodegener Dis* 2008;5(3-4):157-9.

[12] Rhein V, Song X, Wiesner A, Ittner LM, Baysang G, Meier F, et al. Amyloid-beta and tau synergistically impair the oxidative phosphorylation system in triple transgenic Alzheimer's disease mice. *Proc Natl Acad Sci U S A* 2009 Nov 24;106(47):20057-62.

[13] Terry RD, Masliah E, Salmon DP, Butters N, DeTeresa R, Hill R, et al. Physical basis of cognitive alterations in Alzheimer's disease: synapse loss is the major correlate of cognitive impairment. *Ann Neurol* 1991 Oct;30(4):572-80.

[14] Selkoe DJ. Alzheimer's disease is a synaptic failure. *Science* 2002 Oct 25;298(5594):789-91.

[15] Scheff SW, Price DA, Schmitt FA, Scheff MA, Mufson EJ. Synaptic loss in the inferior temporal gyrus in mild cognitive impairment and Alzheimer's disease. *J Alzheimers Dis* 2011;24(3):547-57.

[16] Klein WL. Abeta toxicity in Alzheimer's disease: globular oligomers (ADDLs) as new vaccine and drug targets. *Neurochem Int* 2002 Nov;41(5):345-52.

[17] Klein WL, Stine WB, Jr., Teplow DB. Small assemblies of unmodified amyloid beta-protein are the proximate neurotoxin in Alzheimer's disease. *Neurobiol Aging* 2004 May;25(5):569-80.

[18] Gong Y, Chang L, Viola KL, Lacor PN, Lambert MP, Finch CE, et al. Alzheimer's disease-affected brain: presence of oligomeric A beta ligands (ADDLs) suggests a molecular basis for reversible memory loss. *Proc Natl Acad Sci U S A* 2003 Sep 2;100(18):10417-22.

[19] Townsend M, Shankar GM, Mehta T, Walsh DM, Selkoe DJ. Effects of secreted oligomers of amyloid beta-protein on hippocampal synaptic plasticity: a potent role for trimers. *J Physiol* 2006 Apr 15;572(Pt 2):477-92.

[20] Ono K, Yamada M. Low-n oligomers as therapeutic targets of Alzheimer's disease. *J Neurochem* 2011 Apr;117(1):19-28.

[21] Ono K, Condron MM, Teplow DB. Structure-neurotoxicity relationships of amyloid beta-protein oligomers. *Proc Natl Acad Sci U S A* 2009 Sep 1;106(35):14745-50.

[22] Finder VH. Alzheimer's disease: a general introduction and pathomechanism. *J Alzheimers Dis* 2010;22 Suppl 3:5-19.

[23] Gamba P, Testa G, Sottero B, Gargiulo S, Poli G, Leonarduzzi G. The link between altered cholesterol metabolism and Alzheimer's disease. *Ann N Y Acad Sci* 2012 Jul;1259:54-64.

[24] Blalock EM, Geddes JW, Chen KC, Porter NM, Markesbery WR, Landfield PW. Incipient Alzheimer's disease: microarray correlation analyses reveal major transcriptional and tumor suppressor responses. *Proc Natl Acad Sci U S A* 2004 Feb 17;101(7):2173-8.

[25] Emilsson L, Saetre P, Jazin E. Alzheimer's disease: mRNA expression profiles of multiple patients show alterations of genes involved with calcium signaling. *Neurobiol Dis* 2006 Mar;21(3):618-25.

[26] Brooks WM, Lynch PJ, Ingle CC, Hatton A, Emson PC, Faull RL, et al. Gene expression profiles of metabolic enzyme transcripts in Alzheimer's disease. *Brain Res* 2007 Jan 5;1127(1):127-35.

[27] Dickey CA, Loring JF, Montgomery J, Gordon MN, Eastman PS, Morgan D. Selectively reduced expression of synaptic plasticity-related genes in amyloid precursor protein + presenilin-1 transgenic mice. *J Neurosci* 2003 Jun 15;23(12):5219-26.

[28] Reddy PH, McWeeney S, Park BS, Manczak M, Gutala RV, Partovi D, et al. Gene expression profiles of transcripts in amyloid precursor protein transgenic mice: up-regulation of mitochondrial metabolism and apoptotic genes is an early cellular change in Alzheimer's disease. *Hum Mol Genet* 2004 Jun 15;13(12):1225-40.

[29] Mirnics K, Korade Z, Arion D, Lazarov O, Unger T, Macioce M, et al. Presenilin-1-dependent transcriptome changes. *J Neurosci* 2005 Feb 9;25(6):1571-8.

[30] Song S, Kim SY, Hong YM, Jo DG, Lee JY, Shim SM, et al. Essential role of E2-25K/Hip-2 in mediating amyloid-beta neurotoxicity. *Mol Cell* 2003 Sep;12(3):553-63.

[31] Paratore S, Parenti R, Torrisi A, Copani A, Cicirata F, Cavallaro S. Genomic profiling of cortical neurons following exposure to beta-amyloid. *Genomics* 2006 Oct;88(4):468-79.

[32] Kim JY, Kim H, Lee SG, Choi BH, Kim YH, Huh PW, et al. Amyloid beta peptide (Abeta42) activates PLC-delta1 promoter through the NF-kappaB binding site. *Biochem Biophys Res Commun* 2003 Oct 24;310(3):904-9.

[33] Valerio A, Boroni F, Benarese M, Sarnico I, Ghisi V, Bresciani LG, et al. NF-kappaB pathway: a target for preventing beta-amyloid (Abeta)-induced neuronal damage and Abeta42 production. *Eur J Neurosci* 2006 Apr;23(7):1711-20.

[34] Estus S, Tucker HM, van Rooyen C, Wright S, Brigham EF, Wogulis M, et al. Aggregated amyloid-beta protein induces cortical neuronal apoptosis and concomitant "apoptotic" pattern of gene induction. *J Neurosci* 1997 Oct 15;17(20):7736-45.

[35] Yao M, Nguyen TV, Pike CJ. Beta-amyloid-induced neuronal apoptosis involves c-Jun N-terminal kinase-dependent downregulation of Bcl-w. *J Neurosci* 2005 Feb 2;25(5):1149-58.

[36] Martinez T, Pascual A. Gene expression profile in beta-amyloid-treated SH-SY5Y neuroblastoma cells. *Brain Res Bull* 2007 May 30;72(4-6):225-31.

[37] Malik B, Fernandes C, Killick R, Wroe R, Usardi A, Williamson R, et al. Oligomeric amyloid-beta peptide affects the expression of genes involved in steroid and lipid metabolism in primary neurons. *Neurochem Int* 2012 Aug;61(3):321-33.

[38] Lamsa R, Helisalmi S, Hiltunen M, Herukka SK, Tapiola T, Pirttila T, et al. The association study between DHCR24 polymorphisms and Alzheimer's disease. *Am J Med Genet B Neuropsychiatr Genet* 2007 Oct 5;144B(7):906-10.

[39] Carter CJ. Convergence of genes implicated in Alzheimer's disease on the cerebral cholesterol shuttle: APP, cholesterol, lipoproteins, and atherosclerosis. *Neurochem Int* 2007 Jan;50(1):12-38.

[40] Papassotiropoulos A, Wollmer MA, Tsolaki M, Brunner F, Molyva D, Lutjohann D, et al. A cluster of cholesterol-related genes confers susceptibility for Alzheimer's disease. *J Clin Psychiatry* 2005 Jul;66(7):940-7.

[41] Wolozin B. Cholesterol and the biology of Alzheimer's disease. *Neuron* 2004 Jan 8;41(1):7-10.

[42] Canevari L, Clark JB. Alzheimer's disease and cholesterol: the fat connection. *Neurochem Res* 2007 Apr;32(4-5):739-50.

[43] Grimm MO, Grimm HS, Hartmann T. Amyloid beta as a regulator of lipid homeostasis. *Trends Mol Med* 2007 Aug;13(8):337-44.

[44] Maulik M, Westaway D, Jhamandas JH, Kar S. Role of cholesterol in APP metabolism and its significance in Alzheimer's disease pathogenesis. *Mol Neurobiol* 2013 Feb;47(1):37-63.

[45] Malik B, Currais A, Soriano S. Cell cycle-driven neuronal apoptosis specifically linked to amyloid peptide Abeta1-42 exposure is not exacerbated in a mouse model of presenilin-1 familial Alzheimer's disease. *J Neurochem* 2008 Jul;106(2):912-6.

[46] Malik B, Currais A, Andres A, Towlson C, Pitsi D, Nunes A, et al. Loss of neuronal cell cycle control as a mechanism of neurodegeneration in the presenilin-1 Alzheimer's disease brain. *Cell Cycle* 2008 Mar 1;7(5):637-46.

[47] Williamson R, Usardi A, Hanger DP, Anderton BH. Membrane-bound beta-amyloid oligomers are recruited into lipid rafts by a fyn-dependent mechanism. *FASEB J* 2008 May;22(5):1552-9.

[48] Lambert MP, Barlow AK, Chromy BA, Edwards C, Freed R, Liosatos M, et al. Diffusible, nonfibrillar ligands derived from Abeta1-42 are potent central nervous system neurotoxins. *Proc Natl Acad Sci U S A 1998* May 26;95(11):6448-53.

[49] Krafft GA, Klein WL. ADDLs and the signaling web that leads to Alzheimer's disease. *Neuropharmacology* 2010 Sep;59(4-5):230-42.

[50] Bradford MM. A rapid and sensitive method for the quantitation of microgram quantities of protein utilizing the principle of protein-dye binding. *Anal Biochem* 1976 May 7;72:248-54.

[51] Williamson R, Scales T, Clark BR, Gibb G, Reynolds CH, Kellie S, et al. Rapid tyrosine phosphorylation of neuronal proteins including tau and focal adhesion kinase in response to amyloid-beta peptide exposure: involvement of Src family protein kinases. *J Neurosci 2002* Jan 1;22(1):10-20.

[52] Chomczynski P, Sacchi N. Single-step method of RNA isolation by acid guanidinium thiocyanate-phenol-chloroform extraction. *Anal Biochem* 1987 Apr;162(1):156-9.

[53] Malik B, Nirmalananthan N, Gray AL, La Spada AR, Hanna MG, Greensmith L. Co-induction of the heat shock response ameliorates disease progression in a mouse model of human spinal and bulbar muscular atrophy: implications for therapy. *Brain* 2013 Mar;136(Pt 3):926-43.

[54] Vandesompele J, De PK, Pattyn F, Poppe B, Van RN, De PA, et al. Accurate normalization of real-time quantitative RT-PCR data by geometric averaging of multiple internal control genes. *Genome Biol* 2002 Jun 18;3(7):RESEARCH0034.

[55] Livak KJ, Schmittgen TD. Analysis of relative gene expression data using real-time quantitative PCR and the 2(-Delta Delta C(T)) *Method. Methods* 2001 Dec;25(4):402-8.

[56] Ihaka R, entleman R. R: A language for data analysis and graphics. *J Comput Graph Stat* 1996;5:299-314.

[57] Gentleman RC, Carey VJ, Bates DM, Bolstad B, Dettling M, Dudoit S, et al. Bioconductor: open software development for computational biology and bioinformatics. *Genome Biol* 2004;5(10):R80.

[58] Wilson CL, Miller CJ. Simpleaffy: a BioConductor package for Affymetrix Quality Control and data analysis. *Bioinformatics* 2005 Sep 15;21(18):3683-5.

[59] Smyth GK. Linear models and empirical bayes methods for assessing differential expression in microarray experiments. *Stat Appl Genet Mol Biol* 2004;3:Article3.

[60] Benjamini JA, Hochberg Y. Controlling the False Discovery Rate: a Practical and Powerful Approach to Multiple Testing. *J Roy Stat Soc* 1995;57:289-300.

[61] Storey JD, Tibshirani R. Statistical significance for genomewide studies. *Proc Natl Acad Sci U S A* 2003 Aug 5;100(16):9440-5.

[62] Dennis G, Jr., Sherman BT, Hosack DA, Yang J, Gao W, Lane HC, et al. DAVID: Database for Annotation, Visualization, and Integrated Discovery. *Genome Biol* 2003;4(5):3.

[63] Dahlquist KD, Salomonis N, Vranizan K, Lawlor SC, Conklin BR. GenMAPP, a new tool for viewing and analyzing microarray data on biological pathways. *Nat Genet* 2002 May;31(1):19-20.

[64] Liang WS, Dunckley T, Beach TG, Grover A, Mastroeni D, Ramsey K, et al. Altered neuronal gene expression in brain regions differentially affected by Alzheimer's disease: a reference data set. *Physiol Genomics* 2008 Apr 22;33(2):240-56.

[65] Kanehisa M, Araki M, Goto S, Hattori M, Hirakawa M, Itoh M, et al. KEGG for linking genomes to life and the environment. *Nucleic Acids Res* 2008 Jan;36(Database issue):D480-D484.

[66] Burns MP, Vardanian L, Pajoohesh-Ganji A, Wang L, Cooper M, Harris DC, et al. The effects of ABCA1 on cholesterol efflux and Abeta levels in vitro and in vivo. *J Neurochem* 2006 Aug;98(3):792-800.

[67] Forwood JK, Thakur AS, Guncar G, Marfori M, Mouradov D, Meng W, et al. Structural basis for recruitment of tandem hotdog domains in acyl-CoA thioesterase 7 and its role in inflammation. *Proc Natl Acad Sci U S A* 2007 Jun 19;104(25):10382-7.

[68] Akram A, Schmeidler J, Katsel P, Hof PR, Haroutunian V. Increased expression of cholesterol transporter ABCA1 is highly correlated with severity of dementia in AD hippocampus. *Brain Res* 2010 Mar 8;1318:167-77.

[69] Barry G, Ross IL. Neurosteroids and sporadic Alzheimer's diseas. *Curr Alzheimer Res* 2008 Aug;5(4):367-74.

[70] Matsuzaki T, Sasaki K, Hata J, Hirakawa Y, Fujimi K, Ninomiya T, et al. Association of Alzheimer disease pathology with abnormal lipid metabolism: the Hisayama Study. *Neurology 2011* Sep 13;77(11):1068-75.

[71] Vaya J, Schipper HM. Oxysterols, cholesterol homeostasis, and Alzheimer disease. *J Neurochem* 2007 Sep;102(6):1727-37.

[72] Jeitner TM, Voloshyna I, Reiss AB. Oxysterol derivatives of cholesterol in neurodegenerative disorders. *Curr Med Chem* 2011;18(10):1515-25.

[73] Shobab LA, Hsiung GY, Feldman HH. Cholesterol in Alzheimer's disease. *Lancet Neurol* 2005 Dec;4(12):841-52.

[74] Fonseca AC, Resende R, Oliveira CR, Pereira CM. Cholesterol and statins in Alzheimer's disease: current controversies. *Exp Neurol* 2010 Jun;223(2):282-93.

[75] Suzuki S, Kiyosue K, Hazama S, Ogura A, Kashihara M, Hara T, et al. Brain-derived neurotrophic factor regulates cholesterol metabolism for synapse development. *J Neurosci* 2007 Jun 13;27(24):6417-27.

[76] Zhou B, Teramukai S, Fukushima M. Prevention and treatment of dementia or Alzheimer's disease by statins: a meta-analysis. *Dement Geriatr Cogn Disord* 2007;23(3):194-201.

[77] Di PG, Kim TW. Linking lipids to Alzheimer's disease: cholesterol and beyond. *Nat Rev Neurosci* 2011 May;12(5):284-96.

[78] Pappolla MA, Bryant-Thomas TK, Herbert D, Pacheco J, Fabra GM, Manjon M, et al. Mild hypercholesterolemia is an early risk factor for the development of Alzheimer amyloid pathology. *Neurology* 2003 Jul 22;61(2):199-205.

[79] Kuo YM, Emmerling MR, Bisgaier CL, Essenburg AD, Lampert HC, Drumm D, et al. Elevated low-density lipoprotein in Alzheimer's disease correlates with brain abeta 1-42 levels. *Biochem Biophys Res Commun* 1998 Nov 27;252(3):711-5.

[80] Haag MD, Hofman A, Koudstaal PJ, Stricker BH, Breteler MM. Statins are associated with a reduced risk of Alzheimer disease regardless of lipophilicity. The Rotterdam Study. *J Neurol Neurosurg Psychiatry* 2009 Jan;80(1):13-7.

[81] Sparks DL, Sabbagh MN, Connor DJ, Lopez J, Launer LJ, Browne P, et al. Atorvastatin for the treatment of mild to moderate Alzheimer disease: preliminary results. *Arch Neurol* 2005 May;62(5):753-7.

[82] Sparks DL, Kryscio RJ, Sabbagh MN, Connor DJ, Sparks LM, Liebsack C. Reduced risk of incident AD with elective statin use in a clinical trial cohort. *Curr Alzheimer Res* 2008 Aug;5(4):416-21.

[83] Tong XK, Lecrux C, Rosa-Neto P, Hamel E. Age-dependent rescue by simvastatin of Alzheimer's disease cerebrovascular and memory deficits. *J Neurosci* 2012 Apr 4;32(14):4705-15.

[84] Jick H, Zornberg GL, Jick SS, Seshadri S, Drachman DA. Statins and the risk of dementia. *Lancet* 2000 Nov 11;356(9242):1627-31.

[85] Silva T, Teixeira J, Remiao F, Borges F. Alzheimer's disease, cholesterol, and statins: the junctions of important metabolic pathways. *Angew Chem Int Ed Engl* 2013 Jan 21;52(4):1110-21.

[86] Wolozin B. Cholesterol, statins and dementia. *Curr Opin Lipidol* 2004 Dec;15(6):667-72.

[87] Padala KP, Padala PR, McNeilly DP, Geske JA, Sullivan DH, Potter JF. The effect of HMG-CoA reductase inhibitors on cognition in patients with Alzheimer's dementia: a prospective withdrawal and rechallenge pilot study. *Am J Geriatr Pharmacother* 2012 Oct;10(5):296-302.

[88] Sano M, Bell KL, Galasko D, Galvin JE, Thomas RG, van Dyck CH, et al. A randomized, double-blind, placebo-controlled trial of simvastatin to treat Alzheimer disease. *Neurology* 2011 Aug 9;77(6):556-63.

[89] Ghribi O. Potential mechanisms linking cholesterol to Alzheimer's disease-like pathology in rabbit brain, hippocampal organotypic slices, and skeletal muscle. *J Alzheimers Dis* 2008 Dec;15(4):673-84.

[90] Distl R, Meske V, Ohm TG. Tangle-bearing neurons contain more free cholesterol than adjacent tangle-free neurons. *Acta Neuropathol* 2001 Jun;101(6):547-54.

[91] Refolo LM, Pappolla MA, LaFrancois J, Malester B, Schmidt SD, Thomas-Bryant T, et al. A cholesterol-lowering drug reduces beta-amyloid pathology in a transgenic mouse model of Alzheimer's disease. *Neurobiol Dis* 2001 Oct;8(5):890-9.

[92] Bertram L, Tanzi RE. Thirty years of Alzheimer's disease genetics: the implications of systematic meta-analyses. *Nat Rev Neurosci* 2008 Oct;9(10):768-78.

[93] Shibata N, Kawarai T, Lee JH, Lee HS, Shibata E, Sato C, et al. Association studies of cholesterol metabolism genes (CH25H, ABCA1 and CH24H) in Alzheimer's disease. *Neurosci Lett* 2006 Jan 2;391(3):142-6.

[94] Sundar PD, Feingold E, Minster RL, DeKosky ST, Kamboh MI. Gender-specific association of ATP-binding cassette transporter 1 (ABCA1) polymorphisms with the risk of late-onset Alzheimer's disease. *Neurobiol Aging* 2007 Jun;28(6):856-62.

[95] Wavrant-De VF, Compton D, Womick M, Arepalli S, Adighibe O, Li L, et al. ABCA1 polymorphisms and Alzheimer's disease. *Neurosci Lett* 2007 Apr 12;416(2):180-3.

[96] Tall AR. Cholesterol efflux pathways and other potential mechanisms involved in the athero-protective effect of high density lipoproteins. *J Intern Med* 2008 Mar;263(3):256-73.

[97] Iivonen S, Hiltunen M, Alafuzoff I, Mannermaa A, Kerokoski P, Puolivali J, et al. Seladin-1 transcription is linked to neuronal degeneration in Alzheimer's disease. *Neuroscience* 2002;113(2):301-10.

[98] Sun Y, Yao J, Kim TW, Tall AR. Expression of liver X receptor target genes decreases cellular amyloid beta peptide secretion. *J Biol Chem* 2003 Jul 25;278(30):27688-94.

[99] Wahrle SE, Jiang H, Parsadanian M, Hartman RE, Bales KR, Paul SM, et al. Deletion of Abca1 increases Abeta deposition in the PDAPP transgenic mouse model of Alzheimer disease. *J Biol Chem* 2005 Dec 30;280(52):43236-42.

[100] Koldamova R, Staufenbiel M, Lefterov I. Lack of ABCA1 considerably decreases brain ApoE level and increases amyloid deposition in APP23 mice. *J Biol Chem* 2005 Dec 30;280(52):43224-35.

[101] Williamson R, Sutherland C. Neuronal membranes are key to the pathogenesis of Alzheimer's disease: the role of both raft and non-raft membrane domains. *Curr Alzheimer Res* 2011 Mar;8(2):213-21.

[102] Xiong H, Callaghan D, Jones A, Walker DG, Lue LF, Beach TG, et al. Cholesterol retention in Alzheimer's brain is responsible for high beta- and gamma-secretase activities and Abeta production. *Neurobiol Dis* 2008 Mar;29(3):422-37.

[103] Yip CM, Elton EA, Darabie AA, Morrison MR, McLaurin J. Cholesterol, a modulator of membrane-associated Abeta-fibrillogenesis and neurotoxicity. *J Mol Biol* 2001 Aug 24;311(4):723-34.

[104] Abad-Rodriguez J, Ledesma MD, Craessaerts K, Perga S, Medina M, Delacourte A, et al. Neuronal membrane cholesterol loss enhances amyloid peptide generation. *J Cell Biol* 2004 Dec 6;167(5):953-60.

[105] Schneider A, Schulz-Schaeffer W, Hartmann T, Schulz JB, Simons M. Cholesterol depletion reduces aggregation of amyloid-beta peptide in hippocampal neurons. *Neurobiol Dis* 2006 Sep;23(3):573-7.

[106] Ji SR, Wu Y, Sui SF. Cholesterol is an important factor affecting the membrane insertion of beta-amyloid peptide (A beta 1-40), which may potentially inhibit the fibril formation. *J Biol Chem* 2002 Feb 22;277(8):6273-9.

[107] Ferrera P, Mercado-Gomez O, Silva-Aguilar M, Valverde M, Arias C. Cholesterol potentiates beta-amyloid-induced toxicity in human neuroblastoma cells: involvement of oxidative stress. *Neurochem Res* 2008 Aug;33(8):1509-17.

[108] Abramov AY, Ionov M, Pavlov E, Duchen MR. Membrane cholesterol content plays a key role in the neurotoxicity of beta-amyloid: implications for Alzheimer's disease. *Aging Cell* 2011 Aug;10(4):595-603.

[109] Arispe N, Doh M. Plasma membrane cholesterol controls the cytotoxicity of Alzheimer's disease AbetaP (1-40) and (1-42) peptides. *FASEB J* 2002 Oct;16(12):1526-36.

[110] Cecchi C, Nichino D, Zampagni M, Bernacchioni C, Evangelisti E, Pensalfini A, et al. A protective role for lipid raft cholesterol against amyloid-induced membrane damage in human neuroblastoma cells. *Biochim Biophys Acta* 2009 Oct;1788(10):2204-16.

[111] Martins IC, Kuperstein I, Wilkinson H, Maes E, Vanbrabant M, Jonckheere W, et al. Lipids revert inert Abeta amyloid fibrils to neurotoxic protofibrils that affect learning in mice. *EMBO J* 2008 Jan 9;27(1):224-33.

[112] Pensalfini A, Zampagni M, Liguri G, Becatti M, Evangelisti E, Fiorillo C, et al. Membrane cholesterol enrichment prevents Abeta-induced oxidative stress in Alzheimer's fibroblasts. *Neurobiol Aging* 2011 Feb;32(2):210-22.

[113] Cenini G, Cecchi C, Pensalfini A, Bonini SA, Ferrari-Toninelli G, Liguri G, et al. Generation of reactive oxygen species by beta amyloid fibrils and oligomers involves different intra/extracellular pathways. *Amino Acids* 2010 Apr;38(4):1101-6.

[114] Sponne I, Fifre A, Koziel V, Oster T, Olivier JL, Pillot T. Membrane cholesterol interferes with neuronal apoptosis induced by soluble oligomers but not fibrils of amyloid-beta peptide. *FASEB J* 2004 Jun;18(7):836-8.

[115] Grimm MO, Grimm HS, Patzold AJ, Zinser EG, Halonen R, Duering M, et al. Regulation of cholesterol and sphingomyelin metabolism by amyloid-beta and presenilin. *Nat Cell Biol* 2005 Nov;7(11):1118-23.

[116] Grimm MO, Rothhaar TL, Hartmann T. The role of APP proteolytic processing in lipid metabolism. *Exp Brain Res* 2012 Apr;217(3-4):365-75.

[117] Pierrot N, Tyteca D, D'auria L, Dewachter I, Gailly P, Hendrickx A, et al. Amyloid precursor protein controls cholesterol turnover needed for neuronal activity. *EMBO Mol Med* 2013 Apr;5(4):608-25.

In: Microarrays: Principles, Applications and Technologies ISBN: 978-1-62948-669-7
Editor: James V. Rogers

Chapter 15

Application of Microarrays to Cutaneous Chemical Exposures

James V. Rogers **and Jennifer A. Price**
Battelle, Columbus, Ohio, US

Abstract

The skin is a dynamic organ that serves as the first line of defense against external insults. Profiling gene expression in the skin using microarrays has become a useful approach in many areas of research to enhance the understanding of dermal function, toxicity, diseases, and risk assessment. With respect to cutaneous chemical exposures, there are many factors that influence the transcriptional response from the time at which a chemical comes into contact with the skin through gene activation. The number of microarray studies on chemical-exposed skin is limiting and these often differ in experimental design, utilize different assay and software platforms, and are limiting in the amount of raw data presented. Assessing multiple microarray data sets could provide a basis for identifying potential redundant biological mechanisms or genes associated with cutaneous responses to chemical exposure. Additionally, incorporating inputs from other microarray applications, such as microRNA assessments, can help build a more complete picture of mechanistic biological pathways and processes. As cutaneous chemical exposure models can vary, analyzing multiple microarray data sets could promote the identification of more generalized pathways or genes that could support dermal toxicological assessments and the development of potential target-based therapeutics. This chapter provides an overview of studies and approaches for utilizing microarrays and bioinformatics to understand mechanisms and identify potential biomarkers that could aid in developing therapeutics or treatment regimens for chemical-induced skin damage.

* Correspondence to: James V. Rogers, Ph.D. Battelle, 505 King Avenue, JM-7, Columbus, OH 43201. Tel: (614) 424-5428. Fax: (614) 458-5428. Email: rogersjv@battelle.org.

Introduction

The skin is the largest organ of the body and fulfills many functional and physiological roles that enable communication between the external environment and internal tissue-specific cellular and molecular processes. As a physical barrier, the skin resists desiccation, infection, chemicals, and ultraviolet light. At the physiological level, the skin regulates temperature, provides mechanical support, initiates immunological and pro-inflammatory responses, and provides chemoprotection.

By being the first line of defense and environmental barrier, the skin is susceptible to various types of injuries caused by chemical exposure. Skin injuries resulting from exposure to chemicals ranging from acids, alkali, and halogens to hydrocarbons, metals, and solvents can lead to concerns over potential long-term effects. Therefore, research has focused on chemical-induced burn treatments to prevent or reduce infection, promote healing processes, and reduce scarring. Medical providers have typically used topical or systemic treatments, dressings, or cosmetics to treat the clinical symptoms associated with chemical burns. However, these treatments are sometimes used without clearly understanding or properly targeting the molecular mechanism(s) driving the dermatological condition.

The skin is easily accessible for conducting exposure studies that evaluate many biological and toxicological endpoints through minimally-invasive sampling, such as biopsy punches or tape stripping. From these samples, transcriptomics provides a glimpse into gene expression changes at a single point in time with a goal of establishing a skin-specific transcriptional signature. Genomics studies using microarrays have helped enhance knowledge of skin function and response to homeostatic changes that are typically reflected by increases or decreases in messenger RNA (mRNA) and microRNA (miRNA) as well as subsequent proteins that regulate biological responses [1-11]. Over the years, investigators have used transcriptomics in dermal studies to evaluate diseased skin [12-18], scarring [19, 20], cancer [21-25], irradiation [26-28], chemical exposure [29-49], thermal injury [37, 44, 50, 51], aging [52], hair formation [53, 54], pigmentation [55, 56], wounds [57, 58], and homeostasis [59] using various microarray formats and in vivo models.

Microarray technologies have helped researchers advance their understanding of cellular and molecular events associated with cutaneous chemical burns, enabling the identification of a potential biomarker or biomarker panel that could be exploited as therapeutic targets. Few studies using whole skin have employed transcriptomics to characterize the molecular mechanisms underlying the biological responses following chemical exposure [60]; however, there have been studies in which changes in gene expression patterns are coupled with bioinformatics to identify potential biomarkers for future therapeutic considerations [30, 33, 36-40, 44, 45, 61].

As with all types of biomedical research, one critical component that must be evaluated from the start is the selection of an applicable model system that will provide the best information to address the question(s) being asked. To evaluate specific chemical-induced changes, in vitro models consisting of primary cultures, cell lines, or three-dimensional cultures have been used for measuring different endpoints of interest. These in vitro models offer benefits such as homogeneity of experimental samples, flexibility in the number of replicates, high-throughput, reduced cost, and the ability to monitor the response of a single cell type. Such examples have included the use of transcriptomics to assess the responses of dendritic cells, keratinocytes, fibroblasts, and melanocytes [62-66]. However, the skin functions biochemically as a result of the interplay between multiple cell types and extensive networks that regulate complex gene expression patterns. In vitro models lack such dynamic cell-to-cell interaction, thereby not providing an accurate representation of the responses in whole skin to an external stimulus or chemical exposure.

With chemical-induced cutaneous toxicity, the development and application of post-exposure therapeutics would be intended to treat the resulting pathological changes of the affected tissue rather than a single cell type within the damaged region. Thus, whole skin will provide the most relevent test system for utilizing microarrays and gene expression patterns as a molecular signature in chemically-exposed skin. By applying microarray technology to whole skin studies, the observed changes in mRNA levels derived from a pool of transcripts isolated from static and transient cells would reflect an overall net average in the transcriptional level (i.e., increased or decreased) of each gene. The cumulative reported average expression for each gene as reported by the microarray data should be an important factor when developing a biomarker panel, as this group of specialized genes represents the overall dose-response regulation of mRNAs within the exposed tissue. Subsequent gene ontology, biological response, and canonical pathway analysis of this biomarker panel could aid in the identification of potential therapeutic targets.

Factors Influencing Cutaneous Toxicity

In the fields of pharmacology and toxicology, quantifying and understanding the biological processes associated with chemical penetration into the skin that results in local and systemic perturbations is important. In vivo studies possess advantages over those performed using in vitro models due to intact skin having normal physiological blood flow, active metabolism, nervous system responses, and dynamic immune function. For toxicological studies, humans are the species of primary interest; however, laboratory animals are often used as surrogates when invasive endpoints are needed. The applicability of these results is limited to the experimental design, biological similarity between the test species and humans, and overall correlation of data between species. Therefore, it is critical to understand differences between species for extrapolating laboratory-based data to humans in a manner that supports analytical-based decisions.

The skin resists the movement of foreign objects or chemicals into the body, thereby mitigating the potential for infection as well as local or systemic toxicity. The dead, lipid-rich stratum corneum shields the underlying viable cells of the epidermis and dermis by acting as the initial barrier to the penetration of most chemical substances. If chemicals do penetrate through the stratum corneum, the resulting local toxic responses (e.g., irritation) are assumed to be related to the chemical concentration present in the viable layers of the skin (e.g., epidermis) and not the chemical dose applied.

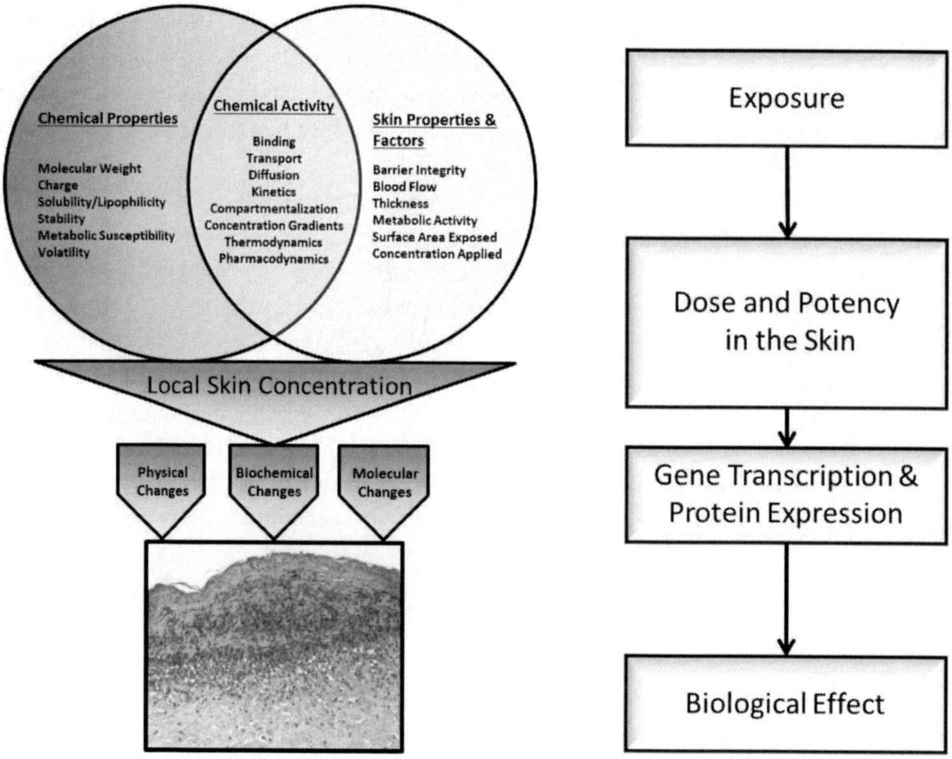

Figure 1. General paradigm of chemical exposure and factors leading to transcriptional and biological responses.

Although the amount of chemical applied to a defined surface exposure area is an important component in toxicological assessments, there are many factors influencing the transcutaneous flux and subsequent chemical concentration in the skin (Figure 1). As the physical properties and thermodynamic forces influence the rate and total amount of chemical penetrating through skin, it can be assumed that the viable epidermal and dermal cells will respond differently to this chemical concentration in terms of varied gene regulation. Chemical dose and potency are two factors that contribute to cutaneous toxicity in which the resulting biological response of the tissue is typically proportional to the chemical, and the magnitude of these biological responses can differ from chemical to chemical. Similarly, various cell types within the epidermis and dermis respond to chemical exposure by releasing pro-inflammatory compounds and/or initiating gene transcription that results in molecular and cellular responses indicative of chemical dose and potency in the skin. Focusing on the specific genes modulated in response to chemical exposure may provide the best approach to identify molecular targets to mitigate the toxic effect of a particular chemical or class of chemicals.

By applying the generalized paradigm in toxicology to microarray studies in which the target tissue dose is the primary determinant of the subsequent response, one could expect that changes in gene expression patterns would behave in a similar manner. That is, a tissue-specific gene expression response following exposure to any chemical could be assumed to be proportionally higher as the tissue chemical concentration increases. Moreover, it could also be expected that two chemicals differing in potency, yet exhibiting similar target tissue doses may induce overall gene expression responses that differ in magnitude. For example, McDougal and Garrett [35] conducted a study to investigate the transcriptional responses in the epidermis of rat skin as a function of chemical concentration and potency of two aromatic and two aliphatic compounds. It was observed that the epidermal chemical concentrations were similar; however, the magnitude of total gene expression levels and pathway-specific genes differed among the four compounds by 10-fold [35].

Microarray Analysis and Bioinformatics

Microarrays provide a snapshot in time of gene transcription, which can result in an overwhelming amount of information to analyze and deduce some type of biological sense. The development and implementation of bioinformatics approaches has paralleled the maturation of microarray platforms and technologies. Blumenberg [2] summarized the bioinformatics activities into four basic steps that can be applied to all of the microarray platforms and technologies, which include (1) data acquisition, (2) database development, (3) data analysis, and (4) data integration.

Statistical reduction enables researchers to condense the volume of raw data into workable lists of significantly-modulated transcripts. Groups of genes identified as being significantly altered within a certain fold-change threshold can then be imported into specialized bioinformatics software. These software programs typically contain resources and vast amounts of information within a knowledge base that helps investigators identify biological functions, signaling pathways, and specific genes of interest. At this stage, data can be presented in a way that provides visualization (e.g., canonical pathway) of the altered biological mechanisms that underscore the observed tissue changes. For microarray transcriptomic studies, combining these data-mining tools, information from the scientific literature, and empirical data enables scientists to focus research efforts into defined areas of interest, thereby eliminating a shotgun approach.

In order to utilize microarray data on a much larger scale to investigate complex biological problems, researchers have performed meta-analyses on multiple microarray datasets. A meta-analysis that combines data from existing studies provides increased statistical power than when using microarray data from a single study. Ramasamy et al. [67] noted that a meta-analysis can increase the reliability and generalization of the results with greater statistical power, thereby enabling a higher degree of confidence in the gene expression changes. However, there are some hurdles to overcome in performing a meta-analysis of microarray data. These include data availability, integration of data derived from different sources, varying microarray platforms, and differences in the data flow process, statistical software, and bioinformatics tools.

In dermatological research, there are few published reports that use a meta-analysis approach to investigate a biological problem. However, meta-analyses have been performed on publicly-available microarray data sets to investigate dermatological diseases and biological processes [68-72]. For example, Tian et al. [69] used data from five microarray studies to develop a meta-analysis derived (MAD) transcriptome of psoriasis using differentially-expressed genes from skin biopsies. This study identified 20 MAD "classifer" genes that could be used for the molecular diagnosis of psoriasis in which six of these genes comprise the top methylation genes discriminating between psoriasis and normal skin and four genes were identified as part of the Residual Disease Genomic Profile or "molecular scar" after anti-TNF treatment [69]. Moreover, the gene encoding kynureninase (KYNU) was suggested to be a potential biomarker due to its increased expression and demethylation state in psoriatic skin [69]. Another study used meta-analysis of microarray datasets to identify a molecular signature characteristic of the cutaneous T-cell lymphoma, mycosis fungoides [68]. These investigators identified gene expression patterns indicative of the mycosis fungoides tumor stage that provided insight into the developing pathogenesis, characterized the immunophenotype of this lymphoma (i.e., IL-32-producing cell lineage), and identified genes such as gametocyte specific factor 1 (GTSF1) and thyroid hormone receptor interactor 13 (TRIP13) that could be used as diagnostic markers or therapeutic targets [68]. Lastly, Liu et al. [70] applied a genome-wide relative significance (GWRS) and genome-wide global significance (GWGS) model to five separate microarray studies and identified a 200 gene signature for melanoma. Within this signature, 12 genes were down-selected as a diagnostic biomarker panel for metastatic melanoma [70].

By searching the open literature, it appears that a meta-analysis has yet to be performed for transcriptomic data obtained from chemically-exposed skin. Based on the approaches taken, such as those described above, it is possible that a meta-analysis could provide a generalized view of chemical-induced gene profiles in skin according to compound class (e.g., irritants) as well as identifying a set of biomarkers (genes or pathways) that could serve as therapeutic targets.

Microarrays and Cutaneous Toxicity

It has been suggested that changes in gene expression, transcriptional analysis, and pathway elucidation may be a more sensitive endpoint than clinical or pathological observations of toxicity [3]. By understanding the transcriptional events that occur with or without observable pathological changes, it is possible to develop customized therapeutics or treatments to mitigate chemical-specific toxicity. The first step in achieving such a goal requires characterization of time- and dose-dependent transcriptional profiles, pathway mapping, and target identification in chemically-exposed skin. Two areas of research using this approach include investigating the cutaneous response to chemical irritants and burns.

Skin Irritants

Contact dermatitis (irritant and allergic) is a skin inflammatory condition following exposure to a hazardous compound and accounts for approximately 90-95% of all occupational skin disease [73]. Irritant contact dermatitis is a non-immune-related response characterized by direct action of a compound on skin tissues and reactions are typically localized to the site of contact [73]. Available data indicate that irritant contact dermatitis represents approximately 80% of all cases of occupational contact dermatitis [73]. Irritant dermatitis can be caused by a wide variety of compounds including solvents, surfactants, hydrocarbons, cutting fluids, and oils, resulting in itching, pain, edema, erythema, small blisters (wheals), and dry, flaking, scaly skin [73]. Although a prevalent disease, the precise underlying mechanisms resulting in skin irritation are still unknown and the severity of the condition depends on factors such as chemical properties, concentration, duration and frequency of exposure, environmental factors, and overall skin condition [73].

One study using oligonucleotide arrays evaluated gene expression changes in skin following exposures to solvents (m-xylene and d-limonene) and a surfactant (sodium lauryl sulfate; SLS) [42]. This study provided some insights into the effects of irritants on skin at the transcriptional level that reflected differences in type of chemical, duration of exposure, and associations with observable histological changes. Four hours after a 1-hour exposure, SLS did not induce observable histological changes, but promoted a higher number of transcriptional changes compared to the solvents. In contrast, m-xylene and d-limonene promoted a greater amount of transcriptional changes after a 1-hour exposure in which time-dependent histological changes progressed through four hours. These results suggest temporal differences in gene expression may exist that varies by the type of chemical and differences in the skin concentrations of SLS, *m*-xylene, and d-limonene may directly affect the number and magnitude of genes changed [42].

Transcriptional modulation associated with irritation induced by individual chemicals as well as mixtures has been investigated. JP-8 (jet propellant-8) is similar to hydrodesulfurized kerosene, but formulated to improve performance in military aviation for use by the US and North Atlantic Treaty Organization militaries [74]. JP-8 is a complex mixture containing aliphatic and aromatic hydrocarbons, including thousands of isomers and up to six performance additives. Due to the properties of JP-8 and high potential for occupational exposures, the skin has been considered an important route of exposure with irritation being a primary concern.

A study by McDougal et al. [34] investigated changes in epidermal gene expression for up to 8 hours after a 1-hour cutaneous exposure. This study demonstrated temporal changes in mRNA patterns over 8 hours in which skin irritation was not observed. Additionally, this study showed that the total number of significantly-changed genes increased over time in which down-regulated mRNA expression predominated by 8 hours [34]. Signaling pathways related to inflammation, apoptosis, cell growth, and proliferation were activated by JP-8 exposure in which 23 genes were commonly shared across all time points, while one gene was consistently decreased at all time periods. Subsequent bioinformatics performed suggested that JP-8-induced gene expression responses were consistent with a "physical" stimulus from the JP-8 interacting with the epidermis, and the "trigger" for inflammation may be due to disruption of the oxidative state, osmotic balance, and changes in membrane integrity [34].

As noted above, McDougal and Garrett [35] demonstrated differences in gene expression as a function of chemical potency and skin concentration for neat exposures to dimethylnaphthalene, trimethylbenzene, undecane, tetradecane, or JP-8 [35]. Although the data sets generated from these two studies [34, 35] have not been directly compared, the results indicate that tumor necrosis factor (TNF) and interleukin-6 (IL-6) could be considered as targets for therapeutic intervention. In skin, both TNF and IL-6 play direct roles in various inflammatory processes, while inducing the expression of additional secondary inflammatory mediators [75-77]. However, the common induction of TNF and IL-6 by JP-8 and some of its components may not be definitive markers of irritation, but could serve as a starting point for potential amelioration of JP-8-induced irritation through therapeutic intervention of TNF and IL-6 activity [35].

A study bu Ku et al. [33] used microarrays and informatics analysis to generate data that could help discriminate between chemical sensitizers and irritants. Mouse ear skin exposed daily for three days to the contact sensitizers 1-chloro-2,4-dinitrobenzene, 2-phenyl-4-ethoxymethylene-5-oxazolone, or toluene 2,4-diisocyanate, or the irritant croton oil [33]. Results showed that histological assessments were unable to differentiate between the sensitizers and irritant and gene expression profiles were similar. However, microarray analysis demonstrated that the three sensitizers changed the allergy-related genes 2'-5' oligoadenylate sythetase-like 2 (Oasl2), Z-DNA binding protein 1 (Zbp1), and chemokine ligands 9 and 10 (Cxcl9 and Cxcl10) by more than 2-fold. The authors suggest that this information could help discriminate skin sensitization from irritation [33].

Chemical Burns

Following cutaneous exposure to hazardous chemicals, it can be difficult to correlate burn severity to applied dose as there may be little to no immediate visible skin damage. This can complicate the means of determining the type of first aid or treatment that should be administered. Furthermore, exposure to hazardous chemicals could lead to incapacitating damage and infection, or long-term sequelae (e.g., scarring). Therefore, the timing and selection of the appropriate burn care treatment is critical to mitigate the level and severity of damage as well as enhance the wound healing process.

Cutaneous toxicity following exposure to the blistering agent sulfur mustard (HD) has been an area of research for decades. Investigators have used microarray analysis to complement pathological assessments to help elucidate underlying molecular responses associated with HD-induced cutaneous injury and subsequent wound healing. A goal of leveraging mRNA expression analysis would be to help streamline the development of therapeutics by utilizing microarrays and informatics analysis to identify potential biomarkers. Dillman et al. [29] monitored transcriptional changes in skin treated with indomethacin, dimercaprol, or octyl homovanillamide following HD exposure. Microarray analysis showed a positive correlation between changes in ear swelling and genes involved with the biological processes of cell cycle control, inflammation, signal transduction, and adhesion [29]. This study showed that gene expression profiling can provide insight into pathways indicative of HD damage for future development of therapeutics [29].

Similarly, Gerecke et al. [30] used gene-expression profiling to focus on HD-induced transcriptional changes in mouse ear skin in the presence or absence of a matrix metalloproteinase (MMP) inhibitor. Skin pre-treatment with MMP-2/MMP-9 inhibitor I prior to HD exposure led to changes in genes associated with metabolic pathways, while HD alone induced changes in pathways for apoptosis, inflammation, repair, and stress-related mechanisms. These gene changes appeared to be more suitable indicators than histological changes, and differences in gene expression patterns may help identify biomarkers specific to HD injury [30].

The similarities between porcine and human skin [78, 79] led to the development of a weanling swine model to assess the damage and subsequent healing of injuries caused by hazardous chemicals. This weanling swine model has been used to assess laser debridement, measure urinary metabolite excretion, evaluate clinical endpoints, monitor wound healing, assess treatment efficacy, and monitor transcriptional activity [36-39, 44, 45, 80-84]. The modification of the weanling swine model to incorporate changes in mRNA levels underlying biological functions and signaling pathways as a means to correlate molecular events with pathological observations has enabled researchers to identify potential gene targets for use in developing or evaluating new treatments that may increase time to healing as well as reduce long-term sequelae. Figure 2 shows an approach using microarrays and informatics analysis to identify these key genes and associated drugs.

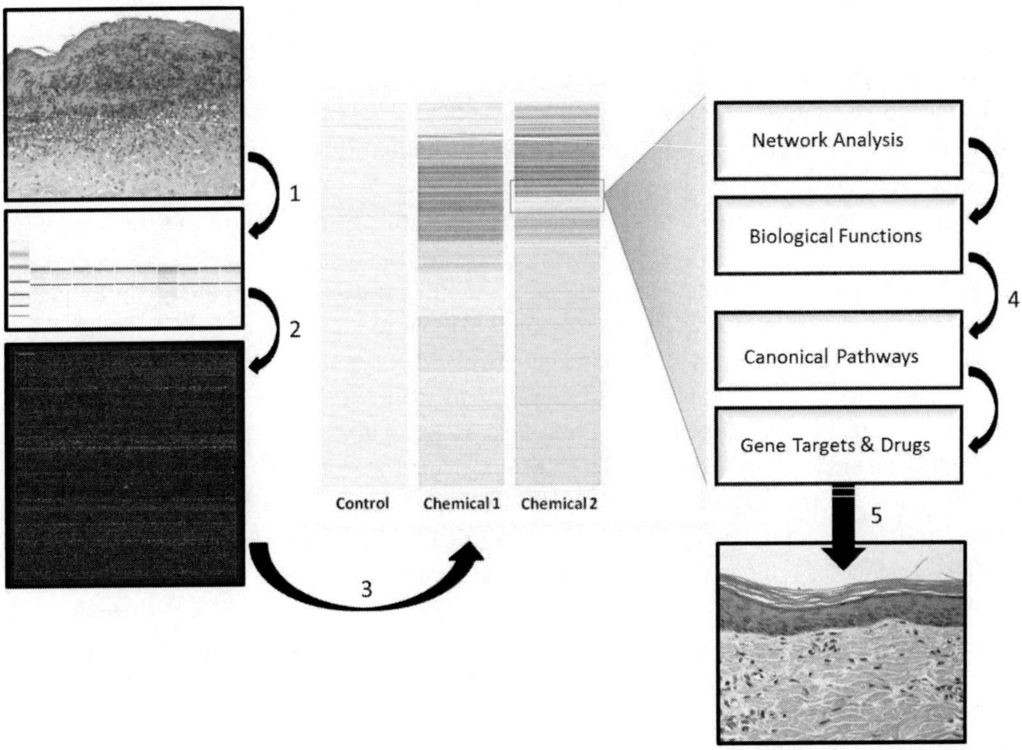

Figure 2. Flow chart highlighting experimental procedures through statistical analysis and informatics. Total RNA isolated from injured skin (1), processed, and hybridized to a microarray (2). Raw data from the microarrays are subjected to statistical analysis to generate lists of significantly-changed transcripts (3). Bioinformatics is then used to sort genes with respect to their associated biological networks, functions, canonical pathways, and subsequent identification of gene targets that could be modulated by drugs (4). Implementation of drug treatments could lead to amelioration of the chemical-induced skin toxicity (5).

The use of microarrays has played a key role in advancing the understanding of HD-induced burns and the toxic effects of other hazardous chemicals [29, 30, 37, 43, 44, 46]. Researchers have utilized the weanling swine model to conduct toxicogenomic analysis following cutaneous exposure to HD [37, 44], liquid bromine [36], bromine vapor [38, 45], and chlorine vapor [39]. These studies led to the identification of key genes that serve as targets for existing drugs for reducing burn severity and improving healing. In skin exposed to HD or bromine (liquid and vapor), the gene encoding heme oxygenase 1 (HMOX1) was significantly up-regulated in response to these hazardous chemicals [36-38, 44, 45]. These results suggest that HMOX1 could be considered a potential molecular target in which the HMOX1 inhibitor, tin mesoporphyrin, could be used to mitigate the severity of burns associated with chemical exposure and improve cutaneous healing. In skin exposed to chlorine vapor, the IL-1 signaling network and the transcripts encoding IL-6, IL-1A and IL-1B were identified as common potential therapeutic targets for Phase II/III clinical trial, or FDA-approved drugs that may serve to decrease the severity of chlorine-induced injury and promote wound healing [39].

MicroRNAs

MicroRNAs (miRNAs) are a major group of small endogenous, non-coding RNAs (~22 nt) known to regulate post-transcriptional gene expression. MicroRNAs play an important role in many biological processes and pathways, including differentiation and development, cell signaling, and response to infection. MicroRNAs are well-conserved in eukaryotic organisms and are thought to be a vital and evolutionarily component of genetic regulation [85].

Although there is substantial evidence indicating that miRNAs play a pivotal role in gene regulation and biological processes, little is known about the fundamental role of miRNAs in the pathogenesis of skin following cutaneous chemical exposure and associated toxicity. As the knowledge base regarding chemical-induced gene expression patterns in skin continues to grow, understanding the function of miRNAs in regulating these underlying molecular transcriptional mechanisms seems to be the logical next step in dermatotoxicity testing. In fact, there are published studies that have focused on the role of miRNAs in skin in which various miRNAs have been recognized as critical regulators of skin morphogenesis, physiology, pathology, cutaneous tissue repair, and wound healing.

With respect to wounds, the global pattern distribution of miRNA expression is unknown; however, specific miRNAs are have been identified for their role in the different phases of wound healing. These include the inflammatory phase (miR-105, miR-140, miR-146a, and miR-125b), proangiogenesis (miR-17-92, miR-126, miR-130a, miR-210, miR-296, and miR-378), antiangiogenesis (miR-92a, miR-17, miR-15b, miR-16, miR-20a, miR-20b, miR-320, miR-221, and miR-222), proliferative phase (miR-184, miR-205, and miR-210), and remodeling (miR-29a, miR-29b, miR-29c, and miR-192) [86]. miR-21 has also been shown to play a role during granulation formation by regulating wound contraction and collagen deposition [87].

MicroRNAs have been shown to be involved in inflammatory diseases such as allergic contact dermatitis and psoriasis [88, 89]. Allergic contact dermatitis is a common occupational and environmental health issue that progresses two stages. An initial phase is acquired after exposure to a specific compound, which is followed by an inflammatory response that results from a repeated exposure to the same chemical. Vennegaard et al. [88] investigated the expression of miRNAs from subjects with allergic responses to diphenylcyclopropenone (DPCP) to determine their involvement in gene expression. The study concluded that miR-21, miR-142-3p, miR-142-5p, and miR-223 were significantly up-regulated in skin from humans sensitized with DPCP, indicating that miRNAs may play a role in the pathogenesis of allergic contact dermatitis [88].

A global assessment of miRNA expression was investigated in canine cervical subcutaneous tissues treated with a matrix metalloproteinase inhibitor (MMPi). miRNAs were identified as being associated with processes indicative of histopathological changes of MMPi-induced fibrodysplasia [90]. This study evaluated temporal miRNA expression in which attenuation of the miR-200 family contributed to the activation of several transcription factors including Sp1, STAT1, and RelA. Moreover, miRNAs were shown to be associated with TGFβ synthesis and regulation of the acute inflammatory response [90].

MicroRNAs have also been shown to play a role in mediating UVB-induced senescence in human diploid fibroblasts [91]. This study evaluated fibroblasts subjected to 8 consecutive UVB treatments over a 4 day period. Genome-wide analysis revealed extensive transcriptional changes in which expression levels of 806 miRNAs in UVB-treated skin were compared to controls and identified several miRNA-mRNA regulatory networks mediating UVB-induced senescence [91]. The results from this study suggested miR-15, miR-20a/b, miR-93, miR-101 and their respective mRNA targets have important roles in which miR-101 and Ezh2 are key players in UVB-induced senescence [91]. An et al. [92] showed that epigallocatechin-3-gallate appears to alter specific miRNAs leading to protection against UVB damage in normal human dermal fibroblasts. Moreover, Xu et al. [93] evaluated the effects of the flavonoid, baicalin, on UVB-induced miRNA expression in mouse skin. These investigators showed that miR-23a was up-regulated in baicalin-treated, UVB-exposed skin, while mmu-miR-181b, mmu-miR-199a-3p, and mmu-miR-378 were down-regulated.

Other miRNAs involved in skin inflammatory-related processes include miR-125b, miR-146a, and miR-203 [89, 94]. In psoriasis, miR-146a is over-expressed, while miR-125b is down-regulated. These two miRNAs appear to regulate the TNF-α pathway in skin and may serve as potential therapeutic targets in the treatment of psoriasis [89]. miR-203 has been shown to be specific to the skin keratinocytes within different squamous epithelia and observed to be over-expressed in psoriasis [94]. miR-203 post-transcriptionally regulates suppressor of cytokine signaling-3, which is a negative regulator of the STAT3 pathway for inflammatory cytokines [94].

As with characterizing mRNA expression patterns, monitoring miRNA-specific responses in skin exposed to chemicals may provide a better understanding towards identifying potential therapeutic targets for the treatment of cutaneous chemical-induced pathologies. The studies investigating the role of miRNAs in inflammatory skin disorders can provide a basis of understanding for or comparison to miRNAs modulated in skin exposed to chemicals.

Conclusion

The skin is easily accessible for evaluating many biological and toxicological endpoints. Transcriptomics of perturbed skin provides a snapshot in time of gene expression changes, which can aid in establishing a *molecular fingerprint when compared to normal skin.* Gene expression studies have enhanced the knowledge of skin function and response to external insults by analyzing regulation in mRNA expression. Scientists have used microarrays in dermal research for years to evaluate underlying mechanisms of disease, cancer, wound healing, UV irradiation, chemical exposure, thermal injury, aging, pigmentation, and hair formation.

This chapter provides a glimpse into the potential of using microarrays to expand the current knowledge of skin function as well as help understand the underlying molecular mechanisms in skin as a result of cutaneous chemical exposure. Advancement in research in the field of chemically-induced dermal toxicology through the use of microarrays can lead to an improvement in understanding the molecular basis of dermal toxicity and risk in humans. Additionally, these molecular approaches can help identify potential genetic variability that may increase the susceptibility of an individual to chemical toxicity.

Identifying biomarkers to aid in the development of therapeutics to reduce chemical damage and enhance cutaneous healing is an elusive, but an important goal in dermatotoxicology. As described in this chapter, microarrays can be used to monitor expression of mRNAs and microRNAs and map these to specific biological functions and signaling pathways. Moreover, in-depth bioinformatics analysis enables researchers to identify specific mRNAs and microRNAs as biomarkers that could be used as therapeutic drug targets. An ultimate goal in using microarray data for cutaneous chemical exposures would be to take a next step toward integrating these data into a useable approach that can promote translational and predictive medicine.

References

[1] Aardema MJ, MacGregor JT. Toxicology and genetic toxicology in the new era of "toxicogenomics": impact of "-omics" technologies. *Mutat Res* 2002; 499(1):13-25.

[2] Blumenberg M. Skinomics: Transcriptional profiling in dermatology and skin biology. *Curr Genomics* 2012; 13:363-368.

[3] Harrill AH, Rusyn I. Systems biology and functional genomics approaches for the identification of cellular responses to drug toxicity. *Expert Opin Drug Metab Toxicol* 2008; 4(11):1379-1389.

[4] Jansen BJ, Schalkwijk J. Transcriptomics and proteomics of human skin. *Brief Funct Genomic Proteomic* 2003; 1(4):326-341.

[5] Kunz M, Ibrahim SM, Koczan D, Scheid S, Thiesen HJ, Gross G. DNA microarray technology and its applications in dermatology. *Exp Dermatol* 2004; 13(10):593-606.

[6] MacGregor JT. The future of regulatory toxicology: impact of the biotechnology revolution. *Toxicol Sci* 2003; 75(2):236-248.

[7] Oberemm A, Onyon L, Gundert-Remy U. How can toxicogenomics inform risk assessment? *Toxicol Appl Pharmacol* 2005; 207(2 Suppl):592-598.

[8] Sellheyer K, Belbin TJ. DNA microarrays: from structural genomics to functional genomics. The applications of gene chips in dermatology and dermatopathology. *J Am Acad Dermatol* 2004; 51(5):681-692.

[9] Wessagowit V, South AP. Dermatological applications of DNA array technology. *Clin Exp Dermatol* 2002; 27(6):485-492.

[10] Mlakar V, Glavac D. DNA microarrays and their use in dermatology. *Acta Dermatovenerol Alp Panonica Adriat* 2007; 16(1):7-12.

[11] Sonkoly E, Ståhle M, Pivarcsi. MicroRNAs: novel regulators in skin inflammation. *Exp Dermatol* 2008; 33:213-215.

[12] Dasu MR, Barrow RE, Hawkins HK, McCauley RL. Gene expression profiles of giant hairy naevi. *J Clin Pathol* 2004; 57(8):849-855.

[13] Merryman-Simpson AE, Wood SH, Fretwell N, Jones PG, McLaren WM, McEwan NA et al. Gene (mRNA) expression in canine atopic dermatitis: microarray analysis. *Vet Dermatol* 2008; 19(2):59-66.

[14] Milano A, Pendergrass SA, Sargent JL, George LK, McCalmont TH, Connolly MK et al. Molecular subsets in the gene expression signatures of scleroderma skin. *PLoS ONE* 2008; 3(7):e2696.

[15] Nomura I, Gao B, Boguniewicz M, Darst MA, Travers JB, Leung DY. Distinct patterns of gene expression in the skin lesions of atopic dermatitis and psoriasis: a gene microarray analysis. *J Allergy Clin Immunol* 2003;112:1195-1202.

[16] Roy S, Khanna S, Rink C, Biswas S, Sen CK. Characterization of the acute temporal changes in excisional murine cutaneous wound inflammation by screening of the wound-edge transcriptome. *Physiol Genomics* 2008;34:162-184.

[17] Wenzel J, Peters B, Zahn S, Birth M, Hofmann K, Kusters D et al. Gene expression profiling of lichen planus reflects CXCL9+-mediated inflammation and distinguishes this disease from atopic dermatitis and psoriasis. *J Invest Dermatol* 2008;128:67-78.

[18] Zhou L, Askew D, Wu C, Gilliam AC. Cutaneous gene expression by DNA microarray in murine sclerodermatous graft-versus-host disease, a model for human scleroderma. *J Invest Dermatol* 2007; 127(2):281-292.

[19] Huang C, Nie F, Qin Z, Li B, Zhao X. A snapshot of gene expression signatures generated using microarray datasets associated with excessive scarring. *Am J Dermatopathol* 2013;35:64-73.

[20] Wu J, Ma B, Yi S, Wang Z, He W, Luo G, Chen X, Wang X, Chen A, Barisoni D. Gene expression of early hypertrophic scar tissue screened by means of cDNA microarrays. *J Trauma* 2004; 57:1276-1286.

[21] Fernandez-Figueras MT, Puig L, Musulen E, Gilaberte M, Lerma E, Serrano S, Ferrandiz C, Ariza A. Expression profiles associated with aggressive behavior in Merkel cell carcinoma. *Mod Pathol* 2007; 20:90-101.

[22] Mendoza-Villanueva D, Diaz-Chavez J, Uribe-Figueroa L, Rangel-Escareao C, Hidalgo-Miranda A, March-Mifsut S, Jimenez-Sanchez G, Lambert P, Gariglio P. Gene expression profile of cervical and skin tissues from human papillomavirus type 16 E6 transgenic mice. *BMC Cancer* 2008;8:347.

[23] Nindl I, Dang C, Forschner T, Kuban RJ, Meyer T, Sterry W, Stockfleth E. Identification of differentially expressed genes in cutaneous squamous cell carcinoma by microarray expression profiling. *Mol Cancer* 2006;5:30.

[24] O'Driscoll L, McMorrow J, Doolan P, McKiernan E, Mehta JP, Ryan E, Gammell P, Joyce H, O'Donovan N, Walsh N, Clynes M. Investigation of the molecular profile of basal cell carcinoma using whole genome microarrays. *Mol Cancer* 2006;5:74.

[25] Riker AI, Enkemann SA, Fodstad O, Liu S, Ren S, Morris C, Xi Y, Howell P, Metge B, Samant RS, Shevde LA, Li W, Eschrich S, Daud A, Ju J, Matta J. The gene expression profiles of primary and metastatic melanoma yields a transition point of tumor progression and metastasis. *BMC Med Genomics* 2008;1:13.

[26] Enk CD, Shahar I, Amariglio N, Rechavi G, Kaminski N, Hochberg M. Gene expression profiling of in vivo UVB-irradiated human epidermis. *Photodermatol Photoimmunol Photomed* 2004; 20(3):129-137.

[27] Lee YS, Choi DK, Kim CD, Im M, Mollah ML, Jang JY, Oh TJ, An S, Seo YJ, Hur GM, Cho MJ, Park JK, Lee JH. Expression profiling of radiation-induced genes in radiodermatitis of hairless mice. *Br J Dermatol* 2006; 154(5):829-838.

[28] Mnich CD, Hoek KS, Virkki LV, Farkas A, Dudli C, Laine E et al. Green tea extract reduces induction of p53 and apoptosis in UVB-irradiated human skin independent of transcriptional controls. *Exp Dermatol* 2009; 18(1):69-77.

[29] Dillman JF, III, Hege AI, Phillips CS, Orzolek LD, Sylvester AJ, Bossone C et al. Microarray analysis of mouse ear tissue exposed to bis-(2-chloroethyl) sulfide: gene expression profiles correlate with treatment efficacy and an established clinical endpoint. *J Pharmacol Exp Ther* 2006; 317(1):76-87.

[30] Gerecke DR, Chen M, Isukapalli SS, Gordon MK, Chang YC, Tong W, Androulakis IP, Georgopoulos PG. Differential gene expression profiling of mouse skin after sulfur mustard exposure: Extended time response and inhibitor effect. *Toxicol Appl Pharmacol* 2009; 234(2):156-165.

[31] Lu J, Xie L, Sylvester J, Wang J, Bai J, Baybutt R, Wang W. Different gene expression of skin tissues between mice with weight controlled by either calorie restriction or physical exercise. *Exp Biol Med (Maywood)* 2007; 232(4):473-480.

[32] Matthews CP, Birkholz AM, Baker AR, Perella CM, Beck GR Jr., Young MR et al. Dominant-negative activator protein 1 (TAM67) targets cyclooxygenase-2 and osteopontin under conditions in which it specifically inhibits tumorigenesis. *Cancer Res* 2007; 67(6):2430-2438.

[33] Ku HO, Jeong SH, Kang HG, Pyo HM, Cho JH, Son SW et al. Gene expression profiles and pathways in skin inflammation induced by three different sensitizers and an irritant. *Toxicol Lett* 2009; 190:231-237.

[34] McDougal JN, Garrett CM, Amato CM, Berberich SJ. Effects of Brief Cutaneous JP-8 Jet Fuel Exposures on Time Course of Gene Expression in the Epidermis. *Toxicol Sci* 2007; 95:495-510.

[35] McDougal JN, Garrett CM. Gene expression and target tissue dose in the rat epidermis after brief JP-8 and JP-8 aromatic and aliphatic component exposures. *Toxicol Sci* 2007; 97:569-581.

[36] Price JA, Rogers JV, McDougal JN, Shaw MQ, Reid FM, Kiser RC, Graham JS. Gene expression analysis of bromine-induced burns in porcine skin. *Toxicol Lett* 2008; 182:69-78.

[37] Price JA, Rogers JV, McDougal JN, Shaw MQ, Reid FM, Graham JS. Transcriptional changes in porcine skin at 7 days following sulfur mustard and thermal burn injury. *Cutan Ocul Toxicol* 2009; 28(3):129-140.

[38] Price JA, Rogers JV, Wendling MQS, Plahovinsak JL, Perry MR, Reid FM, Kiser RC, Graham JS. Temporal effects in porcine skin following bromine vapor exposure. *Cutan Ocul Toxicol* 2011; 30:187-197.

[39] Price JA, Rogers JV, Wendling MQS, Plahovinsak JL, Perry MR, Reid FM, Graham JS. Toxicogenomic analysis of chlorine vapor-induced porcine skin injury. *Cutan Ocul Toxicol* 2012; 31:323-331.

[40] Ridd K, Zhang SD, Edwards RE, Davies R, Greaves P, Wolfreys A, Smith AG, Gant TW. Association of gene expression with sequential proliferation, differentiation and tumor formation in murine skin. *Carcinogenesis* 2006; 27:1556-1566.

[41] Riggs PK, Angel JM, Abel EL, DiGiovanni J. Differential gene expression in epidermis of mice sensitive and resistant to phorbol ester skin tumor promotion. *Mol Carcinog* 2005; 44:122-136.

[42] Rogers JV, Garrett CM, McDougal JN. Gene expression in rat skin induced by irritating chemicals. *J Biochem Mol Toxicol* 2003; 17:123-137.

[43] Rogers JV, Choi YW, Kiser RC, Babin MC, Casillas RP, Schlager JJ, Sabourin CL. Microarray analysis of gene expression in murine skin exposed to sulfur mustard. *J Biochem Mol Toxicol* 2004; 18:289-299.

[44] Rogers JV, McDougal JN, Price JA, Reid FM, Graham JS. Transcriptional responses associated with sulfur mustard and thermal burns in porcine skin. *Cutan Ocul Toxicol* 2008; 27:135-160.

[45] Rogers JV, Price JA, Wendling MQS, Perry MR, Reid FM, Kiser RC, Graham JS. An assessment of transcriptional changes in porcine skin exposed to bromine vapor. *J Biochem Mol Toxicol* 2011; 25:252-262.

[46] Sabourin CL, Rogers JV, Choi YW, Kiser RC, Casillas RP, Babin MC, Schlager JJ. Time- and dose-dependent analysis of gene expression using microarrays in sulfur mustard-exposed mice. *J Biochem Mol Toxicol* 2004; 18(6):300-312.

[47] Schlingemann J, Hess J, Wrobel G, Breitenbach U, Gebhardt C, Steinlein P, Kramer H, Furstenberger G, Hahn M, Angel P, Lichter P. Profile of gene expression induced by the tumour promotor TPA in murine epithelial cells. *Int J Cancer* 2003; 104(6):699-708.

[48] Xie L, Jiang Y, Ouyang P, Chen J, Doan H, Herndon B, Sylvester JE, Zhang K, Molteni A, Reichle M, Zhang R, Haub MD, Baybutt RC, Wang W. Effects of dietary calorie restriction or exercise on the PI3K and Ras signaling pathways in the skin of mice. *J Biol Chem* 2007; 282(38):28025-28035.

[49] Zhang Z, Yao R, Li J, Wang Y, Boone CW, Lubet RA, You M. Induction of invasive mouse skin carcinomas in transgenic mice with mutations in both H-ras and p53. *Mol Cancer Res* 2005; 3(10):563-574.

[50] Feezor RJ, Paddock HN, Baker HV, Varela JC, Barreda J, Moldawer LL, Schultz GS, Mozingo DW. Temporal patterns of gene expression in murine cutaneous burn wound healing. *Physiol Genomics* 2004; 16(3):341-348.

[51] Spies M, Dasu MR, Svrakic N, Nesic O, Barrow RE, Perez-Polo JR, Herndon DN. Gene expression analysis in burn wounds of rats. *Am J Physiol Regul Integr Comp Physiol* 2002; 283(4):R918-R930.

[52] Lener T, Moll PR, Rinnerthaler M, Bauer J, Aberger F, Richter K. Expression profiling of aging in the human skin. *Exp Gerontol* 2006; 41(4):387-397.

[53] Carroll JM, McElwee KJ, King E, Byrne MC, Sundberg JP. Gene array profiling and immunomodulation studies define a cell-mediated immune response underlying the pathogenesis of alopecia areata in a mouse model and humans. *J Invest Dermatol* 2002; 119(2):392-402.

[54] Liu Y, Das S, Olszewski RE, Carpenter DA, Culiat CT, Sundberg JP, Soteropoulos P, Liu X, Doktycz MJ, Michaud EJ, Voy BH. The near-naked hairless (Hr(N)) mutation

disrupts hair formation but is not due to a mutation in the Hairless coding region. *J Invest Dermatol* 2007; 127(7):1605-1614.

[55] Aoki H, Moro O, Tagami H, Kishimoto J. Gene expression profiling analysis of solar lentigo in relation to immunohistochemical characteristics. *Br J Dermatol* 2007; 156(6):1214-1223.

[56] April CS, Barsh GS. Skin layer-specific transcriptional profiles in normal and recessive yellow (Mc1re/Mc1re) mice. *Pigment Cell Res* 2006; 19(3):194-205.

[57] Cole J, Tsou R, Wallace K, Gibran N, Isik F. Early gene expression profile of human skin to injury using high-density cDNA microarrays. *Wound Repair Regen* 2001; 9(5):360-370.

[58] Colwell AS, Longaker MT, Peter LH. Identification of differentially regulated genes in fetal wounds during regenerative repair. *Wound Repair Regen* 2008; 16(3):450-459.

[59] Newkirk KM, MacKenzie DA, Bakaletz AP, Hudson LG, Kusewitt DF. Microarray analysis demonstrates a role for Slug in epidermal homeostasis. *J Invest Dermatol* 2008; 128(2):361-369.

[60] Rogers JV, Price JA, McDougal JN. A review of transcriptomics in cutaneous chemical exposure. *Cutan Ocul Toxicol* 2009; 28:157-170.

[61] Xu H, Cheepala S, McCauley E, Coombes K, Xiao L, Fischer SM et al. Chemoprevention of skin carcinogenesis by phenylretinamides: retinoid receptor-independent tumor suppression. *Clin Cancer* Res 2006; 12:969-979.

[62] Python F, Goebel C, Aeby P. Comparative DNA microarray analysis of human monocyte derived dendritic cells and MUTZ-3 cells exposed to the moderate skin sensitizer cinnamaldehyde. *Toxicol Appl Pharmacol* 2009;239:273-283.

[63] Perez DS, Handa RJ, Yang RS, Campain JA. Gene expression changes associated with altered growth and differentiation in benzo[a]pyrene or arsenic exposed normal human epidermal keratinocytes. *J Appl Toxicol* 2008;28:491-508.

[64] Zheng XH, Watts GS, Vaught S, Gandolfi AJ. Low-level arsenite induced gene expression in HEK293 cells. *Toxicology* 2003;187:39-48.

[65] Ao L, Liu JY, Liu W, Gao LH, Hu R, Fang ZJ et al. Comparison of gene expression profiles in BALB/c 3T3 transformed foci exposed to tumor promoting agents. *Toxicol In Vitro* 2010;24:430-438.

[66] Cheng SL, Huang Liu R, Sheu JN, Chen ST, Sinchaikul S et al. Toxicogenomics of kojic acid on gene expression profiling of a375 human malignant melanoma cells. *Biol Pharm Bull 2006*;29:655-669.

[67] Ramasamy A, Mondry A, Holmes CC, Altman DG. Key issues in conducting a meta-analysis of gene expression microarray datasets. *PLoS Med* 2008;5:e184.

[68] van Kester MS, Borg MK, Zoutman WH, Out-Luiting JJ, Jansen PM, Dreef EJ et al. A meta-analysis of gene expression data identifies a molecular signature characteristic for tumor-stage myosis fungoides. *J Invest Dermatol* 2012;132:2050-2059.

[69] Tian S, Krueger JG, Li K, Jabbari A, Brodmerkel C, Lowes MA et al. Meta-analysis derived (MAD) transcriptome of psoriasis defines the "core" pathogenesis of disease. *PLoS One* 2012;7:e44274.

[70] Liu W, Peng Y, Tobin DJ. A new 12-gene diagnostic biomarker signature of melanoma revealed by integrated microarray analysis. *PeerJ* 2013;1:49.

[71] Noh M, Yeo H, Ko J, Kim HK, Lee CH. MAP17 is associated with T-helper cell cytokine-induced down-regulation of flaggrin transcription in human keratinocytes. *Exp Dermatol* 2010;19:355-362.

[72] Blumenberg M. Profiling and metaanalysis of epidermal keratinocytes responses to epidermal growth factor. *BMC Genomics* 2013;14:85.

[73] NIOSH National Occupational Research Agenda. Allergic and Irritant Dermatitis. http://www.cdc.gov/niosh/topics/skin. Accessed Septermber 22, 2013.

[74] McDougal JN, Rogers JV, Simman R. Understanding systemic and local toxicity of JP-8 after cutaneous exposures. In: Witten ML, Zeiger E, Ritchie GD, eds. *Jet Fuel Toxicology*. Boca Raton:CRC Press, 2010:149-179.

[75] Corsini E, Galli CL. Cytokines and irritant contact dermatitis. *Toxicol Lett* 1998;102-103:277-282.

[76] Luster MI, Simeonova PP, Gallucci R, Matheson J. Tumor necrosis factor alpha and toxicology. *Crit Rev Toxicol 1999*;29:491-511.

[77] Sehgal PB. Interleukin-6: molecular pathophysiology. *J Invest Dermatol* 1990;94:2S-6S.

[78] Dick IP, Scott RC. Pig ear skin as an in-vitro model for human skin permeability. *J Pharm Pharmacol* 1992;44:640-645.

[79] Montagna W, Yun JS. The skin of the domestic pig. *J Invest Dermatol* 1964, 42:11-21.

[80] Graham JS, Reid FM, Smith JR, Stotts RR, Tucker ES, Shumaker SM et al. A cutaneous full-thickness liquid sulfur mustard burn model in weanling swine: clinical pathology and urinary excretion of thiodiglycol. *J Appl Toxicol* 2000;20:S161-S172.

[81] Reid FM, Graham J, Niemuth NA, Singer AW, Janny SJ, Johnson JB. Sulfur mustard-induced skin burns in weanling swine evaluated clinically and histopathologically. *J Appl Toxicol* 2000;20:S153-S160.

[82] Graham JS, Schomacker KT, Glatter RD, Briscoe CM, Braue EH Jr, Squibb KS. Efficacy of laser debridement with autologous split-thickness skin grafting in promoting improved healing of deep cutaneous sulfur mustard burns. *Burns* 2002;28:719-730.

[83] Reid FM, Niemuth NA, Shumaker SM, Waugh JD, Graham JS. Biomechanical monitoring of cutaneous sulfur mustard-induced lesions in the weanling pig model for depth of injury. *Skin Res Technol* 2007;13:217-225.

[84] Graham JS, Stevenson RS, Mitcheltree LW, Simon M, Hamilton TA, Deckert RR, et al. Improved wound healing of cutaneous sulfur mustard injuries in a weanling pig model. *J Burns Wounds* 2006;5:e7.

[85] Kawaji H, Hayashizaki Y. Exploration of small RNAs. *PLoS Genet* 2008;4:e22.

[86] Banerjee J, Chan YC, Sen CK. MicroRNAs in skin and wound healing. *Physiol Genomics* 2011;43:543-556.

[87] Wang T, Feng Y, Sun H, Zhang L, Hao L, Shi C et al. miR-21 regulates skin wound healing by targeting multiple aspects of the healing process. *Am J Pathol* 2012;181:1911-1120.

[88] Vennegaard MT, Bonefeld CM, Hagedorn PH, Bangsgaard N, Løvendorf MB, Odum N et al. Allergic contact dermatitis induces upregulation of identical microRNAs in humans and mice. *Contact Dermatitis* 2012;67:298-305.

[89] Sonkoly E, Ståhle M, Pivarcsi A. MicroRNAs: novel regulators in skin inflammation. *Clin Exp Dermatol* 2008;33:312-315.

[90] Tonge DP, Tugwood JD, Kelsall J, Gant TW. The role of microRNAs in the pathogenesis of MMPi-induced skin fibrodysplasia. *BMC Genomics* 2013;14:338.
[91] Greussing R, Hackl M, Charoentong P, Pauck A, Monteforte R, Cavinato M et al. Identification of microRNA-mRNA functional interactions in UVB-induced senescence of human diploid fibroblasts. *BMC Genomics* 2013;14:224.
[92] An IS, An S, Park S, Lee SN, Bae S. Involvement of microRNAs in epigallocatechin gallate-mediated UVB protection in human dermal fibroblasts. *Oncol Rep* 2013;29:253-253.
[93] Xu Y, Zhou B, Wu D, Yin Z, Luo D. Baicalin modulates microRNA expression in UVB irradiated mouse skin. *J Biomed Res* 2012;26:125-134.
[94] Sonkoly E, Wei T, Janson PC, Sääf A, Lundeberg L, Tengvall-Linder M et al. MicroRNAs: novel regulators involved in the pathogenesis of psoriasis? *PLoS One* 2007;2:e610.

Index

#

A

B

C

D

E

F

G

H

I

J

K

L

M

N

O

P

Q

R

S

T

U

V

W

X

Y

Z

β